Spin-Crossover Complexes

Special Issue Editor
Kazuyuki Takahashi

MDPI • Basel • Beijing • Wuhan • Barcelona • Belgrade

MDPI

Special Issue Editor
Kazuyuki Takahashi
Kobe University
Japan

Editorial Office
MDPI
St. Alban-Anlage 66
Basel, Switzerland

This edition is a reprint of the Special Issue published online in the open access journal *Inorganics* (ISSN 2304-6740) from 2017–2018 (available at: http://www.mdpi.com/journal/inorganics/special_issues/spin_crossover_complexes).

For citation purposes, cite each article independently as indicated on the article page online and as indicated below:

Lastname, F.M.; Lastname, F.M. Article title. *Journal Name* **Year**, *Article number, page range*

First Edition 2018

ISBN 978-3-03842-825-1 (Pbk)
ISBN 978-3-03842-826-8 (PDF)

Table of Contents

About the Special Issue Editor

Kazuyuki Takahashi, received his Doctor of Science degree from the University of Tokyo in 1999, working on the development of novel organic electron acceptors under the supervision of Professor Keiji Kobayashi. After post-doctoral working experience with Prof. Yohji Misaki at Kyoto University, he joined the group of Dr. Osamu Sato at Kanagawa Academy of Science and Technology in 2001, and started to work on photo-responsive transition metal complexes. Kazuyuki Takahshi was a Research Associate in the group of Prof. Hayao Kobayashi at Institute for Molecular Science in 2004, and then moved to the group of Prof. Hatsumi Mori at the Institute for Solid State Physics in 2006, working on the development of functional molecular materials. Since 2011, he has held the position of Associate Professor at Kobe University. His current main interest is studying the development of novel molecular materials exhibiting exotic electronic properties.

inorganics |MDPI|

Editorial
Spin-Crossover Complexes

Kazuyuki Takahashi

Department of Chemistry, Graduate School of Science, Kobe University, 1-1 Rokkodai-cho, Nada-ku, Kobe 657-8501, Japan; ktaka@crystal.kobe-u.ac.jp; Tel.: +81-78-803-5691

Received: 12 February 2018; Accepted: 27 February 2018; Published: 1 March 2018

Spin-crossover (SCO) is a spin-state switching phenomenon between a high-spin (HS) and low-spin (LS) electronic configurations in a transition metal center. The SCO phenomenon is widely recognized as an example of molecular bistability. The SCO compounds most widely studied are six-coordinate first-row transition metal complexes with d^4–d^7 configurations. A relative small enthalpy variation between LS and HS states can be realized by coopetition between ligand field stabilization (LFS) and spin pairing energies, which is illustrated by the Tanabe–Sugano diagrams in common coordination chemistry textbooks. Since an entropy variation in spin multiplicity from LS to HS is always positive, an increase in temperature can induce the transformation in Gibbs free energy from a positive to a negative sign, at which point SCO conversion occurs from the LS to HS state.

Cambi and co-workers' pioneering work on the anomalous magnetic behaviors of mononuclear Fe(III) dithiocarbamate complexes [1] was first recognized as SCO phenomena in the early 1930s. However, progress on SCO complexes awaited the dissemination of ligand field theory into coordination chemistry. The concept of controlling LFS energies by substitution with different field strength ligands resulted in the corroboration of SCO phenomena in some Co(II) and Fe(II) complexes in the early 1960s. Moreover, subsequent findings that pressure [2] and light [3] can induce an SCO phenomenon may attract attention to SCO complexes. In the 1990s the demonstration of device applications using the SCO complex [4] illuminated the potential of SCO complexes in future practical applications in memory, display, and sensing devices. Figure 1 shows the number of published articles per year whose titles or topics contain the words "spin-crossover," "spin equilibrium," or their derivatives. Studies concerning SCO complexes have apparently increased since the 1980s; moreover, the number has more rapidly developed after the 2000s. The fundamentals and applications of SCO complexes have attracted growing interest not only in inorganic coordination chemistry but also in a wide range of relevant research fields.

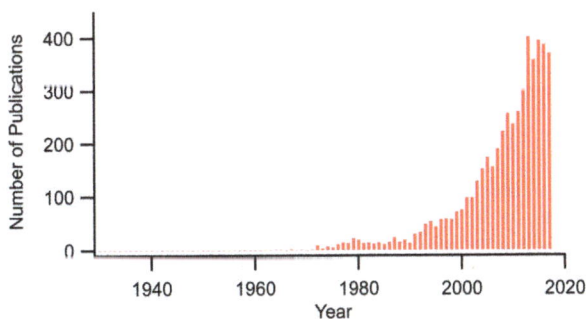

Figure 1. The number of publications per year whose titles or keywords contain "spin-crossover," "spin equilibrium," or their derived words, retrieved from Web of Science.

In the history of SCO complexes, three volumes of earlier research on SCO complexes were edited by Gütlich and Goodwin [5] in 2003. Later, a book edited by Halcrow [6] and a review by Gütlich and co-workers [7] concerning recent advances in SCO complexes were published in 2013. Moreover, a large number of reviews on specific subjects relating to SCO complexes have been published to date. Although considerable knowledge concerning SCO complexes has been accumulated, it is still challenging to design and synthesize an SCO complex that exhibits the target SCO behavior.

The impact on the occurrence of SCO can be considered to be divided into the inner and outer effects on a coordination sphere. The former effect is the prerequisite requirement of the LFS energy of a transition metal complex near the SCO region, as described above, which is strongly dependent on a transition metal center, coordinating ligands, and the coordination geometry. The latter effect arises from various interactions between SCO complexes, counter ions, and solvate molecules, if any. These interactions may affect the coordination structure of an SCO complex and/or SCO cooperativity, which represents the correlation of a spin state between SCO active metal centers. Therefore, further investigation has been undertaken to clarify the impact on SCO complexes.

This Special Issue is devoted to various aspects of recent research on SCO complexes by means of open-access way. Excellent, well-organized, and impressive reviews are contributed by three groups. Kuroiwa [8] reviews the recent advances in supramolecular approaches to SCO complexes, indicating that the successful control of molecular assemblies of SCO complexes provides an opportunity to contribute toward nanoscience for transition metal complexes including SCO complexes. Sugahara and co-workers [9] disclose the outer effects of functional counterions on SCO behaviors for the well-known one-dimensional triazole-coordinated Fe(II) complex. Moreover, they demonstrate that extended X-ray absorption fine structure (EXAFS) is a very useful technique to analyze the coordination structures for non-crystalline SCO complex solids. Banerjee and co-workers [10] give an overview of recent progress in theoretical treatments on SCO complexes and review the applications of the state-of-the-art density functional theory-based calculation technique to microscopic understanding of SCO cooperativity.

This Special Issue also contains 11 original research articles that are devoted to the synthesis and characterization of various molecular systems of SCO complexes. For the investigation of isolated mononuclear SCO complexes, Stock and co-workers [11] report structural dynamics in isolated LS Fe(II) complexes from extended-tripod ligands by means of laser flash photolysis and find that the slowing-down of dynamic exchange from the metastable HS state may arise from the trigonal torsion of a coordination octahedron. With respect to mononuclear SCO complex solids, Kimura and Ishida [12] describe the substitution effect on the pybox ligand in an Fe(II) SCO complex and reveal that the decrease in LFS energy originates from the electron-donating ability of substituents; moreover, the comparison between the solid and solution states may shed light on the possibility of separating the inner and outer effects from a substitution effect. Phonsri and co-workers [13] describe the ethoxy substitution effect on homoleptic and heteroleptic Fe(III) SCO systems and find that the steric effect leads to less cooperative SCO conversions. Nakanishi and co-workers [14] investigate the halogen substitution effect on both the ligand and counter-anion for an Fe(III) SCO complex. They were successful at isolating and characterizing four isostructural SCO complexes and found that the substitution of the ligand may result in an inner electronic effect, whereas substitution of the counter-anion may create an outer chemical pressure effect. As for dinuclear SCO complex solids, Craze and co-workers [15] report the steric effect of bridging groups between two SCO active centers in dinuclear Fe(II) triple helicate complexes and observed slight changes in SCO transition behavior. Hora and Hagiwara [16] also disclose the bridging alkyl length effect in the dinuclear Fe(II) helicate system and discover the widest thermal hysteresis loop among the related dinuclear Fe(II) helicate complexes. For polynuclear SCO complex assemblies, Kosone and co-workers [17] investigate the substitution effect on the axial pyridine ligands in the 2D bilayer Hofmann-type Fe(II) coordination polymer and reveal steric and chemical pressure effects on SCO cooperativity and transition temperature. Iwai and co-workers [18] report a series of Fe(II) assembled complexes from a bidentate bridging ligand with different co-ligands. They indicate that the parallel configuration of coordinated pyridine rings in an FeN_6 coordination core favors the HS state. Imoto and co-workers [19] demonstrate the metal dilution effect on SCO cooperativity for the cyano-bridged Fe(II) SCO network and reveal that the decrease in cooperativity by increasing Co(II) concentration may lead to more gradual SCO transitions and lower transition temperatures. For the development of novel multifunctional SCO hybrid compounds, Kuramochi and co-workers [20] develop Fe(II) SCO hybrid molecular systems with a polyoxometalate (POM) anion, which is known as a multi-functional unit. They clarify the crystal structures and magnetic properties of the SCO–POM hybrids and find that the hydrogen bonding interactions between the Fe(II) cation and POM anion strongly influenced the spin state of the Fe(II) cation. Similarly, Takahashi and co-workers [21] report the crystal structures and physical properties of a new hybrid compound from a well-known Fe(III) SCO complex cation with a π-radical anion. They reveal that the coordination distortion induced by strong π-stacking interactions between an Fe(III) cation and a π-radical anion may suppress the occurrence of SCO. As summarized above, this Special Issue covers a wide range of molecular systems of SCO complexes, and various experimental and theoretical techniques. I am grateful to all the authors for their diverse contributions.

I hope readers will increase their knowledge by engaging with this Special Issue and will go on to contribute to further progress in the research field of SCO complexes. Finally, I am particularly grateful to all the reviewers for their rigorous evaluations and valuable suggestions, which will help to enhance the quality of this Special Issue. In addition, I sincerely thank the editorial staff for their dedicated support in the planning, reviewing, and publishing of this Special Issue.

References

1. Cambi, L.; Szegö, L. Uber die magnetische Susceptibilitat der komplexen Verbindungen. *Ber. Dtsch. Chem. Ges.* **1931**, *64*, 2591–2598. [CrossRef]
2. Ewald, A.H.; Martin, R.L.; Sinn, E.; White, A.H. Electronic equilibrium between the $^6A_1$ and $^2T_2$ states in iron(III) dithio chelates. *Inorg. Chem.* **1969**, *8*, 1837–1846. [CrossRef]
3. Decurtins, S.; Gütlich, P.; Köhler, C.P.; Spiering, H.; Hauser, A. Light-induced excited spin state trapping in a transition-metal complex: The hexa-1-propyltetrazole-iron(II) tetrafluoroborate spin-crossover system. *Chem. Phys. Lett.* **1984**, *105*, 1–4. [CrossRef]
4. Kahn, O.; Kröber, J.; Jay, C. Spin Transition Molecular Materials for Displays and Data Recording. *Adv. Mater.* **1992**, *4*, 718–728. [CrossRef]
5. Gütlich, P.; Goodwin, H.A. (Eds.) *Spin Crossover in Transition Metal Compounds I–III*; Springer: Berlin/Heidelberg, Germany, 2004.
6. Halcrow, M.A. (Ed.) *Spin-Crossover Materials*; John Wiley & Sons, Ltd.: Oxford, UK, 2013.
7. Gütlich, P.; Gaspar, A.B.; Garcia, Y. Spin State Switching in Iron Coordination Compounds. *Beilstein J. Org. Chem.* **2013**, *9*, 342–391. [CrossRef] [PubMed]
8. Kuroiwa, K. Supramolecular Control of Spin Crossover Phenomena Using Various Amphiphiles. *Inorganics* **2017**, *5*, 45. [CrossRef]
9. Sugahara, A.; Kamebuchi, H.; Okazawa, A.; Enomoto, M.; Kojima, N. Control of Spin-Crossover Phenomena in One-Dimensional Triazole-Coordinated Iron(II) Complexes by Means of Functional Counter Ions. *Inorganics* **2017**, *5*, 50. [CrossRef]
10. Banerjee, H.; Chakraborty, S.; Saha-Dasgupta, T. Design and Control of Cooperativity in Spin-Crossover in Metal–Organic Complexes: A Theoretical Overview. *Inorganics* **2017**, *5*, 47. [CrossRef]
11. Stock, P.; Wiedemann, D.; Petzold, H.; Hörner, G. Structural Dynamics of Spin Crossover in Iron(II) Complexes with Extended-Tripod Ligands. *Inorganics* **2017**, *5*, 60. [CrossRef]
12. Kimura, A.; Ishida, T. Pybox-Iron(II) Spin-Crossover Complexes with Substituent Effects from the 4-Position of the Pyridine Ring (Pybox = 2,6-Bis(oxazolin-2-yl)pyridine). *Inorganics* **2017**, *5*, 52. [CrossRef]
13. Phonsri, W.; Darveniza, L.C.; Batten, S.R.; Murray, K.S. Heteroleptic and Homoleptic Iron(III) Spin-Crossover Complexes; Effects of Ligand Substituents and Intermolecular Interactions between Co-Cation/Anion and the Complex. *Inorganics* **2017**, *5*, 51. [CrossRef]

Inorganics **2018**, *6*, 32

14. Nakanishi, T.; Okazawa, A.; Sato, O. Halogen Substituent Effect on the Spin-Transition Temperature in Spin-Crossover Fe(III) Compounds Bearing Salicylaldehyde 2-Pyridyl Hydrazone-Type Ligands and Dicarboxylic Acids. *Inorganics* **2017**, *5*, 53. [CrossRef]
15. Craze, A.R.; Sciortino, N.F.; Badbhade, M.M.; Kepert, C.J.; Marjo, C.E.; Li, F. Investigation of the Spin Crossover Properties of Three Dinulear Fe(II) Triple Helicates by Variation of the Steric Nature of the Ligand Type. *Inorganics* **2017**, *5*, 62. [CrossRef]
16. Hora, S.; Hagiwara, H. High-Temperature Wide Thermal Hysteresis of an Iron(II) Dinuclear Double Helicate. *Inorganics* **2017**, *5*, 49. [CrossRef]
17. Kosone, T.; Kawasaki, T.; Tomori, I.; Okabayashi, J.; Kitazawa, T. Modification of Cooperativity and Critical Temperatures on a Hofmann-Like Template Structure by Modular Substituent. *Inorganics* **2017**, *5*, 55. [CrossRef]
18. Iwai, S.; Yoshinami, K.; Nakashima, S. Structure and Spin State of Iron(II) Assembled Complexes Using 9,10-Bis(4-pyridyl)anthracene as Bridging Ligand. *Inorganics* **2017**, *5*, 61. [CrossRef]
19. Imoto, K.; Takano, S.; Ohkoshi, S.-I. Metal Substitution Effect on a Three-Dimensional Cyanido-Bridged Fe Spin-Crossover Network. *Inorganics* **2017**, *5*, 63. [CrossRef]
20. Kuramochi, S.; Shiga, T.; Cameron, J.M.; Newton, G.N.; Oshio, H. Synthesis, Crystal Structures and Magnetic Properties of Composites Incorporating an Fe(II) Spin Crossover Complex and Polyoxometalates. *Inorganics* **2017**, *5*, 48. [CrossRef]
21. Takahashi, K.; Sakurai, T.; Zhang, W.-M.; Okubo, S.; Ohta, H.; Yamamoto, T.; Einaga, Y.; Mori, H. Spin-Singlet Transition in the Magnetic Hybrid Compound from a Spin-Crossover Fe(III) Cation and π-Radical Anion. *Inorganics* **2017**, *5*, 54. [CrossRef]

inorganics

|MDPI|

Review

Supramolecular Control of Spin Crossover Phenomena Using Various Amphiphiles

Keita Kuroiwa

Department of Nanoscience, Faculty of Engineering, Sojo University, Kumamoto 860-0082, Japan;
keitak@nano.sojo-u.ac.jp; Tel.: +81-96-326-3891

Received: 27 June 2017; Accepted: 12 July 2017; Published: 14 July 2017

Abstract: An aspect of nanochemistry that has attracted significant attention is the formation of nanoarchitectures from the self-assembly of metal complexes, based on the design of compounds having cooperative functionalities. This technique is currently seen as important within the field of nanomaterials. In the present review, we describe the methods that allow tuning of the intermolecular interactions between spin crossover (SCO) complexes in various media. These approaches include the use of lipophilic derivatives, lipids, and diblock copolypeptide amphiphiles. The resulting supramolecular assemblies can enhance the solubility of various SCO complexes in both organic and aqueous media. In addition, amphiphilic modifications of coordination systems can result in metastable structures and dynamic structural transformations leading to unique solution properties, including spin state switching. The supramolecular chemistry of metal complexes is unprecedented in its scope and potential applications, and it is hoped that the studies presented herein will promote further investigation of dynamic supramolecular devices.

Keywords: SCO; nanoarchitecture; self-assembly; metastable; gel; film; nanofiber; nanorod; nanorectangular; supramolecule

---

1. Introduction

The self-assembly and integration of functional metal complexes has attracted significant attention due to the potential for the development of useful molecular systems [1]. Naturally-occurring examples of such systems include the metal complexes that play important roles in various biological functions, such as the transportation of oxygen, gene activation, and the catalytic reactions of enzymes [2,3]. The field of biomimetic chemistry aims both to understand and to utilize the functional reactions and properties of metal complexes, and includes studies of the intermolecular and intramolecular interactions among naturally-occurring metal complexes with specific functions. However, to date, our knowledge of self-assembly on the molecular level has been largely limited to chemical structures composed of only one type of molecule.

Despite this lack of understanding, there are numerous examples of the supramolecular self-assembly of metal complexes, including in molecular crystals [4], colloids [5], monolayers [6], helices [7], grids [8], polymers [9] and metal–organic frameworks (MOFs) [10–12]. Most molecular self-assemblies result from the spontaneous aggregation of molecules under thermodynamic equilibrium conditions into stable, structurally-defined aggregates connected by relatively strong non-covalent forces such as hydrogen bonding and electrostatic interactions [13]. The more complex self-assembly of both metal complexes and organic compounds, which can result in the supramolecular control of various functions, is controlled by stronger interactions such as covalent bonding and coordination and also by weak interactions, including hydrogen bonding, hydrophobic interactions, and van der Waals interactions [13–15]. Of these, the non-covalent interactions are expected to play an important role in the supramolecular control of coordination systems.

Self-assembly depends on the specific properties of the main components of the system, including electronic states, nanostructures, and bulk physical characteristics. As such, the interactions between components can be adjusted by selecting component-specific properties [16]. As an example, moderately-strong interactions and suitable binding constants result in dynamic systems that exhibit reversibility and the potential for self-growth and self-propagation. These assemblies are often referred to as non-equilibrium structures or metastable structures. It is therefore apparent that the self-assembly of molecules will be affected by the properties of the compounds and that these systems may also be responsive to external stimuli.

Spin crossover (SCO) complexes have been investigated within the fields of coordination chemistry and supramolecular chemistry, as well as in other areas of chemistry and physics. The coordination geometry of metal complexes both determines the structure of the assembly and modifies the electronic configuration of the d-electrons of the metal. Strong interactions between ligands and metal ions will result in significant splitting of the d-orbitals, leading to low spin (LS) states. Conversely, weak interactions will give rise to high spin (HS) compounds. If the interaction is of intermediate strength and is responsive to external stimuli, it can be possible to switch from one spin state to the other. Of particular interest are low-dimensional metal complexes [17–48], in which metals are bridged by linear bidentate ligands, and supramolecular three-dimensional crystals [49–54] with cooperative structural transformations, which are sometimes capable of hosting various molecules. Both represent high priorities with regard to the investigation of supramolecular coordination compounds. In addition, low-dimensional coordination polymers and coordination systems have been examined as a means of generating multi-functional materials. In particular, supramolecular assembly–disassembly is a necessary condition for self-integration, self-propagation, and adaptive behavior. Finally, it is also possible to tune the interaction strength over a wide range, from weak to strong. In addition, the change in the spin state can affect both the magnetic and optical properties of the complex.

Since amphiphilic technique was attempted [17], the preparation of supramolecular formations of amphiphiles incorporating SCO complexes using the Langmuir–Blodgett (LB) technique has been reported [18–20]. This approach can produce interesting materials because of its ability to organize molecules in multilayered architectures. In addition, Kurth et al., Kimizuka et al. and Aida et al. have all studied one-dimensional supramolecular assemblies of SCO complexes and demonstrated organogels and films that exhibit aggregation-induced SCO phenomena, providing evidence that such assemblies possess different characteristics compared to the metal complexes in the bulk state [21–23]. Subsequent to reports of these functional self-assembling systems, many chemists and physicists began to examine low-dimensional self-assemblies and low-dimensional coordination systems involving SCO complexes. However, to date, almost all supramolecular systems based on low-dimensional compounds have relied on strong interactions such as covalent bonding and coordination, and the properties of the resulting supramolecules have been similar to those of the original metal complexes in the solid state (Figure 1a).

Recently, our group developed a flexible supramolecular system composed of metal complexes, using both lipophilic and amphiphilic compounds [24–35] (Figure 1b,c). This system exhibited tunable, metastable properties, such as the formation of heat-set gel-like networks [25,26] and supramolecular SCO via adaptive molecular clefts [27], that are not observed in the solid state. The design of supramolecular systems such as these is predicted to lead to the fabrication of flexible, stimuli-responsive supramolecules with unique and specific functions, and is also expected to improve our understanding of multi-functional biomimetic systems.

In this review, we briefly describe the chemical structures and properties that result in flexible supramolecular systems, focusing on the use of lipophilic amphiphiles [27], lipid amphiphiles (Figure 1b) [29], and diblock copolypeptide amphiphiles (Figure 1c) [36,37]. The important roles that flexibility and weak interactions play in the supramolecular control of nanostructure morphologies and in the generation of dynamic, metastable functions involving SCO phenomena are also discussed, based on the most recent findings of our research group.

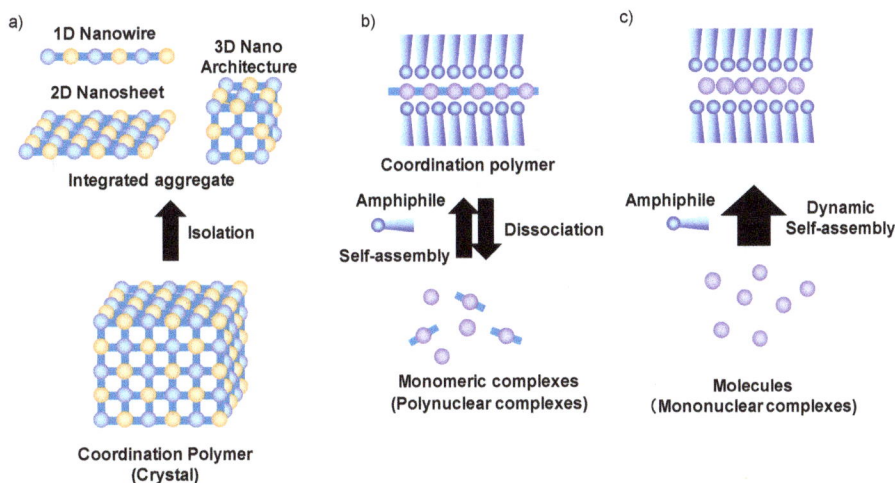

Figure 1. Illustrations showing (**a**) the isolation of various structures from a polymer crystal; (**b**) the self-assembly of a coordination polymer; and (**c**) the integration of amphiphilic molecules, all of which represent means of constructing nanoassemblies of SCO complexes.

2. Self-Assembly of Amphiphilic and Lipophilic Fe^{II} 1,2,4-Triazole Complexes

It is known that 1,2,4-triazoles are able to act as bridging ligands, and linear metal complexes of these compounds have been actively investigated due to their polymeric structures and their magnetic interactions with linearly-aligned metal ions [38–44]. As an example, a series of oligonuclear compounds based on Mn, Fe, Co, Ni and Zn has been prepared, all of which have been found to undergo antiferromagnetic interactions [42–44]. The most interesting feature of these compounds is their characteristic SCO switching between LS (purple, $S = 0$) and HS (colorless, $S = 2$) states [38–44]. In addition, as a result of their magnetic properties, these complexes are obvious candidates for use as information storage materials. In particular, Fe^{II} 1,2,4-triazole complexes have the advantages of ease of synthesis and the ready formation of bridged structures. Despite this, because these compounds are generally only obtained as powders, a remaining challenge is to develop processes to obtain ultrathin films and other delicate structures required for device applications. To date, preliminary studies of triazole complexes have been limited to the study of bulk powder samples, and there has been no general methodology developed for the conversion of these materials to nanostructures. One aspect of the development of such complexes that is theoretically possible is the tuning of their magnetic properties, based on supramolecular control of the spatial arrangements of the metal ions and triazole ligands, resulting in magnetic cooperation between metal complexes.

2.1. 4-Alkylated 1,2,4-Triazole Complexes

Working towards the development of supramolecular triazole complexes, we initially introduced a solvophilic dodecyloxypropyl chain within the ligand [22]. These chains made it possible to dissolve metal complexes containing the lipophilic triazole ligand **1** in organic media. To date, several lipophilic triazole complexes have been reported [17,20,22,23,43–48]. However, it is noteworthy that LS complexes incorporating these ligands are unstable in the solid state or in organogels, since van der Waals interactions between the rigid alkyl chains increase the Fe–to–Fe distance and consequently promote conversion to the HS state. When designing advanced supramolecular systems, it is therefore necessary to consider the chemical composition of the triazole ligands and to carefully select the appropriate media along with suitable external conditions and stimuli, such as temperature, redox reactions, and molecular recognition interactions.

Our own work has focused on the development of flexible, lipophilic, transition metal-triazole complexes in the form of organic solvent solutions [25–28], thin films [27], and liquid crystals [31]. As an example, the flexible ether linkage in triazole ligand **1** enhances the solubility of the metal complexes in organic media and tailors the packing of the alkyl chains (Figure 2). Flexible metal complexes such as these undergo a variety of interactions with organic media, organic molecules, and liquid crystals, and the present review examines the coordination structures, magnetic properties, and morphological dynamism of these complexes.

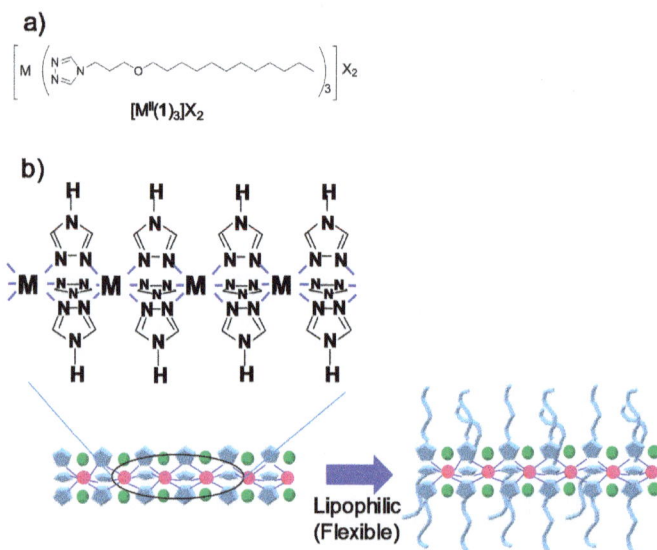

Figure 2. (**a**) Chemical structures of $[Fe^{II}(1)_3]Cl_2$ and (**b**) illustrations of $[Fe^{II}(1)_3]Cl_2$ showing the flexible alkyl chains.

The interactions between alcohols and the surfaces of linear triazole complexes have been employed as a mean of controlling the SCO of the $[Fe^{II}(1)_3]Cl_2$ complex [27]. This material is a purple powder when in its solid form at ambient temperature, which is typical of such complexes in their LS state. However, the complex transitions to a pale yellow organogel (the HS state) when dissolved in chloroform. Such gels result from the formation of nanofiber aggregates, as has been confirmed by transmission electron microscopy (TEM). The casting of a chloroform solution of this type of complex onto solid substrates results in transparent purple films, in which the complex is once again in the LS state. These cast films exhibit sluggish SCO (LS⇌HS) in response to temperature changes, without thermal hysteresis (Figure 3a). In contrast, the co-casting of equimolar quantities of dodecanol or tetradecanol with $[Fe^{II}(1)_3]Cl_2$ forms composite films in which alcohol molecules are bound to the complex by ionic hydrogen bonding between the hydroxyl groups of the alcohols and the chloride ions, as well as by van der Waals interactions. At room temperature, these cast films have regular lamellar structures either with or without doping with alcohol, as determined by wide angle X-ray diffraction (WAXD) measurements. Interestingly, binary films made from $[Fe^{II}(1)_3]Cl_2$ and long chain alcohols (containing 12 or 14 carbons) exhibit reversible and abrupt SCO on heating with thermal hysteresis (Figure 3a). The evident bistability of these films is closely related to dynamic structural transformations between lamellar and hexagonal structures, suggesting a novel supramolecular strategy for controlling the bistability of SCO phenomena (Figure 3b) [27].

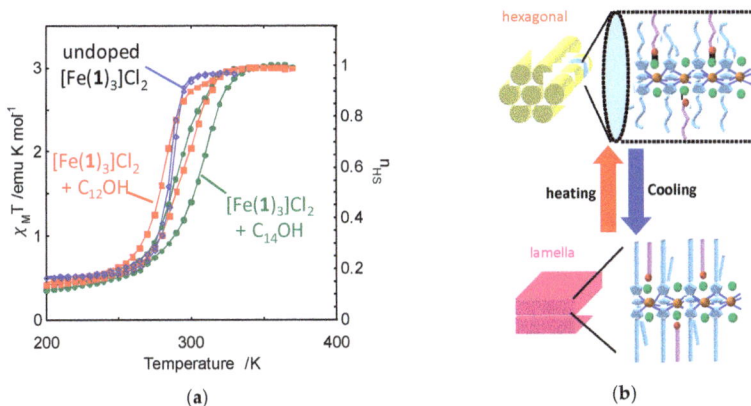

Figure 3. (a) The temperature dependence of the magnetic susceptibility of $[Fe^{II}(1)_3]Cl_2/C_nOH$ samples (n = 12 and 14) and (b) an illustration of the supramolecular unit structures in the cast films. The samples undergo a dynamic structural transformation between lamellar (at 298 K) and hexagonal structures (at 373 K). Adapted with permission from *J. Polym. Sci. A: Polym. Chem.* **2006**, *44*, 5192–5202. Copyright © 2006 Wiley Periodicals, Inc.

In order to derive functional systems through imparting spin-based functionality to guest molecules, we have demonstrated the formation of liquid crystal gels from mixtures of linear Fe^{II}-1,2,4-triazole complexes and nematic liquid crystals [31]. JC–1041XX and JD–1002XX were employed as the liquid crystals, since both display nematic liquid crystal phases over a wide temperature range (T/K of phase transitions: K 291.1 N 365.2 I (JC–1041XX); K 276.8 N 347.9 I (JD–1002XX), Figure 4a). The purple color of the resulting gels indicates that the $[Fe^{II}(1)_3]Cl_2$ complex adopts the LS configuration in either JC–1041XX or JD–1002XX, in contrast to the HS gels formed in chloroform [27]. At elevated temperatures, the macroscopically homogeneous gel structure is preserved, although the color changes from purple to pale yellow. This color change is thermally reversible, as shown by the temperature dependence of reflectance spectra. In addition, the temperature dependence of the magnetic susceptibility demonstrates that the liquid crystal gel composed of $[Fe^{II}(1)_3]Cl_2$ and JC–1041XX exhibits SCO at elevated temperatures, with the appearance of thermal hysteresis. The SCO temperature during the heating cycle (for LS → HS, $T_{sc}\uparrow$, the temperature at which half of the transitioning Fe^{II} changes spin) is approximately 334 K, which is higher than that observed in the cooling cycle (HS → LS, $T_{sc}\downarrow$, 324 K, Figure 4a). Similarly, a combination of $[Fe^{II}(1)_3]Cl_2$ and JD–1002XX exhibits a higher SCO temperature during heating as opposed to cooling (324 K compared to 319 K). These binary $[Fe^{II}(1)_3]Cl_2$/liquid crystal composites therefore undergo sluggish SCO transitions with thermal hysteresis within a higher range of temperatures as compared to pure $[Fe^{II}(1)_3]Cl_2$, for which both $T_{sc}\uparrow$ and $T_{sc}\downarrow$ are 300 K [27]. To date, the behavior of one-dimensional Fe^{II} complexes of 4-substituted 1,2,4-triazoles and their SCO characteristics have been studied with these compounds solely in the solid state or as organogels [45–48]. In contrast, the systems formed from combinations of lipophilic Fe^{II} complexes and hydrophobic liquid crystals have several advantages due to the considerable effects of the liquid crystal environment on either the solvophobic compaction of the N–Fe coordination bonds or the destabilization of the HS state due to the reduction of the Fe Fe distance (Figure 4b) [45]. Composites consisting of liquid crystal and organic (and/or polymeric) molecules have also been reported to have advantageous properties because of the unique interaction that results from the bicontinuous phase separation structures in these mixtures. These bicontinuous structures are formed when the organic components are suitably miscible with the liquid crystals. Molecular assemblies of functional low molecular weight gelators have also been found to form liquid crystal gels. It is anticipated that the incorporation of functional components such as lipophilic triazole complexes in these liquid crystal hybrids will allow the development of intelligent soft materials by imparting spin-based functionalities and solvophobic interactions.

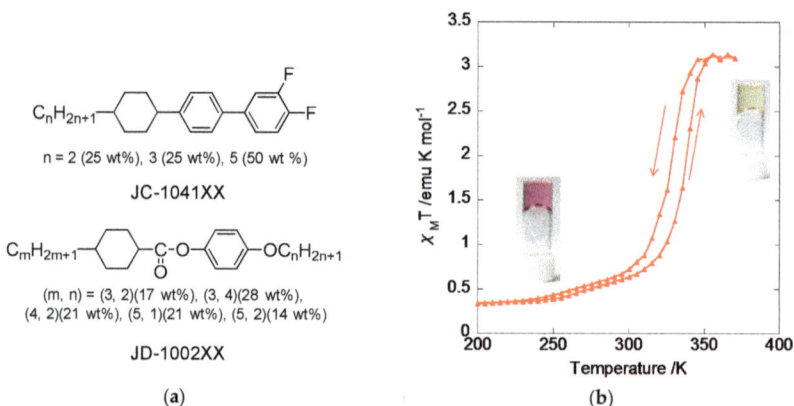

(a) (b)

Figure 4. (a) The chemical structures of the liquid crystals JC–1041XX and JD–1002XX and (b) the temperature dependence of the magnetic susceptibility of a [Fe^{II}(**1**)$_3$]Cl$_2$/JC–1041XX liquid crystal gel. The inset photographs are of the gel at 293 K (left) and 363 K (right). Adapted with permission from *Chem. Commun.* **2010**, *46*, 1229–1231. Copyright © 2010, Royal Society of Chemistry.

2.2. 4R-1,2,4-Triazole Complexes with Lipid Amphiphiles

The lipid-Fe^{II} triazole complexes developed in the present study are shown in Figure 5 [29]. In contrast to the conventional design of 4-alkylated 1,2,4-triazoles, an L-glutamate-derived lipid was introduced as a lipophilic counter anion, and 4-amino-1,2,4-triazole (NH$_2$trz) and 4-(2-hydroxyethyl)-1,2,4-triazole (HOC$_2$trz) were employed as triazole ligands. This noncovalent introduction of a lipophilic moiety is suitable in the case of 1,2,4-triazole-based SCO complexes since the covalent modification of triazole ligands with bulky substituents generally lengthens the Fe–ligand bonds, resulting in the destabilization of LS states [45].

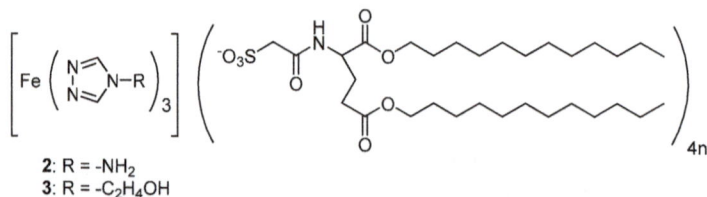

2: R = -NH$_2$
3: R = -C$_2$H$_4$OH

Figure 5. Structures of the lipid-Fe^{II} triazole complexes consisting of an L-glutamate-derived lipid and 4-amino-1,2,4-triazole (NH$_2$trz) (**2**) or 4-(2-hydroxyethyl)-1,2,4-triazole (HOC2trz) (**3**).

The complexes of **2** and **3** were dispersed in toluene with heating at concentrations of 5 or 20 mM. Unexpectedly, toluene dispersions of **2** were purple at room temperature, indicating that the LS state of the complex was maintained in solution (see the photographic image acquired at 298 K in Figure 6a) [29]. Figure 6a also presents the temperature dependence of the absorption spectra obtained from complex **2** (5 unit mM in toluene). The absorption bands around 400 and 800 nm are assigned to d–d transitions of the LS complex ($^1A_1 \rightarrow {}^1T_1$) and the HS complex ($^5T_2 \rightarrow {}^5E$), respectively [45]. The absorbance at 540 nm (associated with the LS state) is decreased upon heating and is accompanied by an increase in the absorption intensity at 800 nm (HS state). These spectral changes were found to be completely reversible upon temperature cycling (Figure 6b). Lipid complex **3** also exhibited reversible thermal spectral changes in a lower temperature range (Figure 6b). The T_{sc} values were ca. 300 K for **2** and 278 K for **3**. The thermally induced changes in spin state observed in the case of the toluene dispersions were compared with those in the solid state, and these same complexes in powder form had T_{sc} values of 280 K for **2** and 170 K for **3** (see the SQUID data in Figure 6b). The T_{sc} value of **2** in toluene was therefore increased by ca. 20 K compared to that of the solid state, indicating stabilization of the LS state in solution. In contrast, **3** in toluene showed a much higher T_{sc} in solution (278 K) compared to that of the solid sample (170 K; Figure 6b). These observations are directly opposite to the typical finding that the LS configuration is destabilized in solution [22,27,28]. In addition, the occurrence of more abrupt spin state changes in solution is extraordinary because, in solution, changes in spin states are governed by spin equilibrium and so spin state transitions tend to occur gradually. This remarkable stabilization of LS states in solution suggests an increase in the ligand-field splitting energies compared to those in the crystalline states. In toluene, these ionic Fe^{II} triazole complexes are dispersed with the help of solvophilic lipid alkyl chains. However, the solvophobic triazole complexes will tend to minimize contact with the nonpolar solvent molecules by contracting Fe–N bonds and/or nearest neighbor Fe–Fe distances (that is, by solvophobic compaction). Evidence for this effect is provided by the dependence of the SCO temperature on solvent polarity. In contrast, these solvophobic interactions do not take place in the solid state, and the packing of the lipid sulfonate groups in crystalline structures may not be dense enough to stabilize the LS complexes [29].

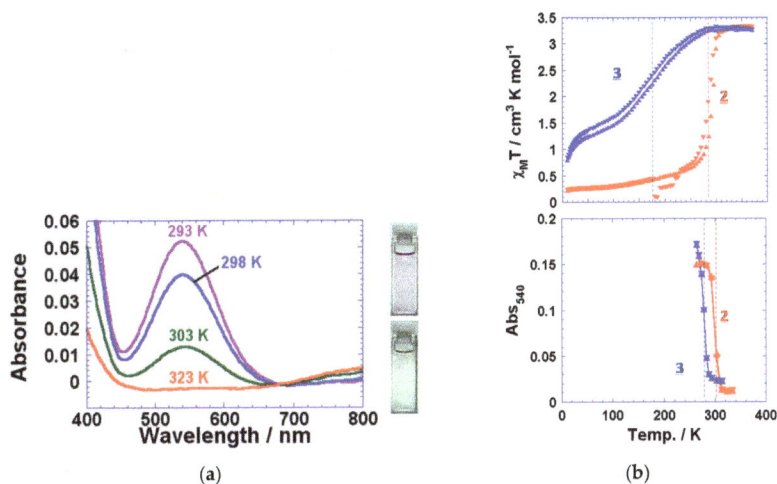

Figure 6. (a) Temperature dependence of the UV/vis absorption spectra of complex **2** (5 unit mM in toluene, on cooling) and photographs of **2** in toluene dispersions (5 unit mM) at 273 K (upper) and 298 K (lower); and (b) temperature dependences of peak absorption intensities at 540 nm (LS complex) for **2** and **3**. Absorbance changes on heating and cooling cycles are plotted on the same curves. Adapted with permission from *J. Am. Chem. Soc.* **2008**, *130*, 5622–5623. Copyright 2008 American Chemical Society.

Atomic force microscopy (AFM) images of complexes **2** and **3** spread from dilute toluene dispersions at various temperatures are provided in Figure 7 [29]. Fibrous nanostructures with widths of 20–30 nm and heights of ca. 7 nm are abundant in the case of **2** in the LS state (Figure 7a). As the bimolecular length of the lipid is ca. 4.4 nm (based on a CPK model), these nanofibers evidently consist of bundled supramolecular complexes. In contrast, aggregates of fragmented structures (Figure 7b) are observed for the HS complex. The presence of nanofibers after cooling of the heat dispersed samples (Figure 7a) and the recovery of LS complexes (Figure 6) clearly indicate the reversible self-assembly of HS fragments into the original LS nanofibers. This effect is further supported by the copolymerization of **2** that occurs after cooling the heat-dissociated mixtures. Because SCO does not change the coordination number of the central atom in the complex, changes in spin state involving the disintegration of coordination bonds would be better described as spin conversion induced by self-assembly. These data demonstrate that lipid-packaged nanowires of Fe^{II} triazole complexes display spin conversion characteristics in organic media. Abrupt spin conversion is observed due to the lipid-assisted solvophobic stabilization of LS complexes, SCO, and the reversible dissociation of coordination main chains. This solvophobic enhancement of ligand–metal interactions and the potential for synergistic control based on self-assembly are expected to expand the possibilities of coordination polymer chemistry.

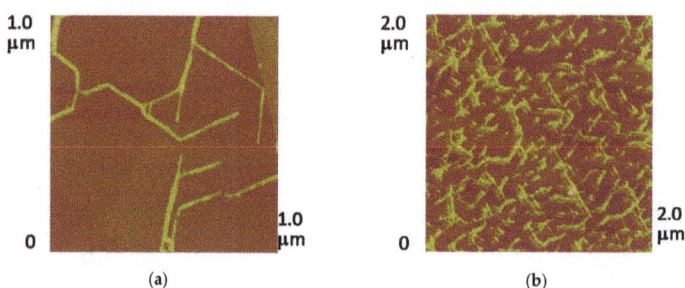

(a) (b)

Figure 7. AFM images of complex **2** cast from toluene dispersions (50 unit μM) on highly oriented pyrolytic graphite (HOPG) (**a**) at ambient temperature (LS state) and (**b**) at 313 K (HS state). Reprinted Adapted with permission from *J. Am. Chem. Soc.* **2008**, *130*, 5622–5623. Copyright 2008 American Chemical Society.

3. Self-Assembly of Metal Complexes with Diblock Copolypeptide Amphiphiles

The SCO of many mononuclear metal complexes also involves a spin change. As an example, it is well-known that the discrete cobalt(II) compounds [Co(terpy)$_2$]X$_2$·nH$_2$O (terpy = 2,2′:6′,2″-terpyridine, X = halide, pseudohalide, BF$_4^-$, NO$_3^-$ or ClO$_4^-$ and n = 0 to 5) have demonstrated gradual SCO behavior. Recently, the high molecular weight alkylated cobalt(II) compounds [Co(R-terpy)$_2$](BF$_4$)$_2$ (R-terpy = 4′-alkoxy-2,2′:6′,2″-terpyridine) have been reported to display a "reverse spin transition" between HS and LS states with a thermal hysteresis loop and triggered by a structural phase transition [49–53]. It has been suggested that the flexibility of the alkyl chains in these complexes plays an important role in determining their unique magnetic properties, intermolecular interactions, and crystal properties. Ideally, the characteristics of such systems would be tunable by controlling the spatial arrangement of the SCO metal complexes, resulting in intermolecular interactions among the metal complexes using supramolecules without covalent or coordinative linkages.

3.1. Co-Terpyridine Complexes with Diblock Copolypeptide Amphiphiles

Our research group has also focused on combining cobalt(II) terpyridine complexes with diblock copolypeptide amphiphiles, and examining the SCO characteristics of the resulting complexes in water [36]. The diblock copolypeptide amphiphiles **4** and **5** and the polypeptide **6** (Figure 8) were synthesized. Compounds **4** and **5** were produced with a specific degree of polymerization intended to enhance the solubility of these compounds in water and also improve the packing of the polypeptides in supramolecular assemblies, resulting in the formation of hydrogels [55–58]. In contrast, polymer **6** was synthesized to a higher degree of polymerization since it was not intended to produce a supramolecular effect. Composites of polypeptides **4–6** and a cobalt(II) terpyridine complex ([Co^{II}(MeO-tery)$_2$]$^{2+}$) were obtained by mixing solutions of the respective compounds, followed by precipitation of the resulting composites, in which the counter anion of the [Co^{II}(MeO-tery)$_2$]$^{2+}$ was replaced by polyglutamates. Elemental analysis indicated the presence of water molecules in the final products and suggested the following formulae: **4**/[Co^{II}(MeO-terpy)$_2$]·H$_2$O, **5**/[Co^{II}(MeO-terpy)$_2$]·3H$_2$O, and **6**/[Co^{II}(MeO-terpy)$_2$]·4H$_2$O.

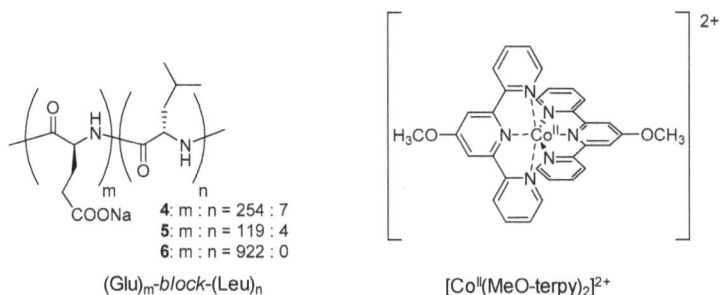

4: m : n = 254 : 7
5: m : n = 119 : 4
6: m : n = 922 : 0

(Glu)$_m$-*block*-(Leu)$_n$

[Co^{II}(MeO-terpy)$_2$]$^{2+}$

Figure 8. The structures of diblock copolypeptide amphiphiles **4** and **5**, polypeptide **6**, and the cobalt(II) terpyridine complex [Co^{II}(MeO-terpy)$_2$]$^{2+}$.

The morphologies of the supramolecular structures generated in dispersions of these complexes were subsequently assessed by TEM. The TEM image of **4**/[Co^{II}(MeO-terpy)$_2$] following transfer to a carbon-coated Cu grid exhibited rectangular structures with widths of 700 nm to 6 mm, while **5**/[Co^{II}(MeO-terpy)$_2$] showed nanostructures with widths ranging from 500 nm to 2 mm.

High-resolution scanning transmission electron microscopy coupled with energy dispersive X-ray spectroscopy (HR-STEM-EDX) also confirmed that the composites consisted of cobalt(II) complexes and polymers. Figure 9 presents STEM-EDX mapping results (Figure 9b, Co; Figure 9c, O) for a nanocomposite of **4** within the boxed area indicated in Figure 9a. These data confirm the formation of nanostructures in which the cobalt complex and the amphiphile are evenly matched. Composite **6**/[Co^{II}(MeO-terpy)$_2$] did not show a specific structure in TEM observations, indicating that combinations of **6** and metal complexes did not form supramolecular structures. Samples of the composites in water were freeze-dried and wide-angle X-ray scattering (WAXS) analysis of the powdered composites was performed, during which no crystalline peaks were observed. This lack of peaks indicated the presence of an amorphous phase in the rectangular supramolecular structures. In addition, small angle X-ray scattering (SAXS) analysis showed the formation of nanostructures of various sizes, which was consistent with the nanostructures having widths of several hundred nm and several μm observed via TEM and dynamic light scattering (DLS).

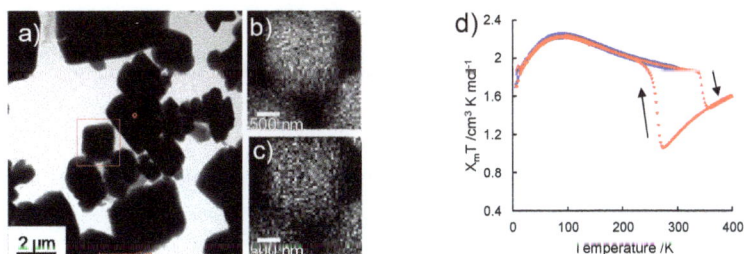

Figure 9. (a) STEM image and (b) Co and (c) O STEM-EDX maps of **4**/[Co^{II}(MeO-terpy)$_2$] within the red square shown in (a); (d) χ_mT versus T for **4**/[Co^{II}(MeO-terpy)$_2$] on warming (▲) and cooling (▼). The blue and red plots indicate the first and second cycles, from 5 K to 300 K and 400 K, respectively. Adapted with permission from *J. Mater. Chem. C* **2015**, *3*, 7779–7783. Copyright © 2015, Royal Society of Chemistry.

The variations in the magnetic susceptibilities of **4**/[Co^{II}(MeO-terpy)$_2$], **5**/[Co^{II}(MeO-terpy)$_2$] and **6**/[Co^{II}(MeO-terpy)$_2$] with temperature were also examined. Composite **4**/[Co^{II}(MeO-terpy)$_2$] was found to remain in the HS state at all temperatures and exhibited a $\chi_m T$ value within the range of 1.69–2.25 cm$^3 \cdot$K$\cdot$mol^{-1} over the temperature range of 5–300 K (see the blue plot in Figure 9d). During further heating up to 400 K, the $\chi_m T$ value was found to decrease at 337 K, which is consistent with the loss of water molecules (the red plot in Figure 9d). However, after annealing, the pre-heated compound displayed markedly different behavior. On cooling, the $\chi_m T$ value gradually decreased from 1.60 cm$^3 \cdot$K$\cdot$mol^{-1} at 400 K to 1.05 cm$^3 \cdot$K$\cdot$mol^{-1} at 274 K, representing normal thermal SCO behavior (the red plot in Figure 9b). Upon further cooling, the $\chi_m T$ value increased abruptly below the $T_{1/2}\downarrow$ value of 260 K, to 1.96 at 222 K. Below this temperature, $\chi_m T$ varied between 1.70 and 2.20 cm^3 K$\cdot$mol^{-1} in the temperature range between 5 and 220 K. With subsequent heating, the $\chi_m T$ value abruptly dropped ($T_{1/2}\uparrow$ = 345 K), indicating a transition from HS to LS. Finally, the $\chi_m T$ value gradually increased between 361 and 400 K. The wide thermal hysteresis loop (ΔT = 85 K) near room temperature was maintained through successive thermal cycles. Thus, reverse spin transition was exhibited by the composite in its solvated state based on intermolecular interactions among the metal complexes [49–53]. Composites **5**/[Co^{II}(MeO-terpy)$_2$] and **6**/[Co^{II}(MeOterpy)$_2$] also showed abnormal reverse spin transitions during heating and cooling cycles (5–300–5–400–5 K). Reversibility between the HS and LS states in the reverse spin transition, however, was dependent on the polymer employed. The diblock copolypeptide amphiphile **4** evidently possessed a suitable degree of polymerization and a balance between hydrophilic and hydrophobic portions, leading to perfect reversibility between HS and LS states, allowing reverse spin transition. The [Co^{II}(MeO-terpy)$_2$]$^{2+}$ complex with BF$_4^-$ anions is typically observed to undergo a gradual SCO transition centered around $T_{1/2}$ = 100–300 K [54]. During this process, water molecules in the solvent have been found to play an important role in the SCO behavior, due to either two-step SCO (involving an H$_2$O-solvated complex) or one-step SCO (via a non-solvated complex). In addition, reverse spin transition can be achieved as the result of an organic-solvated metal complex in solvents such as acetone, in association with a structural phase transition [54]. These prior results suggest that the SCO of aqueous dispersions of solid composites of polymers and a cobalt(II) terpyridine complex should be accompanied by a transition between solvated and non-solvated phases, since the metal complex is dispersed in the amphiphilic polypeptide-induced nanostructure. Therefore, reversible reverse spin transition is thought to result from the balance between the amorphous diblock copolypeptide amphiphile and the loose packing of the cobalt(II) terpyridine complexes (Figure 10) [36].

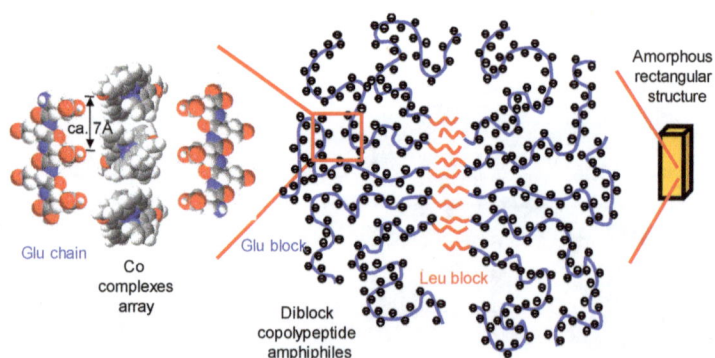

Figure 10. An illustration of the self-assembly of diblock copolypeptide amphiphile/[Co^{II}(MeO-terpy)$_2$] complexes, demonstrating the formation of a rectangular structure from a diblock copolypeptide amphiphile with an amorphous structure.

This work demonstrated that composites consisting of a cobalt(II) terpyridine complex with diblock copolypeptide amphiphiles generate supramolecular structures in water. The formation of these nanostructures results in the evolution of morphologies ranging in size from sub-nanometer to several micrometers. Moreover, these supramolecular composites display reverse spin transition depending on the polypeptide structure. Changes in both the morphology and magnetic properties of these materials can be induced by variations in the temperature and solvated phase.

3.2. Fe-Complexes with Diblock Copolypeptide Amphiphiles

Our group has also demonstrated a new type of nanocomposite based on the iron(II) complex [Fe^{II}(ppi)$_2$(NCS)$_2$] (2-pyridyl-*N*-(phenyl)methylamine (ppi); Figure 11) [59–61] with a low critical solution temperature (LCST), using diblock copolypeptide amphiphiles and SCO complexes. This work demonstrated the possibility of synthesizing multifunctional nanocomposites with metastable properties as well as the formation of cooperative nanocomposites that change spin state switching in solution [37].

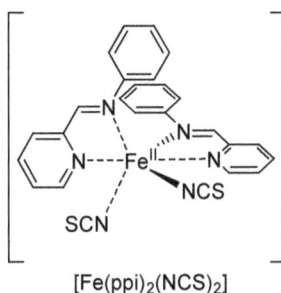

$$[Fe(ppi)_2(NCS)_2]$$

Figure 11. The structure of the iron(II) complex [Fe^{II}(ppi)$_2$(NCS)$_2$].

Composites of polypeptides **4–6** with [Fe^{II}(ppi)$_2$(NCS)$_2$] were synthesized by mixing solutions of the respective compounds, followed by dispersion of hydrophobic [Fe^{II}(ppi)$_2$(NCS)$_2$], and by dissolution of the resulting composites in water. The molar ratio between the metal complex and block copolypeptide amphiphile in each mixture was 250:4, meaning that the molar ratio between the metal complex and the leucine units was ca. 1:2. Although polypeptide **6** was composed of sodium polyglutamate without polyleucine, the hydrophobic metal complex was found to be soluble in an aqueous solution of **6**. This result indicates that the metal complex was able to combine with the charge-neutralized sodium polyglutamate in water at this molar ratio during the preparation stage. Quantities of **4**/[Fe^{II}(ppi)$_2$(NCS)$_2$], **5**/[Fe^{II}(ppi)$_2$(NCS)$_2$] and **6**/[Fe^{II}(ppi)$_2$(NCS)$_2$] were all obtained by lyophilization. Dissolving **4**/[Fe^{II}(ppi)$_2$(NCS)$_2$] in room temperature water at a concentration of 2 mM (on the basis of Fe^{II} ions) at 278 K gave a dark-purple solution (Figure 12a). This result indicated both inclusion of the metal complex and the formation of a homogeneous dispersion of the complex. The **5**/[Fe^{II}(ppi)$_2$(NCS)$_2$] and **6**/[Fe^{II}(ppi)$_2$(NCS)$_2$] composites also dispersed in water at 278 K to give dark-purple solutions, suggesting that these composites were dispersed on the molecular level. Figure 12b shows the UV-vis absorption spectra of **4**/[Fe^{II}(ppi)$_2$(NCS)$_2$] in water at various temperatures. Specific absorptions are evident at 278 K (a peak at 578 nm with a shoulder around 530 nm) that are characteristic of the LS Fe^{II} complex and are attributed to metal-to-ligand charge transfer (MLCT) bands [59]. Without the polypeptide, the hydrophobic, lipophilic [Fe^{II}(ppi)$_2$(NCS)$_2$] cannot be dissolved in water, and therefore transparent, colored aqueous solutions can provide evidence for nanostructures in water. The development of nanostructures requires the supramolecular aggregation resulting from additional intermolecular interactions.

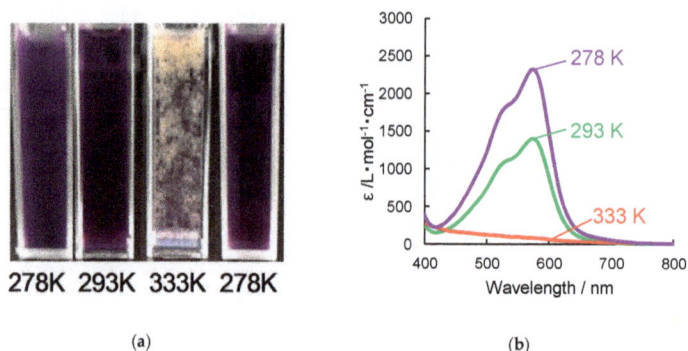

Figure 12. (a) Photographic images of aqueous solutions of **4**/[Fe^{II}(ppi)$_2$(NCS)$_2$] during heating and cooling. These show a dark purple phase at 278 K to 293 K, a yellow phase at 333 K during a heating cycle, and a dark purple phase at 278 K during a cooling cycle; (b) The temperature dependence of the absorption intensity at 578 nm of **4**/[Fe^{II}(ppi)$_2$(NCS)$_2$]. Adapted with permission from *Polymer*, DOI:10.1016/j.polymer.2016.12.079. © 2017 Elsevier Ltd. All rights reserved.

When the dark-purple solution of **4**/[Fe^{II}(ppi)$_2$(NCS)$_2$] is heated above 293 K, the absorption intensity of the LS complex is reduced and a precipitate appears. A yellow precipitate is obtained at 333 K, the color of which indicates the formation of a HS complex via an LCST-type transition (Figure 12a). The precipitation of the HS complexes by heating (Figure 12a) evidently occurs as a consequence of increased junctions (that is, contacts and interactions) between the bundled polypeptide and iron(II) complexes, possibly caused by a hydration-dehydration process. The [Fe^{II}(ppi)$_2$(NCS)$_2$] complex and its derivatives are typically observed to undergo gradual SCO with a transition centered around T_{sc} = ca. 150 K [60,61]. During this process, interactions between iron complexes in the crystal have been found to play an important role in the SCO behavior, due to intermolecular repulsion between ligand molecules [62–64]. In addition, SCO in conjunction with a structural transition can be achieved through the application of physical pressure to crystals [65].

Figure 13 shows TEM images of **4**/[Fe^{II}(ppi)$_2$(NCS)$_2$] transferred from aqueous solution to a carbon-coated Cu grid. At 278 K, various poorly defined structures are seen, including sphere-like structures (Figure 13a). In contrast, at 333 K, well-developed sheet-like morphologies appear (Figure 13b), resulting from the aggregation of the nanocomposite due to passing through the LCST. The thermoresponsive nanocomposite **4**/[Fe^{II}(ppi)$_2$(NCS)$_2$] thus undergoes assembly due to the extension of the polypeptide chain and the hydrophobic interactions of the iron complexes. Below the LCST of the nanocomposite (ca. 313 K), the interactions between polypeptide chains are weak. It is also noteworthy that, above the LCST, two-dimensional sheets with widths of several μm are self-assembled as the temperature increased, in which sheet structures with widths of several micrometers are evident (Figure 13b). The **4**/[Fe^{II}(ppi)$_2$(NCS)$_2$] sheet structures observed in TEM images are several times the length of the diblock copolypeptide amphiphiles. These structures are therefore composed of multiple strands of linear and/or stacked polypeptides combined with the metal complex, leading to SCO in conjunction with an LCST-type transition.

Supramolecular structures on the sub-nanometer to micrometer scale were observed using TEM, DLS, and SAXS, indicating that the hydrophobic interactions of the Leu moieties (Figure 14a) determine the intermolecular interactions of the composites. Nanocomposites formed between metal complexes and polypeptides in water have typically been found to be sensitive to the hydrophilic/hydrophobic balance of the composite. Thus, one possible arrangement among the metal complexes that can be proposed is moderate packing of 1D or 2D sheets (Figure 14b). The close-packed structure of [Fe^{II}(ppi)$_2$(NCS)$_2$] complexes that exhibit SCO phenomena with an LCST-type transition obtained from crystallographic data is a 2D array of metal complexes, which results in a loose packing arrangement. In this structure, the iron(II) complexes are aligned with an average separation between nearest molecules of 6–7 Å, which is consistent with the distance between the neighboring Leu units (ca. 0.7 nm, as estimated by the CPK model; Figure 14b) when polyleucine partially forms a β sheet with an all-trans conformation. Therefore, it is believed that SCO with an LCST-type transition can result from various nanostructures and aggregated structures formed from the amorphous diblock copolypeptide amphiphiles and the loose packing of the $[Fe^{II}(ppi)_2(NCS)_2]$ complexes (Figure 14c) [37].

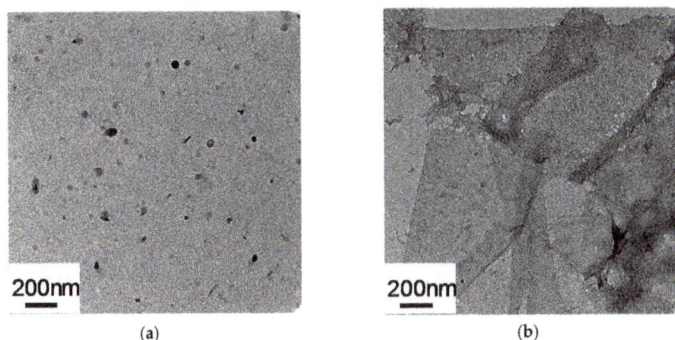

(a) (b)

Figure 13. HR-TEM images of $4/[Fe^{II}(ppi)_2(NCS)_2]$ samples prepared from water dispersions at (a) 278 K and (b) 333 K. Adapted with permission from *Polymer*, DOI:10.1016/j.polymer.2016.12.079. © 2017 Elsevier Ltd. All rights reserved.

The results presented thus far demonstrate that composites consisting of $[Fe^{II}(ppi)_2(NCS)_2]$ complexed with diblock copolypeptide amphiphiles generate supramolecular structures in water. The formation of these nanostructures results in the evolution of morphologies ranging in size from sub-nanometer to several micrometers. Moreover, these supramolecular composites display an unusual SCO phenomenon with an LCST-type transition, depending on the polypeptide structures. Changes in the morphological, spectral, and liquid properties of these materials can be induced by variations in the intermolecular interactions through incorporating hydrophobic parts.

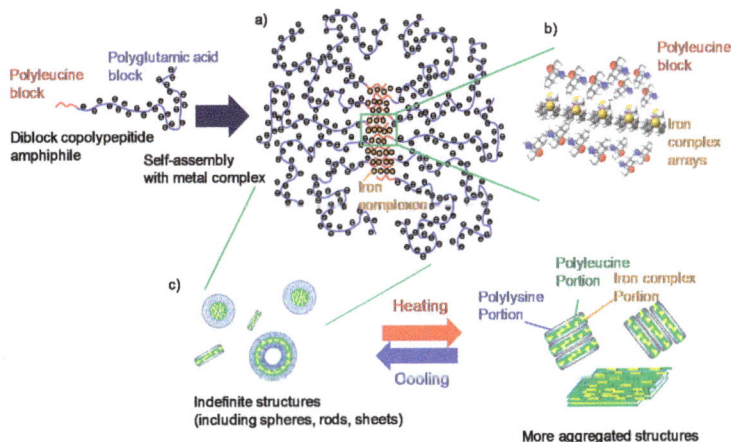

Figure 14. Hierarchical illustration of the self-assembly of the diblock copolypeptide amphiphile complex $4/[Fe^{II}(ppi)_2(NCS)_2]$, showing the manner in which diblock copolypeptide amphiphiles with iron(II) complexes (**a,b**) form nanostructures with LCST-type transitions (**c**).

4. Conclusions

The integration of amphiphile-directed self-assembly and the chemistry of SCO complexes has led to the development of dynamic supramolecular assemblies. Imparting an amphiphilic nature to supramolecules has been found to drive their hierarchical self-assembly. Weak forces such as hydrophobic, hydrogen bonding, van der Waals, and intermolecular interactions evidently play a pivotal role in the determination of the supramolecular architecture, in contrast to reports published during earlier supramolecular chemistry studies. This combinatorial supramolecular approach is also an effective means of developing functional nanomaterials incorporating magnetic metal complexes. The combination of supramolecular nanoarchitectures and metastable engineering should allow new approaches to the design of nanomaterials.

In addition to these nanocomposite techniques, synthetic biomolecules also provide powerful scaffolds for the construction of nanoarchitectures. Therefore, well-designed nanoarchitectures may be fabricated from naturally-occurring lipids, peptides, sugars, and similar compounds. In particular, these materials hierarchically self-assemble into mesoscopic architectures with unique morphologies. This approach is simple and much easier than methods previously devised for supermolecule construction. Supramolecular assemblies having such functional components are anticipated to find increased applications in future.

Finally, it appears that the generation of metastable electronic structures based on self-assembly will be an important issue in nanochemistry. Studies of amphiphilic metal complexes have shown that discrete inorganic complexes can form self-assembling nanoarchitectures in solution, the electronic structures of which are tunable with the aid of amphiphilic molecular assemblies. The dynamic formation of various nanocomposite architectures and networks has been an unexpected result of some of this work. These unique self-assembly properties are not accessible from isolated coordination polymers or thermodynamically stable supramolecular chemistry. The growth of metastable networks in this manner is reminiscent of many central processes that occur in living systems. Thus, we envisage that self-assembling nanoarchitectures will be applied to the design of biomimetic networks that display growth and self-organization in response to external stimuli. These self-assembling systems would contribute to the development of chemical learning systems. They may also provide an opportunity to design molecular (or supramolecular) machines, a new field of molecularly organized chemistry that integrates element blocks [66] to allow transduction, translation, amplification, chemical or physical outputs, dynamic control, and self-propagation.

Acknowledgments: This work was financially supported in part by a Grant-in-Aid for Young Scientists (A) (No. 24685019) and a Grant-in-Aid for Scientific Research on Innovative Areas (New polymeric materials based on element-blocks, #2401) (Nos. 25102547 and 15H00770). Keita Kuroiwa is also grateful for the financial support of The Canon Foundation (No. K16-0146) and for a research grant from Sojo University (No. RT02000001).

Conflicts of Interest: The authors declare no conflicts of interest.

References

1. Gispert, J.R. *Coordination Chemistry*; Wiley-VCH: Weinheim, Germany, 2008.
2. Lippard, S.J.; Berg, J.M. *Principles of Bioinorganic Chemistry*; University of Science Books: Mill Valley, CA, USA, 1994.
3. Anastassopoulou, J.; Theophanaides, T. The role of metal ions in biological systems and medicine. In *Bioinorganic Chemistry: An Inorganic Perspective of Life*; Nato Science Series C; Kessissoglou, D.P., Ed.; Springer: Berlin, Germany, 1995; pp. 209–218.
4. Anthony, A.; Desiraju, G.R.; Jetti, R.K.R.; Kuduva, S.S.; Madhavi, N.N.L.; Nangia, A.; Thaimattam, R.; Thallad, V.R. Crystal Engineering: Some Further Strategies. *Cryst. Eng.* **1998**, *1*, 1–18. [CrossRef]
5. Evans, D.F.; Wennerström, H. *The Colloidal Domain: Where Physics, Chemistry, Biology, and Technology Meet*; John Wiley & Sons: New York, NY, USA, 1999.
6. Jones, M.N.; Chapman, D. *Micelles, Monolayers and Biomembranes*; John Wiley & Sons: New York, NY, USA, 1995.

7.   Lehn, J.M. Toward self-organization and complex matter. *Science* **2002**, *295*, 2400–2403. [CrossRef] [PubMed]
8.   Hanan, G.S.; Volkmer, D.; Schubert, U.S.; Lehn, J.M.; Baum, G.; Fenske, D. Coordination Arrays: Tetranuclear Cobalt(II) Complexes with [2 × 2]-Grid Structure. *Angew. Chem. Int. Ed.* **1997**, *36*, 1842–1844. [CrossRef]
9.   Lehn, J.M. Supramolecular polymer chemistry—Scope and perspectives. *Polym. Int.* **2002**, *51*, 825–839. [CrossRef]
10. Fujita, M.; Yazaki, J.; Ogura, K. Preparation of a macrocyclic polynuclear complex, [(en)Pd(4,4′-bpy)]$_4$(NO$_3$)$_8$ (en = ethylenediamine, bpy = bipyridine), which recognizes an organic molecule in aqueous media. *J. Am. Chem. Soc.* **1990**, *112*, 5645–5647. [CrossRef]
11. Kondo, M.; Yoshitomi, T.; Seki, K.; Matsuzaka, H.; Kitagawa, S. Three-dimensiona framework with channeling cavities for small molecules: {[M$_2$(4,4′-bpy)$_3$(NO$_3$)$_4$]·xH$_2$O}$_n$ (M = Co, Ni, Zn). *Angew. Chem. Int. Ed.* **1997**, *36*, 1725–1727. [CrossRef]
12. Li, H.; Eddaoudi, M.; Groy, T.L.; Yaghi, O.M. Establishing microporosity in open metal–organic frameworks: Gas sorption isotherms for Zn(BDC) (BDC = 1,4-benzenedi carboxylate). *J. Am. Chem. Soc.* **1998**, *120*, 8571–8572. [CrossRef]
13. Whitesides, G.M.; Mathias, J.P.; Seto, C.T. Molecular self-assembly and nanochemistry: A chemical strategy for the synthesis of nanostructures. *Science* **1991**, *254*, 1312–1319. [CrossRef] [PubMed]
14. Lehn, J.M. *Supramolecular Chemistry: Concept and Perspectives*; Wiley-VCH: Weinheim, Germany, 1995.
15. Awod, J.L.; Davies, J.E.D.; MacNicol, D.M.; Vögtle, F.; Lehn, J.M. *Comprehensive Supramolecular Chemistry*; Pergamon: Oxford, UK, 1996; Volume 9.
16. Whitesides, G.M.; Grzybowski, B. Self-assembly at all scales. *Science* **2002**, *295*, 2418–2421. [CrossRef] [PubMed]
17. Armand, F.; Badoux, C.; Bonville, P.; Ruaudel-Teixier, A.; Kahn, O. Langmuir–Blodgett Films of Spin Transition Iron(II) Metalloorganic Polymers. 1. Iron(II) Complexes of Octadecyl-1,2,4-triazole. *Langmuir* **1995**, *11*, 3467–3472. [CrossRef]
18. Soyer, H.; Dupart, E.; Gómez-García, C.J.; Mingotaud, C.; Delhaès, P. First Magnetic Observation of a Spin Crossover in a Langmuir–Blodgett Film. *Adv. Mater.* **1999**, *11*, 382–384. [CrossRef]
19. Soyer, H.; Mingotaud, C.; Boillot, M.L.; Delhaes, P. Spin Crossover of a Langmuir–Blodgett Film Based on an Amphiphilic Iron(II) Complex. *Langmuir* **1998**, *14*, 5890–5895. [CrossRef]
20. Roubeau, O.; Agricole, B.; Clérac, R.; Ravaine, S. Triazole-Based Magnetic Langmuir–Blodgett Films: Paramagnetic to Spin-Crossover Behavior. *J. Phys. Chem. B* **2004**, *108*, 15110–15116. [CrossRef]
21. Kurth, D.G.; Lehmann, P.; Schütte, M. A route to hierarchical materials based on complexes of metallosupramolecular polyelectrolytes and amphiphiles. *Proc. Natl. Acad. Sci. USA* **2000**, *97*, 5704–5707. [CrossRef] [PubMed]
22. Shibata, T.; Kimizuka, N.; Kunitake, T. Construction of mesoscopic supramolecular assemblies consisted of hydrophobic triazole derivative and Fe(II) ion. In Proceedings of the 76th CSJ National Meeting, Yokohama, Japan, 28 March 1999; CSJ: Tokyo, Japan, 1999.
23. Fujigaya, T.; Jiang, D.L.; Aida, T. Spin crossover properties of self-assembled iron(II) complexes with alkyl-tethered triazole ligands. *J. Am. Chem. Soc.* **2003**, *125*, 14690–14691. [CrossRef] [PubMed]
24. Kuroiwa, K.; Oda, N.; Kimizuka, N. Supramolecular solvatochromism. Effect of solvents on the self-assembly and charge transfer absorption characteristics of lipid-packaged, linear mixed valence platinum complexes. *Sci. Tech. Adv. Mater.* **2006**, *7*, 629–634. [CrossRef]
25. Kuroiwa, K.; Shibata, T.; Takada, A.; Nemoto, N.; Kimizuka, N. Heat-set gel-like networks of lipophilic Co(II) triazole complexes in organic media and their thermochromic structural transitions. *J. Am. Chem. Soc.* **2004**, *126*, 2016–2021. [CrossRef] [PubMed]
26. Kuroiwa, K.; Kimizuka, N. Electrochemically controlled self-assembly of lipophilic Fe^{II} 1,2,4-triazole complexes in organic media. *Chem. Lett.* **2010**, *39*, 790–791. [CrossRef]
27. Kuroiwa, K.; Shibata, T.; Sasaki, S.; Ohba, M.; Takahara, A.; Kunitake, T.; Kimizuka, N. Supramolecular control of spin crossover phenomena in lipophilic Fe(II) 1,2,4-triazole complexes. *J. Polym. Sci. A Polym. Chem.* **2006**, *44*, 5192–5202. [CrossRef]
28. Kume, S.; Kuroiwa, K.; Kimizuka, N. Photo-responsive molecular wires of Fe(II) triazole complexes in organic media and light-induced morphological transformations. *Chem. Commun.* **2006**, *42*, 2442–2444. [CrossRef] [PubMed]

29. Matsukizono, H.; Kuroiwa, K.; Kimizuka, N. lipid-packaged linear iron(II) triazole complexes in solution: Controlled spin conversion via solvophobic self-assembly. *J. Am. Chem. Soc.* **2008**, *130*, 5622–5623. [CrossRef] [PubMed]

30. Matsukizono, H.; Kuroiwa, K.; Kimizuka, N. Self-assembly-directed spin conversion of iron(II) 1,2,4-triazole complexes in solution and their effect on photorelaxation processes of fluorescent counter ions. *Chem. Lett.* **2008**, *37*, 446–447. [CrossRef]

31. Kuroiwa, K.; Kikuchi, H.; Kimizuka, N. Spin crossover characteristics of nanofibrous Fe^{II}-1,2,4-triazole complexes in liquid crystals. *Chem. Commun.* **2010**, *46*, 1229–1231. [CrossRef] [PubMed]

32. Noguchi, T.; Chikara, C.; Kuroiwa, K.; Kaneko, K.; Kimizuka, N. Controlled morphology and photoreduction characteristics of polyoxometalate(POM)/lipid complexes and the effect of hydrogen bonding at molecular interfaces. *Chem. Commun.* **2011**, *47*, 6455–6457. [CrossRef] [PubMed]

33. Kuroiwa, K.; Kimizuka, N. Self-assembly and functionalization of lipophilic metal–triazole complexes in various media. *Polym. J.* **2013**, *45*, 384–390. [CrossRef]

34. Kuroiwa, K.; Yoshida, M.; Masaoka, S.; Kaneko, K.; Sakai, K.; Kimizuka, N. Self-assembly of tubular microstructures from mixed-valence metal complexes and their reversible transformation via external stimuli. *Angew. Chem. Int. Ed.* **2012**, *51*, 656–659. [CrossRef] [PubMed]

35. Kuroiwa, K. Dynamic self-assembly from mixed-valence metal complexes and their reversible transformations by external stimuli. *Kobunshi Ronbunshu* **2012**, *69*, 485–492. [CrossRef]

36. Kuroiwa, K.; Arie, T.; Sakurai, S.; Hayami, S.; Deming, T.J. Supramolecular control of reverse spin transitions in cobalt(II) terpyridine complexes with diblock copolypeptide amphiphiles. *J. Mater. Chem. C* **2015**, *3*, 7779–7783. [CrossRef]

37. Arie, T.; Otsuka, S.; Maekawa, T.; Takano, R.; Sakurai, S.; Deming, T.J.; Kuroiwa, K. Development of hybrid diblock copolypeptide amphiphile/magnetic metal complexes and their spin crossover with lower-critical-solution-temperature(LCST)-type transition. *Polymer* **2017**. [CrossRef]

38. Kahn, O.; Martinez, C.J. Spin-transition polymers: From molecular materials toward memory devices. *Science* **1998**, *279*, 44–48. [CrossRef]

39. Kahn, O. Chemistry and physics of supramolecular magnetic materials. *Acc. Chem. Res.* **2000**, *33*, 647–657. [CrossRef] [PubMed]

40. Kahn, O.; Köber, J.; Jay, C. Spin transition molecular materials for displays and data recording. *Adv. Mater.* **1992**, *4*, 718–728. [CrossRef]

41. Köber, J.; Codjovi, E.; Kahn, O.; Groliére, F.; Jay, C. A spin transition system with a thermal hysteresis at room temperature. *J. Am. Chem. Soc.* **1993**, *115*, 9810–9811. [CrossRef]

42. Haasnoot, J.G. Mononuclear, oligonuclear and polynuclear metal coordination compounds with 1,2,4-triazole derivatives as ligands. *Coord. Chem. Rev.* **2000**, *200–202*, 131–185. [CrossRef]

43. Haasnoot, J.G. 1,2,4-Triazoles as ligands for iron(II) high spin-low spin crossovers. In *Magnetism: Supramolecular Function*; Kahn, O., Ed.; Kluwer Academic Publications: Dordrecht, Netherlands, 1995; pp. 299–321.

44. Vos, G.; le Fêbre, L.A.; de Graaff, R.A.G.; Haasnoot, J.G.; Reedijk, J. Unique highspin-low-spin transition of the central ion in a linear, trinuclear iron(II) triazole compound. *J. Am. Chem. Soc.* **1983**, *105*, 1682–1683. [CrossRef]

45. Roubeau, O.; Gomez, M.A.; Balskus, E.; Kolnaar, J.J.A.; Haasnoot, J.G.; Reedijk, J. Spin-transition behaviour in chains of Fe(II) bridged by 4-substituted 1,2,4-triazoles carrying alkyl tails. *New J. Chem.* **2001**, *25*, 144–150. [CrossRef]

46. Fujigaya, T.; Jiang, D.L.; Aida, T. Spin-crossover dendrimers: Generation numberdependent cooperativity for thermal transition. *J. Am. Chem. Soc.* **2005**, *127*, 5484–5489. [CrossRef] [PubMed]

47. Fujigaya, T.; Jiang, D.L.; Aida, T. Spin-crossover physical gels: A quick thermoreversible response assisted by dynamic self-organization. *Chem. Asian J.* **2007**, *2*, 106–113. [CrossRef] [PubMed]

48. Seredyuk, M.; Gaspar, A.B.; Ksenofontov, V.; Reiman, S.; Galyametdinov, Y.; Haase, W.; Rentschler, E.; Gütlich, P. Room temperature operational thermochromic liquid crystals. *Chem. Mater.* **2006**, *18*, 2513–2519. [CrossRef]

49. Hayami., S.; Shigeyoshi, Y.; Akita, M.; Inoue, K.; Kato, K.; Osaka, K.; Takata, M.; Kawajiri, R.; Mitani, R.; Maeda, Y. Reverse spin transition triggered by a structural phase transition. *Angew. Chem. Int. Ed.* **2005**, *44*, 4899–4903. [CrossRef] [PubMed]

50. Hayami, S.; Moriyama, R.; Shuto, A.; Maeda, Y.; Ohta, K.; Inoue, K. Spin transition at the mesophase transition temperature in a cobalt(II) compound with branched alkyl chains. *Inorg. Chem.* **2007**, *46*, 7692–7694. [CrossRef] [PubMed]

51. Hayami, S.; Murata, K.; Urakami, D.; Kojima, Y.; Akita, M.; Inoue, K. Dynamic structural conversion in a spin-crossover cobalt(II) compound with long alkyl chains. *Chem. Commun.* **2008**, *48*, 6510–6512. [CrossRef] [PubMed]

52. Hayami, S.; Kato, K.; Komatsu, Y.; Fuyuhiro, A.; Ohba, M. Unique spin transition and wide thermal hysteresis loop for a cobalt(II) compound with long alkyl chain. *Dalton Trans.* **2011**, *40*, 2167–2169. [CrossRef] [PubMed]

53. Komatsu, Y.; Kato, K.; Yamamoto, Y.; Kamihata, H.; Lee, Y.H.; Akita, F.; Kawata, S.; Hayami, S. Spin-crossover behaviors based on intermolecular interactions for cobalt(II) complexes with long alkyl chains. *Eur. J. Inorg. Chem.* **2012**, *16*, 2769–2775. [CrossRef]

54. Hayami, S.; Nakaya, M.; Ohmagari, H.; Alao, A.S.; Nakamura, M.; Ohtani, R.; Yamaguchi, R.; Kuroda-Sowa, T.; Clegg, J.K. Spin crossover behaviors in solvated cobalt(II) compounds. *Dalton Trans.* **2015**, *44*, 9345–9348. [CrossRef] [PubMed]

55. Pochan, D.J.; Pakstis, L.; Ozbas, B.; Nowak, A.P.; Deming, T.J. SANS and cryo-TEM study of self-assembled diblock copolypeptide hydrogels with rich nano-through microscale morphology. *Macromolecules* **2002**, *35*, 5358–5360. [CrossRef]

56. Novak, A.P.; Breedveld, V.; Pakstis, L.; Ozbas, B.; Pine, D.J.; Pochan, D.; Deming, T.J. Rapidly recovering hydrogel scaffolds from self-assembling diblock copolypeptide amphiphiles. *Nature* **2002**, *417*, 424–428.

57. Holowka, E.P.; Sun, V.Z.; Kamei, D.T.; Deming, T.J. Polyarginine segments in block copolypeptides drive both vesicular assembly and intracellular delivery. *Nat. Mater.* **2007**, *6*, 52–57. [CrossRef] [PubMed]

58. Yang, C.Y.; Song, B.; Ao, Y.; Nowak, A.P.; Abelowitz, R.B.; Korsak, R.A.; Havton, L.A.; Deming, T.J.; Sofroniew, M.V. Biocompatibility of amphiphilic diblock copolypeptide hydrogels in the central nervous system. *Biomaterials* **2009**, *30*, 2881–2898. [CrossRef] [PubMed]

59. Barth, P.; Schemauss, G.; Specker, H. Complexes of iron (II) with substituted 2-pyridinalphenylimines. *Z. Naturforsch* **1972**, *27*, 1149–1154.

60. Wei, H.H.; Kao, S.P.; Jean, Y.C. The effect of methyl substitution on the spinstate in solid bis(Methylsubstituted 2-pyridinalphenylimine) di(thiocyanato) iron(II). *Transit. Met. Chem.* **1986**, *11*, 405–408. [CrossRef]

61. Yu, Z. Light-induced excited spin-state trapping of the complex. *Transit. Met. Chem.* **1996**, *21*, 472–473. [CrossRef]

62. Letard, J.F.; Guionneau, P.; Codjovi, E.; Lavastre, O.; Bravic, G.; Chasseau, D.; Kahn, O. Wide thermal hysteresis for the mononuclear spin-crossover compound cis-bis(thiocyanato)bis[*N*-(20-pyridylmethylene)-4-(phenylethynyl) anilino]iron(II). *J. Am. Chem. Soc.* **1997**, *119*, 10861–10862. [CrossRef]

63. Oso, Y.; Ishida, T. Spin-crossover transition in a mesophase iron(II) thiocyanate complex chelated with 4-hexadecyl-*N*-(2-pyridylmethylene)aniline. *Chem. Lett.* **2009**, *38*, 604–605. [CrossRef]

64. Djukic, B.; Seda, T.; Gorelsky, S.I.; Lough, A.J.; Lemaire, M.T. p-Extended and sixcoordinate iron(II) complexes: Structures, magnetic properties, and the electrochemical synthesis of a conducting iron(II) metallopolymer. *Inorg. Chem.* **2011**, *50*, 7334–7343. [CrossRef] [PubMed]

65. Ksenofontov, V.; Levchenko, G.; Spiering, H.; Gütlich, P.; Letard, J.F.; Bouhedja, Y.; Kahn, O. Spin crossover behavior under pressure of Fe(PM-L)$_2$(NCS)$_2$ compounds with substituted 20-pyridylmethylene 4-anilino ligands. *Chem. Phys. Lett.* **1998**, *294*, 545–553. [CrossRef]

66. Chujo, Y.; Tanaka, K. New polymeric materials based on element-blocks. *Bull. Chem. Soc. Jpn.* **2015**, *88*, 633–643. [CrossRef]

inorganics

Review

Design and Control of Cooperativity in Spin-Crossover in Metal–Organic Complexes: A Theoretical Overview

Hrishit Banerjee [1], Sudip Chakraborty [2] and Tanusri Saha-Dasgupta [1,*]

[1] Department of Condensed Matter Physics and Material Sciences, S N Bose National Centre for Basic Sciences, JD Block, Sector III, Salt Lake, Kolkata 700106, India; h.banerjee10@gmail.com
[2] Materials Theory Division, Department of Physics and Astronomy, Uppsala University, Box 516, 75120 Uppsala, Sweden; sudiphys@gmail.com
* Correspondence: t.sahadasgupta@gmail.com; Tel.: +91-33-2335-5707

Received: 20 June 2017; Accepted: 14 July 2017; Published: 20 July 2017

Abstract: Metal organic complexes consisting of transition metal centers linked by organic ligands, may show bistability which enables the system to be observed in two different electronic states depending on external condition. One of the spectacular examples of molecular bistability is the spin-crossover phenomena. Spin-Crossover (SCO) describes the phenomena in which the transition metal ion in the complex under the influence of external stimuli may show a crossover between a low-spin and high-spin state. For applications in memory devices, it is desirable to make the SCO phenomena cooperative, which may happen with associated hysteresis effect. In this respect, compounds with extended solid state structures containing metal ions connected by organic spacer linkers like linear polymers, coordination network solids are preferred candidates over isolated molecules or molecular assemblies. The microscopic understanding, design and control of mechanism driving cooperativity, however, are challenging. In this review we discuss the recent theoretical progress in this direction.

Keywords: Spin-Crossover (SCO); stimulus-induced spin transition; crystal engineering of SCO and related complexes

1. Introduction

The phenomena of Spin-Crossover (SCO) has caught the attention of scientific community for ages and has shown growing interest in recent time due to various different application possibilities in the information technology [1], as sensors [2], optical switches [3], etc. The essence of the SCO phenomena is described in the following. SCO may take place in transition metal complexes, specially those consisting of transition metal ions and flexible organic ligands, wherein the spin state of the metal ion changes between low spin (LS) and high spin (HS) configuration under the influence of external perturbation [4]. Though this process, in principle, can be observed in any octahedrally or tetrahedrally coordinated transition metal complexes with transition metal ions in d^4–d^7 or d^3–d^6 electronic configurations, the most commonly observed cases are that of octahedrally coordinated iron(II) complexes with Fe^{2+} ions in 3 d^6 electronic configuration [5–7]. The transition between the paramagnetic high spin (S = 2) and the diamagnetic low spin (S = 0) state of Fe^{2+} can be triggered by several different possibilities such as temperature, pressure, light [8]. The SCO phenomenon deserves attention due to accompanying changes in magnetic and optical properties. To be of use in device applications, it is important to induce cooperativity in the SCO phenomena implying spin transition rather than spin crossover, which may happen with associated hysteresis effect as shown in Figure 1. The issue of cooperativity and associated hysteresis is important as it is expected to confer memory effect to the system.

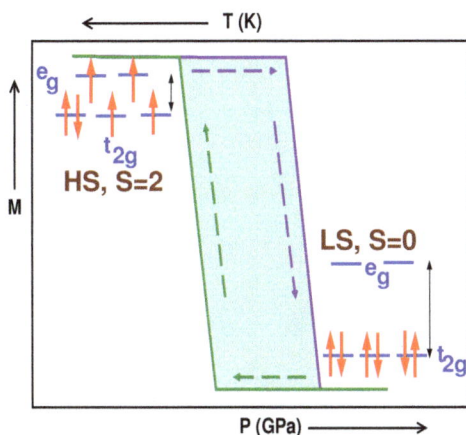

Figure 1. Schematic diagram demonstrating the cooperativity associated with spin crossover phenomena in Fe(II) complexes. The y axis shows the magnetic moment (M) and x axis represents temperature (T) or pressure (P) increasing in two different directions as shown in the figure. Upon increasing T or decreasing P, a transition happens from LS S = 0 state where all spins are paired up in the t_{2g} states octahedrally coordinated d^6 Fe, to HS S = 2 state where two of the electrons from t_{2g} states are promoted to e_g states. Due to the cooperativity between Fe(II) centres in the extended solid state geometry, hysteresis loop appears between the heating and cooling cycles, or between cycles of exertion and release of pressure.

In this context, in comparison to molecular assemblies or crystals with isolated molecular units connected by weak, van der Waals or hydrogen bonding, the coordination polymeric compounds with repeating coordination entities having extended solid structures are better choices. The presence of chemical bridges, linking the SCO sites to each other, as in coordination polymers are expected to propagate the interaction between SCO centers more efficiently than that in molecular crystals. Admitting the suitability of such compounds in exhibiting cooperativity there are several issues that need attention. The key questions are, (a) understanding the microscopic mechanism, i.e., what is the driving force for the cooperativity and the hysteresis; and (b) how the cooperativity can be tuned or modified to suit the specific need. These understandings are expected to provide an advancement of the field in terms of possible commercialization of this technologically important property which relies on critical parameters of cross-over being close to ambient condition, and a large enough hysteresis width.

The above two issues are also intimately connected to materials issue, namely what are the materials to look for cooperative SCO. As mentioned above, 1, 2 or 3-dimensional coordination polymers, which are materials with repeating array of coordination entities, are the suitable choices. The dimensionality of a coordination polymer is defined by the number of directions in space the array extends to. Most studied SCO materials showing cooperativity, so far are linear 1-dimensional coordination polymers which are compounds extending through repeating coordination entities in 1-dimension forming chain like structures, with weak links between individual chains [9], as shown in left panel of Figure 2. The other possibilities are coordination network solids [9], which are compounds extending through repeating coordination entities in 2 or even 3 dimensions. Strategic crystal engineering that makes use of multidentate ligands, connected by spacers, facilitates to increase the dimensionality from 1-d to 2-d or 3-d. Pressure-induced LS–HS transition in 2-d net was first reported for [Fe(btr)$_2$(NCS)$_2$]·H$_2$O [10]. The compound consisted of Fe(II) ions linked by btr in two directions producing infinite layers which were connected by means of van der Waals or weak H bonds. [Fe(btr)$_3$][(ClO$_4$)$_2$] [11] represents the first 3-d SCO coordination polymer. For a concise review on 1-d, 2-d and 3-d Fe(II) polymers, please see Reference [11].

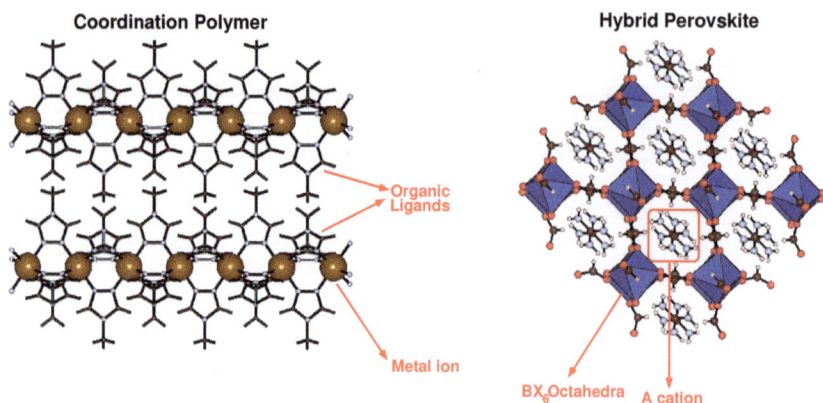

Figure 2. Schematic representation of linear, 1-d coordination polymer (**left** panel) and hybrid perovskite (**right** panel). In case of linear, 1-d coordination polymer, chains of metals are linked by organic ligands. In the ABX_3 structure of hybrid perovskite, BX_6 forms octahedra with A cation sitting in the voids. B cation is a metal ion and A cation is an organic cation. BX_6 octahedra are also linked by organic linkers.

For the sake of description of theoretical studies, we will restrict ourselves to only linear or 1-d coordination polymers, and not extending our discussion to 2-d or 3-d polymers of type $[Fe(btr)_2(NCS)_2] \cdot H_2O$ or $[Fe(btr)_3][(ClO_4)_2]$. Rather we will focus in a new class of materials in context of SCO, namely hybrid perovskites. A subclass of coordination network solids are metal organic frameworks (MOFs) which are coordination networks with organic ligands containing potential voids [9], and thus can be labeled as porous coordination polymers. Most of the research on MOFs are related to porosity of the systems, which have been explored for applications in gas storage [12], chemical sensing [13], drug delivery [14], bio mimetic mineralization [14] and catalysis [15]. Recently, however, the attention is also given to dense MOFs with limited porosity which show potential for applications in other areas like optical devices, batteries and semiconductors [16]. Hybrid perovskites are a class of compounds with general formula ABX_3 having long range connectivity that form a subclass of dense MOFs. The extended, 3-dimensional connectivity with limited void space, together with possibility of synthesizing hybrid perovskites containing transition metal ions have made these compounds also probable candidates for exhibiting cooperative SCO. While linear coordination polymers have already been explored to a large extend in search of cooperative SCO, ABX_3 type hybrid perovskites which though in attention in recent time has not been explored for cooperative SCO. In the following we give brief description of each of these classes of compounds, namely linear chain compounds, and hybrid perovskites.

Among the linear coordination polymers, or 1-d chain compounds 4R-1,2,4-triazole based Fe(II) chain compounds have been in focus both in early studies and in recent developments. $[Fe(4R-1,2,4-triazole)_3]A_2 \cdot solv$, where A is the counterion and *solv* denotes the solvent molecule are made up of linear chains in which the adjacent Fe(II) ions in the chain are linked by three triazole ligands. The coordination linkers, which are 1,2,4-triazole blocks form efficient chemical bonds to transmit cooperative effect, leading to hysteresis loop of width ranging $\approx$ 2–20 K. [17] Sometimes these hysteresis loops are also found to be centered at room temperature. [18] Bimetallic 1-d chain compounds like $Fe(aqin)_2(\mu_2\text{-}M(CN)_4)$, M = Ni(II) or Pt(II), have been recently synthesized which were found to show abrupt HS–LS SCO. [19] Novel 1-d Fe(II) SCO coordination polymers with 3,3′-azopyridine as axial ligand has been synthesized which were found to show kinetic trapping effects and spin transition above room temperature. [20] Combination of rigid links and a hydrogen bond network between 1-d Fe(II) chains has been recently shown as a promising tool to trigger SCO with hysteresis loops having widths as large as $\approx$43 K. [21]

Hybrid perovskites [22,23] having general formula of ABX_3, on the other hand, are counterparts of inorganic perovskites. In case of hybrid perovskites, while the B cation is a metal ion as in inorganic perovskites, both the A cation as well as the ligand can be organic. Lead halides hybrid perovskite family [23,24] having $[AmH]MX_3$ composition, where AmH^+ is the protonated amine part; M is either Sn^{2+} or Pb^{2+} and X is the halogen part (Cl^-, Br^-, or I^-) have been shown to demonstrate high performance and efficiency in applications relating to design of mesostructural (and/or nanostructural) solar cells and other photovoltaic devices [25–27]. Easy processing techniques like spin-coating, dip-coating and vapor deposition techniques have been known to be of advantage in this case [28,29]. There are also transition metal formate based hybrid perovskites, which are of interest in the present context, having formula $[AmH]M(HCOO)_3$ (M = Mn, Cu, Ni, Fe, Co). [30] In the crystal structure of these compounds, as shown schematically in right panel of Figure 2, formate bridges act as linkers that connect the MO_6 octahedra, with the protonated amine cations situated at the hollow spaces formed by the linked octahedra. These hollow spaces act as pseudo-cubic ReO_3 type cavities. Organic ligands like formate being simple enough have been mostly studied with varied A cations. This class of materials has been shown to exhibit curious properties, of which multiferroicity seems to be an intriguing one [31–33]. Ferroelectricity and especially multiferroicity in these materials has been extensively studied by Stroppa and coworkers mostly from a DFT based first principles perspective and at times combined with experimental studies [34–40]. Structural details and effects due to structural phase transitions, strain tuning of various effects like polarisation, and magnetic structure has also been studied by the same groups [41–45]. As mentioned already, the presence of transition metal in these compounds together with its octahedral environment makes them suited also for exhibiting SCO behavior and possibly also cooperativity due to dense nature of framework, which would provide another dimension to functionality of these interesting class of compounds. This aspect of this interesting class of compounds has remained unexplored, apart from a very recent theoretical proposal [46].

The plan of the review is the following. We will start with a brief summary of the experimental investigations on cooperative SCO, as presented in Section 2. With this background, we will discuss in Section 3 different theoretical models that have been proposed to describe cooperative SCO. In Section 4, we will take up the material specific studies, where density functional theory (DFT) based electronic structure calculations have been employed to investigate cooperative SCO phenomena in well studied case of linear coordination polymer and proposed case of hybrid perovskite. We will conclude the review with a summary and outlook in Section 5, where the future prospective and challenges that needs to be overcome will be discussed.

2. Experimental Evidences of Cooperativity

During the past few decades, there has been a growing number of experimental studies focusing on spin crossover phenomena in general. Numerous studies have been carried out on various materials ranging from molecules to molecular assemblies to coordination polymers, linear coordination polymers being the most studied case in latter category. In case of linear coordination polymers, often a significant hysteresis effect is observed, primarily under the influence of temperature, though there exists studies showing the existence of hysteresis in SCO under pressure. There have been also studies involving light-induced spin-crossover transitions. In the following we provide a brief over-view. Given the vastness of the experimental study on SCO, the over-view presented in the following is far from being complete and should be considered as representative. In the first part of the brief over-view, we provide description of experimental studies of SCO phenomena in general, while in the later part we specifically focus on cooperative SCO.

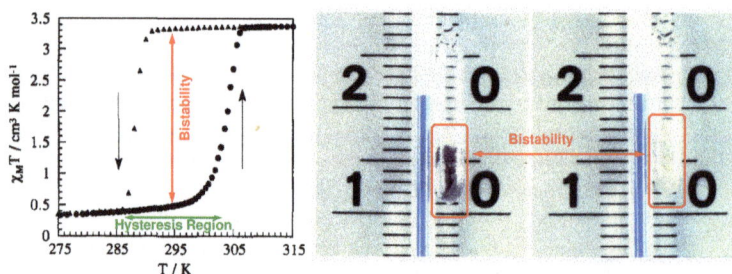

Figure 3. Experiment evidence of cooperativity in Fe-triazole, as reported by Kahn et al. [18] (Adapted with permission from Krober, J.; Codjovi, E.; Kahn, O.; Groliere, F.; Jay, C. A spin transition system with a thermal hysteresis at room temperature. *J. Am. Chem. Soc.* **1993**, *115*, 9810–9811). The **left** panel shows the variation of susceptibility with increasing and decreasing temperature. A large hysteresis loop and corresponding bistability is observed confirming the existence of cooperativity; The **right** panel demonstrates the bistability in the system, reflected in two different colors of the same material at the same temperature, being on cooling or heating path.

SCO phenomena was first reported in the pioneering work by Cambi et al. [47] as early as in 1931, in a series of Fe based metal–organic compounds viz. iron (III) tris-dithiocarbamate. In their study, the authors observed a transition from $S = 1/2$ to $S = 3/2$ spin state. Since then there have been subsequent experiments which demonstrate SCO in other materials, which are mostly Fe based compounds, but also in Co, Ni, Cr, and Mn based compounds [5–7,48]. König et al. [49] in 1961 first observed a $S = 0$ to $S = 2$ transition in a iron (II) thiocyanide material which later became a popular material of study in the area of spin crossover. Various experimental techniques have been employed to study the phenomena of spin crossover, the most popular one being magnetic susceptibility measurement. X-ray diffraction or Extended X-ray absorption Fine Structure (EXAFS) has been employed to measure the structural properties of SCO compounds. [50,51] Mössbauer Spectroscopy has been also employed to detect SCO in iron complexes. [49,52,53] Raman spectroscopy has been used to identify spin state of SCO complexes. [54] Fourier Transform Infrared Spectroscopy (FT-IR) has been employed to plot the area under the curves of the absorption peaks as a function of temperature which have been used to map out the SCO. [55] UV-vis spectroscopy has also been an important tool for detecting the changes of electronic states in SCO materials. [56] Electron Paramagnetic Resonance (EPR) [57], Muon Spin Relaxation (μSR), Positron Annihilation Spectroscopy (PAS), Nuclear Resonant Scattering of synchroton radiation (NRS) are among other techniques which have also been used for experimental study of SCO [58]. While for most experimental studies, the temperature or pressure has been used as external perturbation, there have been also several experimental studies on light induced spin crossover or as it is known Light induced spin state trapping (LIESST). This was first reported by Decurtins and coworkers. [59] In the work by Decurtins and coworkers, [59] the Fe(II) based compound was converted from stable LS state to metastable HS state at low temperature by irradiating the sample with green light. In a subsequent work, Hauser [60] reported the back conversion to LS state by irradiation of red light. It was found [59] that very low temperatures are required for trapping the system in the metastable HS state. The systems relaxes to the LS state above a critical temperature which was defined as T_C(LIESST) [61]. Estimation of the T_C(LIESST) value was found to give an idea of the capability of the material to store the light induced HS state. Superconducting quantum interference device (SQUID) magnetometer measurements were carried out in a study by Kahn et al. [61] to characterize LIESST and determine T_C values for the thermally induced SCO for a variety of Fe SCO materials. In this work correlation between the photomagnetic properties and the thermal properties of a variety of SCO materials was drawn. A plot of the T_C(LIESST) vs the spin crossover temperature for a collection of materials was found to show a correlation between the two. It was concluded that the positioning of a given compound in plot is governed by the cooperativity of the compound.

Other than the case of molecular mononuclear moieties, a large number of studies involve one, two and three-dimensional coordination polymers with multiple SCO centers, exhibiting intriguing hysteresis effects. In a seminal experimental study by Kahn et al. [18] thermal hysteresis at room temperature was demonstrated in case of Fe-triazole compound, in particular [Fe(Htrz)$_{3-3x}$,(4-NH$_2$trz)$_{3x}$](ClO$_4$)$_2$·nH$_2$O with x = 0.05. As shown in Figure 3 the temperature dependence of measured susceptibility showed a spin state transition accompanied by a wide hysteresis loop close to room temperatures, as is desirable for commercial applications. As seen in the figure, the transition is also accompanied by a dramatic change of color. The two spin states correspond to two different electronic states exhibiting two different colors. Depending on whether the observation is being made in the heating or the cooling cycle, the sample at the same temperature falling within the hysteresis width exhibits two different colors, establishing the bistability of the two spin states. The molecular system described in this particular study was probably the first one of its kind exhibiting such a bistability at room temperature. This paved the way for future experimental and simulation based studies on cooperativity.

The experimental study by Linares et al. [62] which was also backed by simulation, reported the existence of hysteresis in SCO phenomena induced thermally as well as through pressure. Their experimental and simulation studies on [Fe$_x$Co$_{1-x}$(btr)$_2$(NCS)$_2$]·H$_2$O with x = 0.847 showed large hysteresis loops under heating and cooling cycle confirming cooperativity in these materials. Their study on [Fe$_x$Ni$_{1-x}$(btr)$_2$(NCS)$_2$]·H$_2$O with x = 0.66 showed large hysteresis in pressure induced SCO.

Baldé et al. [63] considered coordination polymers with the general formula [FeL$_{eq}$(L$_{ax}$)]·solvent where L$_{eq}$ = 3,3′-[1,2-phenylene bis(iminomethylidyne)]bis(2,4-pentanedionato)(2-)-N,N′,O^2,O′^{2} and 3 different components for L$_{ax}$ such as 4,4′-bipyridine(bipy), 1,2-bis(4-pyridyl)ethane (bpea), and 1,3-bis(4-pyridyl)propane (bppa). Cooperative SCO was previously [64–66] reported for the bipy linker, which showed thermal hysteresis and bpea and bppa based compounds showed multi-step transitions. In the study by Baldé et al. [63] the authors also investigated the LIESST effects and estimated the T$_C$(LIESST) values by irradiating the samples in a SQUID. The thermal measurements showed hysteresis in bipy linker based compound with a 2 step transition in bpea based compound and a 2 step transition accompanied by hysteresis in bppa based compound. Evaluation of the T$_C$(LIESST) values associated with the LIESST effect demonstrated the stability of the photoinduced HS state. Experiments on [Fe$_x$Ni$_{1-x}$(btr)$_2$(NCS)$_2$]·H$_2$O with x = 0.5 by Linares et al. [62] under constant light exposure allowed the investigation of the competition between photoexcitation and subsequent relaxation of the metastable state, and a thermal hysteresis under constant light irradiation.

Numerous other studies on SCO compounds, apart from the ones described above, have been carried out which recorded hysteresis in thermal, pressure and light induced spin state transitions. For the sake of brevity, we will not get into the details of all the studies. The few studies described above provide conclusive evidences of experimental observation of cooperativity in SCO complexes.

3. Theoretical Models on Cooperativity

Several theoretical studies have been reported in literature with an aim to provide microscopic understanding of the origin of spin state transition and its cooperativity in metal organic complexes. While the spin crossover in molecular systems with a single SCO center happens due to competition between the crystal field splitting and Hund's rule coupling, in case of extended polymeric systems, interactions between different SCO centers may drive the cooperativity in SCO observed at molecular level. It is important to understand the microscopic origin of this interaction. This understanding is expected to enable designing of extended solid state structures with desired cooperative properties.

The most prevailing concept in this context, is that the interaction is given by phononic excitations due to coupling between local lattice distortions at each molecular unit, setting up a long range elastic wave. Two different model Hamiltonian approaches have been followed to deal with elastic interactions between SCO centers to calculate thermodynamic quantities like HS fraction. In the first category of calculations, Ising type Hamiltonians, $H = \sum_{ij} J_{ij}\sigma_i\sigma_j$ are used with LS or HS state described by pseudo-spin operators $\sigma_i = -1(+1)$ for LS(HS) that interact via nearest neighbor elastic coupling. For example, the study by Boukheddaden et al. [67] considered one-dimensional spin-phonon model, which is expected to model the linear polymeric compounds. In particular the model consisted of assembly of two-level system with elastic interaction between them. The HS state, represented by pseudo-spin $\sigma = 1$ was assumed to be n_H fold degenerate, while the LS state represented by $\sigma = -1$ was assumed to be n_L fold degenerate. The elastic interaction linking the sites i and $i+1$, was denoted by $e_{i,i+1}$. Volume of the molecule changes upon the spin state transition, and thus the elastic interactions between two successive sites was assumed to depend on their spin states. It was conjectured that $e_{i,i+1} = e_{++}$ for $\sigma_i = \sigma_{i+1} = 1$, $e_{i,i+1} = e_{+-}$ for $\sigma_i = -\sigma_{i+1}$, and $e_{i,i+1} = e_{--}$ for $\sigma_i = \sigma_{i+1} = -1$. The constructed model was solved in the framework of classical statistical mechanics using transfer matrix technique. The effective interaction turned out to be ferroelastic for $e_{+-} > \sqrt{e_{++} \times e_{--}}$ and antiferroelastic for $e_{+-} < \sqrt{e_{++} \times e_{--}}$, which in turn was found to be determining factor for the lattice of SCO centers to exhibit cooperativity or not. It was found that ferroelastic interaction favors cooperativity with sharp spin state transition while the antiferroelastic interaction makes the transition much smoother. This indicated that in presence of ferroelastic interaction, first-order phase transition along with hysteresis might take place in higher dimensions. The authors thus introduced inter chain interactions, justified by the fact that in SCO polymeric samples, interchain steric and electrostatic interactions arise due to presence of non-coordinated molecules or counter anions. The mean field treatment of interchain interactions indeed was found to reproduce the first-order nature of HS–LS transition together with hysteresis for ferroelastic intrachain interactions.

In the second category of calculations, free energy of SCO systems were calculated based on anisotropic sphere model describing the volume and shape changes of the lattice at the transition [68–70]. Though the anisotropic models were found to describe the spin crossover phenomena successfully, these models have not been very successful in describing the cooperativity with the exception of a recent work by Spiering et al. [70] where transitions both with and without hysteresis have been achieved. In the model by Spiering et al. [70], an anharmonic lattice having SCO molecules having a certain misfit to the lattice was considered, which were assumed to interact via elastic interactions by the change of shape and volume.

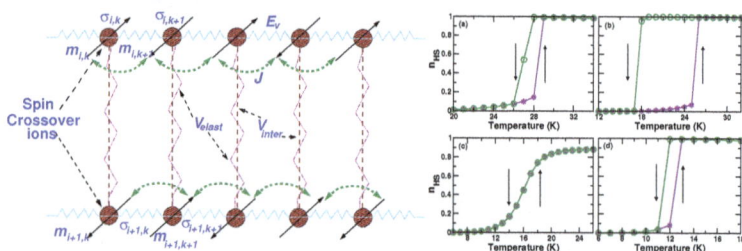

Figure 4. Figure showing schematic diagram of various interactions present in linear SCO polymeric systems along with their respective effects on the system. The left panel shows all the interactions present in the system. The red balls denote the SCO ions. The light blue springs denote the intra chain spin dependent elastic interaction. The black arrows denote the magnetic spins at the SCO sites with the green bent arrows denoting the magnetic exchange between the spins. The pink dashed lines show the inter chain elastic interaction and the brown dashed lines show the interchain superexchange interaction. The right panel shows 4 different graphs showing the effect of (a) ferroelastic intra chain interaction in absence of magnetic interaction (b) ferroelastic in presence of magnetic intra chain interaction (c) anti ferroelastic intra chain interaction in absence of magnetic interaction and (d) anti-ferroelastic in presence of magnetic intra chain interaction. Figure has been adapted from Banerjee et al. [71].

The above described approaches of considering the elastic interaction as the sole driving mechanism, however completely disregards the importance of the long range magnetic interaction that may build up between transition metal centers via superexchange interaction mediated through the organic linkers connecting the metal centers. In order to address this issue, in the study by Timm [72], the collective properties of spin-crossover chains were investigated taking into account both elastic interaction and Ising-like magnetic interaction, and the ground state phase diagram was mapped out. This calculation though ignored the spin-phonon coupling, considering the elastic interactions to be spin independent, and thus were unable to distinguish between ferro and antiferroelastic situations. In a recent study, Banerjee et al. [71] have studied extensively the microscopic mechanism giving rise to cooperativity in spin crossover considering both the effect of magnetic super-exchange interaction and elastic interactions together with spin-phonon coupling in a coupled 1-d chain model. A model Hamiltonian was setup in the basis of pseudo-spins, corresponding to the elastic part of the interaction, and the actual spin quantum number, describing the magnetic superexchange interaction. The system consisted of connected chains of spin-crossover ions, as shown in left panel of Figure 4. The SCO ions in a chain were assumed to be connected by intrachain elastic interactions that depend on their spin states given by $E_v(\sigma_{i,k}, \sigma_{i,k+1})$, where σ denotes fictitious spins or pseudo spins as described above, and $E_v(\sigma_{i,k}, \sigma_{i,k+1}) = \frac{e_{k,k+1}}{2} q_i^2$ wherein q_i denotes the small displacements. The spin dependence of $e_{k,k+1}$ was considered to be same as in the model by Boukheddaden et al. [67]. Magnetism was accounted for by actual spin S, where $m_{i,k} = S_{i,k}$ with $m_{i,k} = 0$ for $\sigma_{i,k} = -1$ (LS state) and $m_{i,k} = -2,...,2$ for $\sigma_{i,k} = 1$, corresponding to the HS state S = 2. Thus the constructed model Hamiltonian was given by,

$$\mathcal{H} = -\sum_k E_v \sigma_{i,k} \sigma_{i,k+1} + \Delta \sum_{i,k} \sigma_{i,k} + J \sum_k m_{i,k} m_{i,k+1}$$

$$+ \sum_{i \neq k} [V_\perp \sigma_{i,k} \sigma_{i+1,k} + V_\parallel (\sigma_{i,k} + \sigma_{i+1,k})]$$

where Δ denoted the difference in energy between the HS and the LS states, $V_\perp$ and $V_\parallel$ represented the interchain interactions arising out of electrostatic and steric interactions, respectively, as discussed earlier. The third term of the Hamiltonian describes the magnetic exchange, which involves Ising-like antiferromagnetic superexchange interaction J acting between two Fe^{2+} ions with spin S = 2.

The Monte Carlo study of this above model Hamiltonian established the important role of magnetic super-exchange interaction acting between the SCO ions, in the cooperativity of the spin transition. They turn out to be equally important as the elastic interaction. Depending on the nature of the spin-dependent elastic interaction, which depends on the nature of the spin-phonon coupling, the magnetic super-exchange was found to contribute in development of cooperativity in a quantitative or qualitative manner. In case of ferrotype elastic interaction, the super-exchange interaction was found to enhance the hysteresis effect, setup by the elastic interaction. In the case of antiferro nature of elastic interaction, the magnetic exchange interaction was found to play a decisive role in driving the hysteresis in the system, putting the elastic coupling to a back seat. As is seen from the right panel of Figure 4 in presence of antiferro type intra chain elastic interaction alone, the system did not show any hysteresis. Only when the magnetic superexchange interaction J was introduced, a hysteresis loop was observed in the HS fraction. It was thus concluded that the magnetic superexchange interaction is of prime importance in driving cooperativity in the system if the intra chain elastic interaction is of antiferro type.

In another study a vibronic model for the SCO transition has been proposed [73], where the coupling between spin state and molecular geometry was considered, which causes quantum mixing between LS and HS states resulting from higher order spin-orbit coupling. It was concluded from such study that non-adiabaticity does not play an important role in case of cooperativity.

4. Ab Initio Studies on Materials Showing Cooperativity

These materials of technological importance, however, are challenges for material specific, predictive theoretical descriptions through ab-initio calculations. Challenges arise due to (a) strong electron-electron correlation in open d shell of the transition metal SCO centers; (b) complex geometry of the materials with large number of atoms in the unit cell; and (c) externally stimulated changes in electronic and structural properties. Both wavefunction based quantum chemical approaches [74–79] and density based Density functional theory (DFT) [80–83] calculations have been undertaken to study the phenomena of SCO in general. However it appears that the wavefunction based approaches require quite a large basis set and even though various techniques like complete active space self-consistent field (CASSCF) [75] or CASPT2 method [76], the coupled-cluster (CC) theory or its variation CCSD(T) method [79], (S and D denote inclusion of single and double excitations, respectively and (T) represents a single/triple correct term) have been used, these methods are computationally quite expensive and application of these methods to 2 or 3 dimensional connectivity of metal–organic frameworks is not feasible. We shall thus primarily discuss and describe applications of DFT based methods which have been used to study the cooperativity in SCO systems. The DFT based methods can handle a large number of atoms in the periodic systems quite efficiently, and can also take advantage of DFT+U methods [84] with supplemented onsite Hubbard U parameter to handle strong correlation effect at transition metal sites. As established, the influence of external perturbation like pressure on crystal structure and electronic structure can be described rather accurately through DFT calculations. Therefore, DFT calculations incorporate the metal-ligand bond length change that happens during SCO, which has been pointed out as an important ingredient driving SCO [85]. The handling of temperature effect within DFT is more challenging which is achievable in recent time through the progress made in ab-initio molecular dynamics (AIMD) simulations [86]. AIMD simulations solve electronic Kohn–Sham equation and Newton's equation of motion for ions, coupled via Hellmann-Feynmann theorem for forces. In the following we describe in detail application of such state-of-art techniques to two representative cases, one of linear coordination polymer and another of hybrid perovskite. There also exists two other DFT studies, cyanide-based bimetallic 3-d coordination polymers [81,83]. The study by Tarafder et al. [83] also studied the hysteresis effect in the pressued induced spin state transitions in cyanide-based Fe–Nb 3-d coordination polymer.

4.1. Linear or 1-d Coordination Polymers

Considering the example of a linear coordination polymeric compound, Fe-triazole complex, the existence of hysteresis through ab initio molecular dynamics (AIMD) simulation was shown in a recent study by Banerjee et al. [71] As shown in left panel of Figure 5, a simple computer designed model of Fe-triazole complex like Fe[(hyetrz)$_3$](4-chlorophenylsulphonate)$_2$ ·3H$_2$O where hyetrz stands for 4-(2'-hydroxyethyl)-1,2,4-triazole was considered in this study. The polymeric Fe(II) triazole compound mentioned above, consists of Fe(II) ions with neighboring iron ions connected through three pyrazole bridges, forming a chain like structure where the counterions and non bonded water molecules form the links between the chains. Even though this compound shows SCO with associated hysteresis, there is hardly any available crystal structure, owing to the fact that the samples were only available in the form of powder, insoluble in water and organic solvents. Absence of accurate crystal structure data pose a problem for computational study. The SCO property in Fe-triazole compounds has been seen for variety of choices of counterions and the residues. A simplified model crystal structure was thus used in the study by Banerjee et al. [71] , where the local environment around Fe(II) ions was kept same as in real compounds, and the distant environment was simplified to a large extend. In the simplified model the counterions were replaced by F$^-$ with the substituent being chosen as CH$_3$ and non-bonded water molecules were removed for the sake of simplification (cf. top panel of Figure 5).

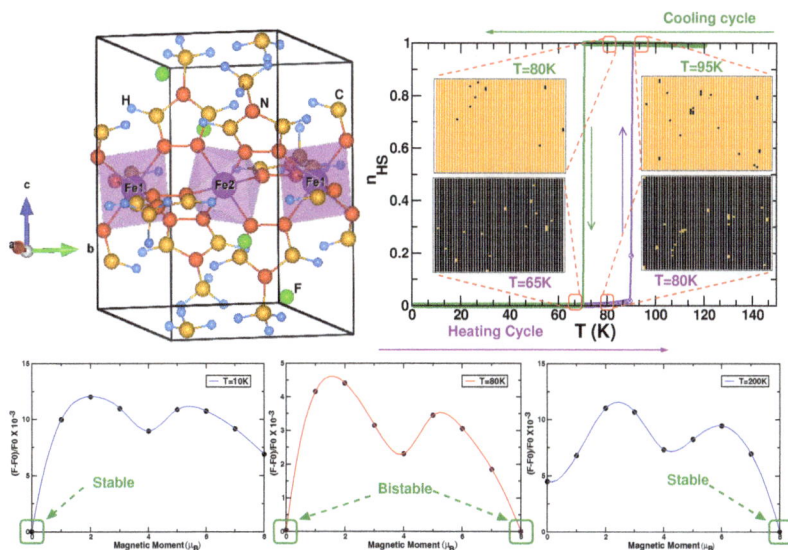

Figure 5. Figure showing crystal structure of Fe-triazole complex along with AIMD and DFT based model Hamiltonian MC study on Fe-triazole. The top left panel shows the simplified computer designed model of Fe[(hyetrz)$_3$](4-chlorophenylsulphonate)$_2$·3H$_2$O. The medium sized golden and red balls represent N and C, respectively. The small sky blue balls represent the H atoms, while the green medium sized balls denote counterion F$^-$. The top right panel shows HS fraction as a function of increasing and decreasing temperature, as given in MC study with DFT based elastic and magnetic interactions. The insets are the snapshots of pseudo-spin configurations in a 50 × 50 lattice with LS(HS) state shown in black (yellow), at T = 65 K and 80 K in the heating cycle and at 80 K and 95 K in the cooling cycle. At T = 80 K, in the HS and LS states are seen to coexist demonstrating the bistability in the system. The bottom panel shows the free energy profiles at three different temperatures T = 10 K, 80 K, and 200 K. Figure has been adapted from Banerjee et al. [71].

The AIMD calculations were carried out within the framework of DFT implemented on a plane wave basis set in Vienna Ab-initio Simulation package (VASP) [87]. The exchange correlation functional was approximated by Perdew-Burke-Ernzerhof [88] (PBE)+U functional. In contrast to classical molecular dynamics where the ground state potential energy surface is represented in the force field, in case of Ab initio molecular dynamics the electronic behavior is obtained in a first principles manner by using a quantum mechanical method, like DFT as is the case under discussion. Born-Oppenheimer AIMD was carried out which is a MD on a Born-Oppenheimer surface. Calculations were carried out using the canonical ensemble and a thermostat was used to exchange the energy of endothermic and exothermic processes. Corresponding to the thermal expansion of Fe-triazole at the various studied temperatures the atomic positions were relaxed at 0 K at various chosen volumes, until the maximum forces on atoms were less than 0.01 eV/A^0. The temperature was then increased within the NVT ensemble from T = 0 K to the desired final temperature using a Berendsen thermostat with a time step of 1 fs for each MD step. At the final step the system was thermalized for a time duration of 1 ps using the Nosé-Hoover thermostat. At the very end, a microcanonical NVE ensemble thermalization was carried out to evaluate the free energy of the system at the final temperature.

The AIMD calculations were carried out at temperatures of 10 K, 80 K, and 200 K, the choice of temperatures being dictated by the experimental measure of susceptibility on Fe-triazole, [82] for which a transition is found to occur at ∼80 K with a hysteresis width of ∼20 K. Free energy profiles were created by plotting the Free energy as a function of the fixed magnetic moments and it was observed that for 10 K the LS state S = 0 was the lowest energy state while for 200 K the HS state was the lowest energy state with S = 2. However a clear bistability was observed at 80 K (cf. bottom panel in Figure 5) proving the capability of the method to capture the cooperativity developed in the system.

Obtaining the intra chain elastic interaction parameters were shown to be imperative in case of Fe-triazole since it is the combination of intra chain elastic parameters that determines whether the effective elastic interaction is ferro- or antiferro- type. In this case the spin dependent intra chain elastic interactions, as defined in the previous section setting the small displacement q = 0.005 turn out to be as follows: $E_{v(++)} = 72$ K, $E_{v(--)} = 101$ K, and $E_{v(+-)} = 80$ K. This clearly shows that the intra chain elastic interaction is of antiferro type. It was thus inferred that the observed bistability and consequently hysteresis in this linear polymeric SOC system is driven by the magnetic exchange interaction.

In a previous work by Jeschke et al. [82] the magnetic exchanges in Fe-triazole complex was computed from DFT. This paper is particularly notable since from ab-initio perspective, this was first study to point out that magnetic exchange might be as important as elastic interaction in giving rise to the cooperativity in these compounds. The magnetic superexchange interactions between neighboring Fe ions were calculated by carrying out DFT+U total energy calculations of FM and AFM orientations of Fe spins considering the predominance of the HS states between the ions. The energy difference between the two spin orientations gave rise to a antiferromagnetic superexchange interaction of magnitude 18 K.

Plugging in the DFT estimated values of material specific constants with good approximations for parameters like Δ which are not easy to calculate using DFT methods, MC simulations on first-principles derived model Hamiltonian carried out by Banerjee et al. [71]. The results are shown in the top right panel of Figure 5. The calculated LS-HS transition temperature was found to be $T_C = 80$ K with the large hysteresis, and the width of the hysteresis was found to be $\Delta T = 20$ K, both in excellent agreement with that obtained from the measured susceptibility reported for the real compound, [82] proving the effectiveness of first-principles approach. Top right panel of Figure 5, also shows the snapshots of the pseudo-spin configurations of the lattice at different temperatures, $T < T_C$, $T > T_C$ and $T = T_C$ during the heating and the cooling cycles of the MC simulation. The snapshot configuration of pseudo-spins at $T = T_C$ show sites with primarily HS configurations for the cooling cycle and LS configurations for the heating cycle, establishing signature of bistability in this material, in conformity with the results obtained from rigorous AIMD simulations.

4.2. Hybrid Perovskites

While the hybrid perovskites have been discussed heavily in literature for many different potential applications, as discussed in introduction, they have not been explored as possible SCO materials. The pressure-driven cooperative spin crossover in metal–organic hybrid perovskites, apparently unconventional materials for SCO, was proposed and investigated by Banerjee et al. [46] using DFT based electronic structure calculations. By considering two popular hybrid perovskite materials, viz., Dimethylammonium Iron Formate (DMAFeF) and Hydroxylamine Iron Formate (HAFeF), [46] crystal structures of which are shown in the left panel of Figure 6, it was demonstrated that application of external pressure can drive spin state transition along with associated hysteresis in these compounds. Interestingly the width of hysteresis turned out to be rather different for the two compounds with two different cations DMA and HA, indicating possible tuning of cooperativity by chemical means.

The mechanical strengths of DMAFeF and HAFeF were calculated using DFT+U calculations in plane wave basis to understand the microscopic origin of the quantitative differences in response to the applied pressure of the two perovskite hybrids. It was argued that due to the very different strength and number of H bonding in the two compounds, the lattice for HAFeF is more rigid compared to that of DMAFeF. The difference in the H bondings in the two cases was understood in terms of enhanced polarity of O–H bond compared to the N–H bond due to the differences in electronegativity of N and O with O being more electronegative compared to N. The magnitudes of the calculated bulk moduli of the two systems confirmed this fact. The bulk modulus turned out to be 21.55 GPa for DMAFeF and

24.27 GPa for HAFeF respectively. The critical pressure required to induce spin-switching through change in Fe–O bond-length was found to be significantly larger in more rigid lattice of HAFeF compared to that in less rigid lattice of DMAFeF. This was reflected in different values of $P_{C\uparrow}$ (critical pressure in increasing pressure path) in the two compounds as seen from Figure 6. A significant hysteresis was found in both compounds in terms of differences between $P_{C\uparrow}$ and $P_{C\downarrow}$ (critical pressure in increasing and decreasing pressure path, respectively).

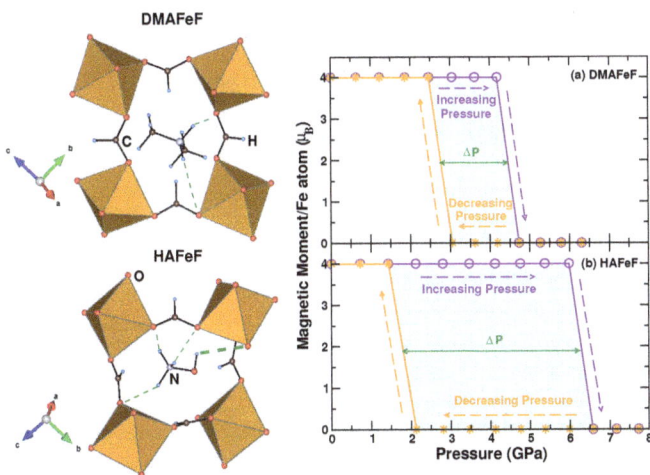

Figure 6. Figure showing the effect of exerting and releasing pressure on DMAFeF and HAFeF. The **left** panel shows the crystal structures of DMAFeF and HAFeF. In the crystal structure the golden octahedra denote the Fe–O octahedra. C, N, H and O are marked in the figure. The N–H··· O H bonding is marked with a thin dashed green line whereas the O–H···O H bonding is marked with a thick dashed green line to show the relative difference of their strengths arising due to the difference in polarity of N–H and O–H bonds; The **right** panel shows comparatively the plots of Magnetic Moment/Fe atom as a function of pressure for both DMAFeF (**top**) and HAFeF (**bottom**). Figure has been adapted from Banerjee et al. [46].

As discussed in Section 3, the interplay between elastic and magnetic interaction in building up cooperativity relies significantly on the spin-dependent rigidity of the lattice. In the study by Banerjee et al. [46] the authors found DMAFeF to be very weakly ferroelastic with $e_{+-} \simeq \sqrt{e_{++}e_{--}}$ while HAFeF was found to be strongly ferroelastic with $e_{+-} > \sqrt{e_{++}e_{--}}$. The effective elastic constants for DMAFeF and HAFeF were found to be 3.52 K and 8.93 K respectively thus making it evident that the contribution to cooperativity and hence the enhanced hysteresis width in HAFeF comes primarily from elastic interaction. The calculated magnetic superexchanges turned out to be of similar values for DMAFeF and HAFeF (3.19 K and 2.85 K) and negative indicating antiferromagnetic nature of magnetic exchanges. This lead to the conclusion that while the magnetic exchanges in the two compounds are of same sign and similar strengths, the spin-dependent elastic interactions are of ferroelastic nature with significantly larger strength for HAFeF compared to DMAFeF. The latter is responsible for HAFeF having a sufficiently larger hysteresis loop width compared to DMAFeF. It was thus concluded that the primary factor responsible in driving cooperativity in these formate frameworks is the spin-dependent lattice effect, which gets assisted by the magnetic exchange. This aspect in hybrid perovskites is very different from the previously described case of linear coordination polymers in which magnetic superexchange was found to have a dominating effect on cooperativity.

5. Summary and Outlook

In this review, we focused on the issue of cooperativity in SCO properties exhibited by coordination polymers with extended solid state structures, as in linear coordination polymers, or hybrid perovskites. This topic needs attention from the perspective of understanding of microscopic mechanisms driving cooperativity, and the use of the obtained understanding in designing of suitable materials aimed at improving the performance of the devices based on them. In this article, we have discussed different model studies proposed for microscopic understanding of cooperativity. Out of these studies, it emerged that the cooperativity-assisted hysteresis is primarily driven by magnetic super-exchanges in case of systems with elastic interactions of antiferroelastic type. In case of systems described by ferroelastic interactions, on the other hand, it is the elastic interaction that builds up the cooperativity-assisted hysteresis with magnetic super-exchange playing a secondary role. It was further demonstrated through rigorous first-principles calculations, considering specific examples, that while the chain polymeric compounds with SCO centers connected by chemical bonds belong to first category, the hybrid perovskites where H-bondings are important belong to the second category. We would like to stress at this point that the work described here on hybrid perovskites is theoretical proposal for which experimental validation needs to be carried out. So far this has remained as an unexplored territory for hybrid perovskites which deserves attention.

This review focused solely on theoretical studies of temperature and pressure induced cooperative SCO. A much less studied and worth exploring area from theoretical point of view would be study of cooperativity in light induced spin state trapping. While empirical theories have been proposed within the non-adiabatic multiphonon framework [89] for mononuclear compounds in terms of ΔE^0_{HL}, the energy difference between the lowest vibrational levels of HS and LS states, and the change of metal–ligand bond length Δr_{HL}, the ab-initio description of the complete process is lacking, apart from few quantum-chemical calculations [90,91] studying the electronic structure of excited state geometries. More importantly, extension to multinuclear systems with possible cooperative effect is non-existent. The study by Létard et al. [61] through irradiation of the sample at low temperature with laser coupled to an optical fiber within a SQUID cavity showed that the temperature dependence of the photomagnetic properties of mononuclear and multinuclear systems to be very different. To the best of our knowledge no microscopic theory has been developed to explain this difference.

This review also focused on Fe(II) based systems, which are so far the most popular ones. However, in addition to Fe or Co based systems, there are few examples of SCO in Mn(III) [92,93]. Mn(III) is a particularly interesting candidate for SCO as it should exhibit a significant Jahn–Teller effect in its HS state. It is an interesting question to ask how the Jahn–Teller distortion effects lattice contribution to cooperativity in terms of influencing the spin-lattice coupling. This may result into different profiles of SCO, like double transition, compared to those observed for Fe(II). This issue demands future attention.

Finally, hybrid perovskites which are the new candidates proposed for observing cooperative SCO effect, should be also explored for LIESST effect both from experimental and theoretical point of view. Given the observation of multiferroicity, already reported in literature for hybrid perovskites, it will be also worth to investigate any exotic magnetic ordering of HS Fe(II)'s leading to breaking of inversion symmetry, and thus resulting into magneto-electric coupling.

Acknowledgments: Hrishit Banerjee and Tanusri Saha-Dasgupta acknowledge the support of Department of Science and Technology (DST) India, Nanomission for funding through Thematic Unit of Excellence on Computational Materials Science.

Author Contributions: The literature survey was carried out by Hrishit Banerjee and Sudip Chakraborty. The manuscript was written by Hrishit Banerjee and Tanusri Saha-Dasgupta. The project was coordinated by Tanusri Saha-Dasgupta.

Conflicts of Interest: The authors declare no conflict of interest.

Inorganics **2017**, *5*, 47

Abbreviations

The following abbreviations are used in this manuscript:

SCO	Spin crossover
MOF	Metal organic framework
LS	Low Spin
HS	High Spin
EXAFS	Extended X-ray absorption Fine Structure
FT-IR	Fourier Transform Infrared Spectroscopy
UV-vis	Ultraviolet and visible
EPR	Electron Paramagnetic Resonance
μSR	Muon Spin Relaxation
PAS	Positron Annihilation Spectroscopy
NRS	Nuclear Resonant Scattering of synchroton radiation
LIESST	Light induced spin state trapping
SQUID	Superconducting quantum interference device
trz	triazole
btr	bis-triazole
bipy	4,4'-bipyridine
bpea	1,2-bis(4-pyridyl)ethane
bppa	1,3-bis(4-pyridyl)propane
hyetrz	4-(2'-hydroxyethyl)-1,2,4-triazole
MC	Monte Carlo
DFT	Density functional theory
CASSCF	Complete active space self consistent field
CASPT2	Complete active space with second order perturbation theory
CC	Couple cluster theory
CCSD(T)	Coupled Cluster single-double and perturbative triple
VASP	Vienna Ab-initio Simulation Package
PBE	Perdew, Burke and Ernzerhof
AIMD	Ab-initio Molecular dynamics
DMAFeF	Dimethyl ammonium Iron Formate
HAFeF	Hydroxyl ammonium Iron Formate

References

1. Létard, J.-F.; Guionneau, P.; Goux-Capes, L. *Spin Crossover in Transition Metal Compounds I-III*; Gütlich, P., Goodwinpp, H., Eds.; Springer: Berlin, Germany, 2004; pp. 221–249.
2. Linares, J.; Codjovi, E.; Garcia, Y. Pressure and Temperature Spin Crossover Sensors with Optical Detection. *Sensors* **2012**, *12*, 4479–4492.
3. Cobo, S.; Molnár, G.; Real, J.A.; Bousseksou, A. Multilayer Sequential Assembly of Thin Films that Display Room-Temperature Spin Crossover with Hysteresis. *Angew. Chem. Int. Ed.* **2006**, *45*, 5786–5789.
4. Saha-Dasgupta, T.; Oppeneer, P.M. Computational design of magnetic metal–organic complexes and coordination polymers with spin-switchable functionalities. *MRS Bull.* **2014**, *39*, 614–620.
5. Brooker, S.; Kitchen, J.A. Nano-magnetic materials: Spin crossover compounds vs. single molecule magnets vs. single chain magnets. *Dalton Trans.* **2009**, 7331–7340, doi:10.1039/B907682D.
6. Ohkoshi, S.I.; Imoto, K.; Tsunobuchi, Y.; Takano, S.; Tokoro, H. Light-induced spin-crossover magnet. *Nat. Chem.* **2011**, *3*, 564–569.
7. Halder, G.J.; Kepert, C.J.; Moubaraki, B.; Murray, K.S.; Cashion, J.D. Guest-Dependent Spin Crossover in a Nanoporous Molecular Framework Material. *Science* **2002**, *298*, 1762–1765.
8. Olguín, J.; Brooker, S. Spin crossover active iron(II) complexes of selected pyrazole-pyridine/pyrazine ligands. *Coord. Chem. Rev.* **2011**, *255*, 203–240.

9. Batten, S.R.; Champness, N.R.; Chen, X.M.; Garcia-Martinez, J.; Kitagawa, S.; Öhrström, L.; O'Keeffe, M.; Suh, M.P.; Reedijk, J. Terminology of metal–organic frameworks and coordination polymers (IUPAC Recommendations 2013). *Pure Appl. Chem.* **2013**, *85*, 1715–1724.

10. Ozarowski, A.; Shunzhong, Y.; McGarvey, B.R.; Mislankar, A.; Drake, J.E.   EPR and NMR study of the spin-crossover transition in bis(4,4'-bi-1,2,4-triazole)bis(thiocyanato)iron hydrate and bis(4,4'-bi-1,2,4-triazole)bis(selenocyanato)iron hydrate.   X-ray structure determination of Fe(4,4'-bi-1,2,4-triazole)$_2$(SeCN)$_2$·H$_2$O. *Inorg. Chem.* **1991**, *30*, 3167–3174, doi:10.1021/ic00016a013.

11. Yann Garcia, N.N.A.; Naik, A.D. Crystal Engineering of Fe-II Spin Crossover Coordination Polymers Derived from Triazole or Tetrazole Ligands. *Chimia* **2013**, *67*, 411–418, doi:10.2533/chimia.2013.411.

12. Morris, R.; Wheatley, P. Gas Storage in Nanoporous Materials. *Angew. Chem. Int. Ed.* **2008**, *47*, 4966–4981.

13. Kreno, L.E.; Leong, K.; Farha, O.K.; Allendorf, M.; Van Duyne, R.P.; Hupp, J.T. Metal–Organic Framework Materials as Chemical Sensors. *Chem. Rev.* **2012**, *112*, 1105–1125, doi:10.1021/cr200324t, PMID:22070233.

14. Horcajada, P.; Chalati, T.; Serre, C.; Gillet, B.; Sebrie, C.; Baati, T.; Eubank, J.F.; Heurtaux, D.; Clayette, P.; Kreuz, C.; et al. Porous metal–organic-framework nanoscale carriers as a potential platform for drug delivery and imaging. *Nat. Mater.* **2010**, *9*, 172–178.

15. Czaja, A.U.; Trukhan, N.; Muller, U. Industrial applications of metal–organic frameworks. *Chem. Soc. Rev.* **2009**, *38*, 1284–1293.

16. Cheetham, A.K.; Rao, C.N.R. There's Room in the Middle. *Science* **2007**, *318*, 58–59.

17. Garcia, Y.; Niel, V.; Muñoz, M.C.; Real, J.A. Spin Crossover in 1D, 2D and 3D Polymeric Fe(II) Networks. In *Spin Crossover in Transition Metal Compounds I*; Gütlich, P., Goodwin, H., Eds.; Springer: Berlin/Heidelberg, Germany, 2004; pp. 229–257.

18. Krober, J.; Codjovi, E.; Kahn, O.; Groliere, F.; Jay, C. A spin transition system with a thermal hysteresis at room temperature. *J. Am. Chem. Soc.* **1993**, *115*, 9810–9811, doi:10.1021/ja00074a062.

19. Setifi, F.; Milin, E.; Charles, C.; Thétiot, F.; Triki, S.; Gómez-García, C.J. Spin Crossover Iron(II) Coordination Polymer Chains: Syntheses, Structures, and Magnetic Characterizations of [Fe(aqin)$_2$(μ_2-M(CN)$_4$)] (M = Ni(II), Pt(II), aqin = Quinolin-8-amine). *Inorg. Chem.* **2014**, *53*, 97–104, doi:10.1021/ic401721x.

20. Schonfeld, S.; Lochenie, C.; Thoma, P.; Weber, B. 1D iron(II) spin crossover coordination polymers with 3,3[prime or minute]-azopyridine-kinetic trapping effects and spin transition above room temperature. *CrystEngComm* **2015**, *17*, 5389–5395.

21. Bauer, W.; Lochenie, C.; Weber, B. Synthesis and characterization of 1D iron(II) spin crossover coordination polymers with hysteresis. *Dalton Trans.* **2014**, *43*, 1990–1999.

22. Manser, J.S.; Christians, J.A.; Kamat, P.V. Intriguing Optoelectronic Properties of Metal Halide Perovskites. *Chem. Rev.* **2016**, *116*, 12956–13008, doi:10.1021/acs.chemrev.6b00136, PMID:27327168.

23. Mitzi, D.B.; Feild, C.A.; Harrison, W.T.A.; Guloy, A.M. Conducting tin halides with a layered organic-based perovskite structure. *Nature* **1994**, *369*, 467–469.

24. Náfrádi, B.; Szirmai, P.; Spina, M.; Lee, H.; Yazyev, O.V.; Arakcheeva, A.; Chernyshov, D.; Gibert, M.; Forró, L.; Horváth, E. Optically switched magnetism in photovoltaic perovskite CH$_3$NH$_3$(Mn:Pb)I$_3$. *Nat. Commun.* **2016**, *7*, 13406.

25. Kojima, A.; Teshima, K.; Shirai, Y.; Miyasaka, T. Organometal Halide Perovskites as Visible-Light Sensitizers for Photovoltaic Cells. *J. Am. Chem. Soc.* **2009**, *131*, 6050–6051, doi:10.1021/ja809598r, PMID:19366264.

26. Zhou, H.; Chen, Q.; Li, G.; Luo, S.; Song, T.B.; Duan, H.S.; Hong, Z.; You, J.; Liu, Y.; Yang, Y. Interface engineering of highly efficient perovskite solar cells. *Science* **2014**, *345*, 542–546.

27. Heo, J.H.; Song, D.H.; Han, H.J.; Kim, S.Y.; Kim, J.H.; Kim, D.; Shin, H.W.; Ahn, T.K.; Wolf, C.; Lee, T.W.; et al. Planar CH$_3$NH$_3$PbI$_3$ Perovskite Solar Cells with Constant 17.2% Average Power Conversion Efficiency Irrespective of the Scan Rate. *Adv. Mater.* **2015**, *27*, 3424–3430.

28. Burschka, J.; Pellet, N.; Moon, S.J.; Humphry-Baker, R.; Gao, P.; Nazeeruddin, M.K.; Gratzel, M. Sequential deposition as a route to high-performance perovskite-sensitized solar cells. *Nature* **2013**, *499*, 316–319.

29. Liu, M.; Johnston, M.B.; Snaith, H.J. Efficient planar heterojunction perovskite solar cells by vapour deposition. *Nature* **2013**, *501*, 395–398.

30. Wang, Z.; Hu, K.; Gao, S.; Kobayashi, H. Formate-Based Magnetic Metal–Organic Frameworks Templated by Protonated Amines. *Adv. Mater.* **2010**, *22*, 1526–1533.

31. Jain, P.; Dalal, N.S.; Toby, B.H.; Kroto, H.W.; Cheetham, A.K. Order-Disorder Antiferroelectric Phase Transition in a Hybrid Inorganic–Organic Framework with the Perovskite Architecture. *J. Am. Chem. Soc.* **2008**, *130*, 10450–10451, doi:10.1021/ja801952e, PMID:18636729.

32. Jain, P.; Ramachandran, V.; Clark, R.J.; Zhou, H.D.; Toby, B.H.; Dalal, N.S.; Kroto, H.W.; Cheetham, A.K. Multiferroic Behavior Associated with an Order-Disorder Hydrogen Bonding Transition in Metal–Organic Frameworks (MOFs) with the Perovskite ABX_3 Architecture. *J. Am. Chem. Soc.* **2009**, *131*, 13625–13627, doi:10.1021/ja904156s, PMID:19725496.

33. Weng, D.F.; Wang, Z.M.; Gao, S. Framework-structured weak ferromagnets. *Chem. Soc. Rev.* **2011**, *40*, 3157–3181.

34. Stroppa, A.; Jain, P.; Barone, P.; Marsman, M.; Perez-Mato, J.M.; Cheetham, A.K.; Kroto, H.W.; Picozzi, S. Electric Control of Magnetization and Interplay between Orbital Ordering and Ferroelectricity in a Multiferroic Metal–Organic Framework. *Angew. Chem. Int. Ed.* **2011**, *50*, 5847–5850.

35. Di Sante, D.; Stroppa, A.; Jain, P.; Picozzi, S. Tuning the Ferroelectric Polarization in a Multiferroic Metal–Organic Framework. *J. Am. Chem. Soc.* **2013**, *135*, 18126–18130, doi:10.1021/ja408283a, PMID:24191632.

36. Stroppa, A.; Barone, P.; Jain, P.; Perez-Mato, J.M.; Picozzi, S. Hybrid Improper Ferroelectricity in a Multiferroic and Magnetoelectric Metal–Organic Framework. *Adv. Mater.* **2013**, *25*, 2284–2290.

37. Jain, P.; Stroppa, A.; Nabok, D.; Marino, A.; Rubano, A.; Paparo, D.; Matsubara, M.; Nakotte, H.; Fiebig, M.; Picozzi, S.; et al. Switchable electric polarization and ferroelectric domains in a metal–organic-framework. *Npj Quant. Mater.* **2016**, *1*, 16012.

38. Tian, Y.; Stroppa, A.; Chai, Y.; Yan, L.; Wang, S.; Barone, P.; Picozzi, S.; Sun, Y. Cross coupling between electric and magnetic orders in a multiferroic metal–organic framework. *Sci. Rep.* **2014**. *4*, 6062.

39. Tian, Y.; Stroppa, A.; Chai, Y.S.; Barone, P.; Perez-Mato, M.; Picozzi, S.; Sun, Y. High-temperature ferroelectricity and strong magnetoelectric effects in a hybrid organic–inorganic perovskite framework. *Phys. Status Solidi Rapid Res. Lett.* **2015**, *9*, 62–67.

40. Zhao, W.P.; Shi, C.; Stroppa, A.; Di Sante, D.; Cimpoesu, F.; Zhang, W. Lone-Pair-Electron-Driven Ionic Displacements in a Ferroelectric Metal–Organic Hybrid. *Inorg. Chem.* **2016**, *55*, 10337–10342, doi:10.1021/acs.inorgchem.6b01545, PMID:27676140.

41. Gómez-Aguirre, L.C.; Pato-Doldán, B.; Stroppa, A.; Yáñez-Vilar, S.; Bayarjargal, L.; Winkler, B.; Castro-García, S.; Mira, J.; Sánchez-Andújar, M.; Señarís-Rodríguez, M.A. Room-Temperature Polar Order in [NH4][Cd(HCOO)3]—A Hybrid Inorganic–Organic Compound with a Unique Perovskite Architecture. *Inorg. Chem.* **2015**, *54*, 2109–2116, doi:10.1021/ic502218n, PMID:25664382.

42. Kamminga, M.E.; Stroppa, A.; Picozzi, S.; Chislov, M.; Zvereva, I.A.; Baas, J.; Meetsma, A.; Blake, G.R.; Palstra, T.T.M. Polar Nature of $(CH_3NH_3)_3Bi_2I_9$ Perovskite-Like Hybrids. *Inorg. Chem.* **2017**, *56*, 33–41, doi:10.1021/acs.inorgchem.6b01699, PMID:27626290.

43. Ptak, M.; Maczka, M.; Gagor, A.; Sieradzki, A.; Stroppa, A.; Di Sante, D.; Perez-Mato, J.M.; Macalik, L. Experimental and theoretical studies of structural phase transition in a novel polar perovskite-like $[C_2H_5NH_3][Na_{0.5}Fe_{0.5}(HCOO)_3]$ formate. *Dalton Trans.* **2016**, *45*, 2574–2583.

44. Ghosh, S.; Di Sante, D.; Stroppa, A. Strain Tuning of Ferroelectric Polarization in Hybrid Organic Inorganic Perovskite Compounds. *J. Phys. Chem. Lett.* **2015**, *6*, 4553–4559, doi:10.1021/acs.jpclett.5b01806, PMID:26512946.

45. Mazzuca, L.; Cañadillas-Delgado, L.; Rodríguez-Velamazán, J.A.; Fabelo, O.; Scarrozza, M.; Stroppa, A.; Picozzi, S.; Zhao, J.P.; Bu, X.H.; Rodríguez-Carvajal, J. Magnetic Structures of Heterometallic M(II)–M(III) Formate Compounds. *Inorg. Chem.* **2017**, *56*, 197–207, doi:10.1021/acs.inorgchem.6b01866, PMID:27935298.

46. Banerjee, H.; Chakraborty, S.; Saha-Dasgupta, T. Cationic Effect on Pressure Driven Spin-State Transition and Cooperativity in Hybrid Perovskites. *Chem. Mater.* **2016**, *28*, 8379–8384, doi:10.1021/acs.chemmater.6b03755.

47. Cambi, L.; Szegö, L. Über die magnetische Susceptibilität der komplexen Verbindungen. *Berichte der Deutschen Chemischen Gesellschaft (A B Ser.)* **1931**, *64*, 2591–2598.

48. Stoufer, R.C.; Busch, D.H.; Hadley, W.B. Unusual Magnetic Properties of Some Six-Coördinate Cobalt(II) Complexes 1—Electronic Isomers. *J. Am. Chem. Soc.* **1961**, *83*, 3732–3734, doi:10.1021/ja01478a051.

49. Madeja, K.; König, E. Zur frage der bindungsverhältnise in komplexverbindungen des eisen(II) mit 1,10-phenanthrolin. *J. Inorg. Nucl. Chem.* **1963**, *25*, 377–385.

50. Kojima, N.; Murakami, Y.; Komatsu, T.; Yokoyama, T. EXAFS study on the spin-crossover system, $[Fe(4-NH_2trz)_3](R-SO_3)_2$. *Synth. Met.* **1999**, *103*, 2154.

51. Okabayashi, J.; Ueno, S.; Wakisaka, Y.; Kitazawa, T. Temperature-dependent EXAFS study for spin crossover complex: Fe(pyridine)$_2$Ni(CN)$_4$. *Inorg. Chim. Acta* **2015**, *426*, 142–145.

52. Koenig, E.; Madeja, K.   5T$_2$-1A$_1$ Equilibriums in some iron(II)-bis(1,10-phenanthroline) complexes. *Inorg. Chem.* **1967**, *6*, 48–55, doi:10.1021/ic50047a011.

53. Gütlich, Philipp, G.H.A. *Spin Crossover in Transition Metal Compounds I*; Springer: Berlin, Germany, 2004.

54. Molnár, G.; Kitazawa, T.; Dubrovinsky, L.; McGarvey, J.J.; Bousseksou, A.   Pressure tuning Raman spectroscopy of the spin crossover coordination polymer Fe(C$_5$H$_5$N)$_2$[Ni(CN)$_4$]. *J. Phys. Condens. Matter* **2004**, *16*, S1129.

55. Tuchagues, J.P.; Bousseksou, A.; Molnár, G.; McGarvey, J.J.; Varret, F. The Role of Molecular Vibrations in the Spin Crossover Phenomenon. In *Spin Crossover in Transition Metal Compounds III*; Springer: Berlin/Heidelberg, Germany, 2004; pp. 84–103.

56. Hauser, A. Light-Induced Spin Crossover and the High-Spin→Low-Spin Relaxation. In *Spin Crossover in Transition Metal Compounds II*; Springer: Berlin/Heidelberg, Germany, 2004; pp. 155–198.

57. Krivokapic, I.; Zerara, M.; Daku, M.L.; Vargas, A.; Enachescu, C.; Ambrus, C.; Tregenna-Piggott, P.; Amstutz, N.; Krausz, E.; Hauser, A. Spin-crossover in cobalt(II) imine complexes. *Coord. Chem. Rev.* **2007**, *251*, 364–378.

58. Gütlich, P.; Gaspar, A.B.; Garcia, Y. Spin state switching in iron coordination compounds. *Beilstein J. Org. Chem.* **2013**, *9*, 342–391.

59. Decurtins, S.; Gütlich, P.; Köhler, C.; Spiering, H.; Hauser, A. Light-induced excited spin state trapping in a transition-metal complex: The hexa-1-propyltetrazole-iron(II) tetrafluoroborate spin-crossover system. *Chem. Phys. Lett.* **1984**, *105*, 1–4.

60. Hauser, A.   Reversibility of light-induced excited spin state trapping in the Fe(ptz)$_6$(BF$_4$)$_2$, and the Zn$_{1-x}$Fe$_x$(ptz)$_6$(BF$_4$)$_2$ spin-crossover systems. *Chem. Phys. Lett.* **1986**, *124*, 543–548.

61. Létard, J.F.; Capes, L.; Chastanet, G.; Moliner, N.; Létard, S.; Real, J.A.; Kahn, O. Critical temperature of the LIESST effect in iron(II) spin crossover compounds. *Chem. Phys. Lett.* **1999**, *313*, 115–120.

62. Enachescu, C.; Machado, H.; Menendez, N.; Codjovi, E.; Linares, J.; Varret, F.; Stancu, A. Static and light induced hysteresis in spin-crossover compounds: Experimental data and application of Preisach-type models. *Phys. B Condens. Matter* **2001**, *306*, 155–160.

63. Baldé, C.; Bauer, W.; Kaps, E.; Neville, S.; Desplanches, C.; Chastanet, G.; Weber, B.; Létard, J.F. Light-Induced Excited Spin-State Properties in 1D Iron(II) Chain Compounds. *Eur. J. Inorg. Chem.* **2013**, *2013*, 2744–2750.

64. Weber, B.; Tandon, R.; Himsl, D. Synthesis, Magnetic Properties and X-ray Structure Analysis of a 1-D Chain Iron(II) Spin Crossover Complex with wide Hysteresis. *Zeitschrift für Anorganische und Allgemeine Chemie* **2007**, *633*, 1159–1162.

65. Bauer, W.; Scherer, W.; Altmannshofer, S.; Weber, B. Two-Step versus One-Step Spin Transitions in Iron(II) 1D Chain Compounds. *Eur. J. Inorg. Chem.* **2011**, *2011*, 2803–2818.

66. Weber, B.; Kaps, E.S.; Desplanches, C.; Létard, J.F. Quenching the Hysteresis in Single Crystals of a 1D Chain Iron(II) Spin Crossover Complex. *Eur. J. Inorg. Chem.* **2008**, *2008*, 2963–2966.

67. Boukheddaden, K.; Miyashita, S.; Nishino, M. Elastic interaction among transition metals in one-dimensional spin-crossover solids. *Phys. Rev. B* **2007**, *75*, 094112.

68. Willenbacher, N.; Spiering, H. The elastic interaction of high-spin and low-spin complex molecules in spin-crossover compounds. *J. Phys. C Solid State Phys.* **1988**, *21*, 1423–1439.

69. Spiering, H.; Willenbacher, N.   Elastic interaction of high-spin and low-spin complex molecules in spin-crossover compounds. II. *J. Phys. Condens. Matter* **1989**, *1*, 10089–10105.

70. Spiering, H.; Boukheddaden, K.; Linares, J.; Varret, F. Total free energy of a spin-crossover molecular system. *Phys. Rev. B* **2004**, *70*, 184106.

71. Banerjee, H.; Kumar, M.; Saha-Dasgupta, T.   Cooperativity in spin-crossover transition in metalorganic complexes: Interplay of magnetic and elastic interactions. *Phys. Rev. B* **2014**, *90*, 174433.

72. Timm, C. Collective effects in spin-crossover chains with exchange interaction. *Phys. Rev. B* **2006**, *73*, 014423.

73. D'Avino, G.; Painelli, A.; Boukheddaden, K. Vibronic model for spin crossover complexes. *Phys. Rev. B* **2011**, *84*, 104119.

74. Miralles, J.; Castell, O.; Caballol, R.; Malrieu, J.P. Specific CI calculation of energy differences: Transition energies and bond energies. *Chem. Phys.* **1993**, *172*, 33–43.

75. Roos, B.O.; Taylor, P.R.; Siegbahn, P.E. A complete active space SCF method (CASSCF) using a density matrix formulated super-CI approach. *Chem. Phys.* **1980**, *48*, 157–173.

76. Fouqueau, A.; Mer, S.; Casida, M.E.; Daku, L.M.L.; Hauser, A.; Mineva, T.; Neese, F. Comparison of density functionals for energy and structural differences between the high-$[5T_{2g}:(t_{2g})4(e_g)2]$ and low-$[1A_{1g}:(t_{2g})6(e_g)0]$ spin states of the hexaquoferrous cation $[Fe(H_2O)_6]^{2+}$. *J. Chem. Phys.* **2004**, *120*, 9473–9486, doi:10.1063/1.1710046.

77. Ordejón, B.; de Graaf, C.; Sousa, C. Light-Induced Excited-State Spin Trapping in Tetrazole-Based Spin Crossover Systems. *J. Am. Chem. Soc.* **2008**, *130*, 13961–13968, doi:10.1021/ja804506h.

78. Pierloot, K.; Vancoillie, S. Relative energy of the high-$(T_{2g}5)$ and low-$(A_{1g}1)$ spin states of $[Fe(H_2O)_6]^{2+}$, $[Fe(NH_3)_6]^{2+}$, and $[Fe(bpy)_3]^{2+}$: CASPT2 versus density functional theory. *J. Chem. Phys.* **2006**, *125*, 124303, doi:10.1063/1.2353829.

79. Watts, J.D.; Gauss, J.; Bartlett, R.J. Coupled cluster methods with noniterative triple excitations for restricted open shell Hartree Fock and other general single determinant reference functions. Energies and analytical gradients. *J. Chem. Phys.* **1993**, *98*, 8718–8733, doi:10.1063/1.464480.

80. Maldonado, P.; Kanungo, S.; Saha-Dasgupta, T.; Oppeneer, P.M. Two-step spin-switchable tetranuclear Fe(II) molecular solid: Ab initio theory and predictions. *Phys. Rev. B* **2013**, *88*, 020408.

81. Sarkar, S.; Tarafder, K.; Oppeneer, P.M.; Saha-Dasgupta, T. Spin-crossover in cyanide-based bimetallic coordination polymers-insight from first-principles calculations. *J. Mater. Chem.* **2011**, *21*, 13832–13840.

82. Jeschke, H.O.; Salguero, L.A.; Rahaman, B.; Buchsbaum, C.; Pashchenko, V.; Schmidt, M.U.; Saha-Dasgupta, T.; Valentí, R. Microscopic modeling of a spin crossover transition. *New J. Phys.* **2007**, *9*, 448.

83. Tarafder, K.; Kanungo, S.; Oppeneer, P.M.; Saha-Dasgupta, T. Pressure and Temperature Control of Spin-Switchable Metal–Organic Coordination Polymers from Ab Initio Calculations. *Phys. Rev. Lett.* **2012**, *109*, 077203.

84. Dudarev, S.L.; Botton, G.A.; Savrasov, S.Y.; Humphreys, C.J.; Sutton, A.P. Electron-energy-loss spectra and the structural stability of nickel oxide: An LSDA+U study. *Phys. Rev. B* **1998**, *57*, 1505–1509.

85. Alvarez, S. Distortion Pathways of Transition Metal Coordination Polyhedra Induced by Chelating Topology. *Chem. Rev.* **2015**, *115*, 13447–13483, doi:10.1021/acs.chemrev.5b00537.

86. Kresse, G.; Hafner, J. Ab initio molecular-dynamics simulation of the liquid-metal-amorphous-semiconductor transition in germanium. *Phys. Rev. B* **1994**, *49*, 14251–14269.

87. Kresse, G.; Furthmüller, J. Efficient iterative schemes for ab initio total-energy calculations using a plane-wave basis set. *Phys. Rev. B* **1996**, *54*, 11169–11186.

88. Perdew, J.P.; Burke, K.; Ernzerhof, M. Generalized Gradient Approximation Made Simple. *Phys. Rev. Lett.* **1996**, *77*, 3865–3868.

89. Buhks, E.; Navon, G.; Bixon, M.; Jortner, J. Spin conversion processes in solutions. *J. Am. Chem. Soc.* **1980**, *102*, 2918–2923, doi:10.1021/ja00529a009.

90. Hauser, A.; Guetlich, P.; Spiering, H. High-spin .fwdarw. low-spin relaxation kinetics and cooperative effects in the hexakis(1-propyltetrazole)iron bis(tetrafluoroborate) and $[Zn_{1-x}Fe_x(ptz)_6](BF_4)_2$ (ptz = 1-propyltetrazole) spin-crossover systems. *Inorg. Chem.* **1986**, *25*, 4245–4248, doi:10.1021/ic00243a036.

91. Létard, J.F.; Guionneau, P.; Rabardel, L.; Howard, J.A.K.; Goeta, A.E.; Chasseau, D.; Kahn, O. Structural, Magnetic, and Photomagnetic Studies of a Mononuclear Iron(II) Derivative Exhibiting an Exceptionally Abrupt Spin Transition. Light-Induced Thermal Hysteresis Phenomenon. *Inorg. Chem.* **1998**, *37*, 4432–4441, doi:10.1021/ic980107b, PMID:11670580.

92. Morgan, G.G.; Murnaghan, K.D.; Müller-Bunz, H.; McKee, V.; Harding, C.J. A Manganese(III) Complex that Exhibits Spin Crossover Triggered by Geometric Tuning. *Angew. Chem.* **2006**, *118*, 7350–7353.

93. Martinho, P.N.; Gildea, B.; Harris, M.M.; Lemma, T.; Naik, A.D.; Müller-Bunz, H.; Keyes, T.E.; Garcia, Y.; Morgan, G.G. Cooperative Spin Transition in a Mononuclear Manganese(III) Complex. *Angew. Chem.* **2012**, *124*, 12765–12769.

inorganics

MDPI

Article

Synthesis, Crystal Structures and Magnetic Properties of Composites Incorporating an Fe(II) Spin Crossover Complex and Polyoxometalates

Satoshi Kuramochi [1], Takuya Shiga [1], Jamie M. Cameron [1], Graham N. Newton [2] and Hiroki Oshio [1,*]

[1]   Graduate School of Pure and Applied Sciences, University of Tsukuba, Tennodai 1-1-1,
      Tsukuba 305-8571, Japan; kurmochi@dmb.chem.tsukuba.ac.jp (S.K.); shiga@chem.tsukuba.ac.jp (T.S.);
      jamie.c@chem.tsukuba.ac.jp (J.M.C.)
[2]   School of Chemistry, University of Nottingham, University Park, Nottingham NG7 2RD, UK;
      graham.newton@nottingham.ac.uk
*   Correspondence: oshio@chem.tsukuba.ac.jp; Tel.: +81-29-853-4238

Received: 30 June 2017; Accepted: 19 July 2017; Published: 22 July 2017

Abstract: $[Fe(dppOH)_2]^{2+}$ (dppOH = 2,6-di(pyrazol-1-yl)-4-(hydroxymethyl)pyridine) is known to show spin crossover (SCO) behavior and light-induced excited spin state transitions (LIESST). Here, we show that the SCO properties of the $[Fe(dppOH)_2]^{2+}$ complex can be altered by a crystal engineering approach employing counter anion exchange with polyoxometalate (POM) anions. Using this strategy, two new composite materials $(TBA)[Fe(dppOH)_2][PMo_{12}O_{40}]$ (**1**) and $[Fe(dppOH)_2]_3[PMo_{12}O_{40}]_2$ (**2**) (TBA = tetra-n-butylammonium) have been isolated and studied by single crystal X-ray diffraction and magnetic susceptibility measurements. **1** was found to be in a high spin state at 300 K and showed no spin crossover behavior due to a dense packing structure induced by hydrogen bonding between the hydroxyl group of the dppOH ligands and the POM anions. Conversely, **2** contains two crystallographically unique Fe centers, where one is in the low spin state whilst the other is locked in a high spin state in a manner analogous to **1**. As a result, **2** was found to show partial spin crossover behavior around 230 K with a decrease in the $\chi_m T$ value of 1.9 emu·mol^{-1}·K. This simple approach could therefore provide a useful method to aid in the design of next generation spin crossover materials.

Keywords: spin crossover; polyoxometalate; crystal engineering; magnetic properties; iron

---

1. Introduction

Spin crossover (SCO) materials have been studied extensively due to their ability to reversibly switch spin states in response to external stimuli, allowing their potential application in molecular electronic devices [1,2]. Spin crossover behavior is often shown in hexacoordinated Fe(II) [3], Fe(III) [4] and Co(II) [5] complexes. Among them, Fe(II) complexes which have dpp ligands (dpp = 2,6-di(pyrazol-1-yl)pyridine) have been widely studied by Halcrow, Howard and co-workers [6–13]. In addition, $[Fe(dpp)_2]^{2+}$ complexes show light-induced excited spin state transitions (LIESST) [8,10], where the spin state can be controllably switched in response to visible light. Significantly, the spin state of the Fe(II) ion is directly affected by the coordination environment and distortions in the plane of the dpp ligand [11,12].

Polyoxometalates (POMs) are large anions in which transition metal cations in their highest oxidation state are bridged by oxo-anions, leading to a huge variety of unique, nano-sized structures [14–16]. POMs are also of interest due to their appealing properties, including reversible, multi-electron redox behavior, electrocatalytic activity [17], photo-oxidizing properties [18,19] and proton conductivity [20,21], and show great promise for advanced applications. POMs can also act as efficient spacers to magnetically isolate paramagnetic species: for example, Miyasaka, Hayashi and co-workers have recently used POMs as magnetic spacers in Mn^{3+} dimers of a single molecule magnet (SMM) [22].

In this work, the Fe(II) spin crossover complex, $[Fe(dppOH)_2]^{2+}$ (dppOH = 2,6-di(pyrazol-1-yl)-4-(hydroxymethyl)pyridine; Figure 1) and a Keggin-type POM, $[PMo_{12}O_{40}]^{3-}$, are combined to make hybrid composite materials. Specifically, the tridentate ligand dppOH has a pendant hydroxyl group which was expected to show favorable H-bonding interactions with neighboring polyoxoanions. Controlled spin state conversion and modification of the electronic state of $[Fe(dppOH)_2]^{2+}$ were targeted by employing POM anions as crystallographic spacers with potential to display proton acceptor and photo-oxidative properties. Two hybrid materials $(TBA)[Fe(dppOH)_2][PMo_{12}O_{40}]$ (**1**) and $[Fe(dppOH)_2]_3[PMo_{12}O_{40}]_2$ (**2**) (TBA = tetra-n-butylammonium), were successfully isolated using a simple synthetic approach. The spin state of the Fe(II) ions in **1** are shown to be high spin across the entire temperature range measured, which is likely due to its highly distorted coordination geometry. Conversely, **2** has three Fe(II) centers, one of which is shown to be stabilized in a low spin configuration and displays moderate partial spin crossover behavior between 100 and 300 K.

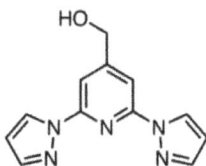

Figure 1. The dppOH (2,6-di(pyrazol-1-yl)-4-(hydroxymethyl)pyridine) ligand.

2. Results

2.1. Synthesis

The tridentate dppOH ligand (dppOH = 2,6-di(pyrazol-1-yl)-4-(hydroxymethyl)pyridine) and the Fe(II) complex, $[Fe(dppOH)_2](BF_4)_2$ were prepared according to previously reported methods [13]. $(TBA)[Fe(dppOH)_2][PMo_{12}O_{40}]\cdot 2CH_3NO_2$ (**1**) and $[Fe(dppOH)_2]_3[PMo_{12}O_{40}]_2\cdot 15H_2O$ (**2**) were synthesized by slow diffusion of the Fe(II) complex into $(TBA)_3[PMo_{12}O_{40}]$ and $H_3PMo_{12}O_{40}$ solutions, respectively. A solution of the POM in 1 mL nitromethane (10 mM) was placed in a glass tube (ϕ = 8 mm), after which a buffer layer of nitromethane and acetonitrile (1:1 v/v, 0.5 mL) was layered on top. Finally, a solution of $[Fe(dppOH)_2](BF_4)_2$ in 1 mL acetone/acetonitrile (2:1 v/v, 10 mM) was carefully layered on top of the middle buffer layer. After around 1 week, crystals suitable for X-ray analysis were obtained.

2.2. Crystal Structures

The structures of **1**·$2CH_3NO_2$ and **2**·$2CH_3NO_2$ were obtained by single crystal X-ray diffraction at 100 K (Figures 2 and 3). The crystal system of **1** is monoclinic and the space group is $P2_1/c$. The asymmetric unit contains one $[Fe(dppOH)_2]^{2+}$ cation, a TBA cation, a $[PMo_{12}O_{40}]^{3-}$ anion, and two nitromethane solvent molecules. Considering charge balance and BVS (bond valence sum) calculations, the valence of the molybdenum ions can be estimated as 6+, in which the overall charge on the $[PMo_{12}O_{40}]$ anion should be 3−.

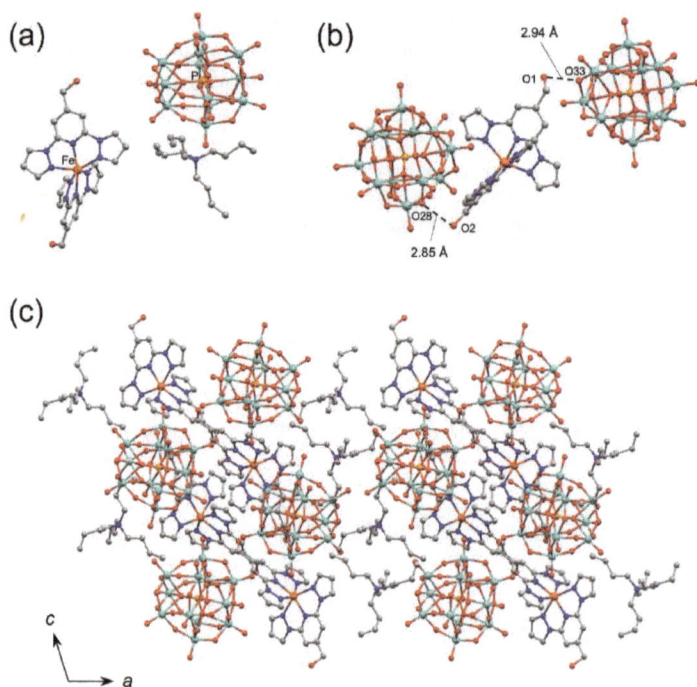

Figure 2. (a) X-ray structure of complex **1**; (b) hydrogen bonding interactions between [Fe(dppOH)$_2$]$^{2+}$ and [PMo$_{12}$O$_{40}$]$^{3-}$ and (c) packing structure of **1** viewed along the crystallographic *b*-axis. Color code: C: gray, N: blue, O: red, P: orange, Fe: brown, Mo: light blue (hydrogen atoms and solvent molecules omitted for clarity).

The Fe(II) ions in **1** were found to be in a distorted octahedral coordination geometry where the angle of *trans*-N(pyridine)–Fe–N(pyridine) (ϕ) and the sum difference of the twelve possible *cis*-N–Fe–N angles (α) from the perpendicular (Σ, where $\Sigma = \Sigma |90 - \alpha|$) [11,12] were found to be $\phi = 158.1°$ and $\Sigma = 151°$, respectively, with an average Fe–N bond length of 2.2 Å. These values are highly typical of high spin Fe(II) and **1** can therefore be assigned as assuming a high spin configuration. In the packing structure of **1**, both pendant hydroxyl groups on the dppOH ligands hydrogen bond to a bridging μ_2-oxo group on the [PMo$_{12}$O$_{40}$]$^{3-}$ anion (Figure 2b), where the distance between the oxygen atoms of the hydroxyl groups and the μ_2-oxo bridges ($d_{O1 \ldots O33}$ and $d_{O2 \ldots O28}$) are 2.94 Å and 2.85 Å, respectively. The dense packing arrangement observed in **1** leads to distorted coordination structure, exerting a strong influence on the coordination environment parameters of the Fe(II) centers and stabilizes its high-spin state.

The crystal system of **2**·2CH$_3$NO$_2$, which was synthesized with the free acid of the POM ([H$_3$PMo$_{12}$O$_{40}$]) as a starting material, is monoclinic with space group $C2/c$. The asymmetric unit contains one and a half [Fe(dppOH)$_2$]$^{2+}$ cationic units and one full [PMo$_{12}$O$_{40}$]$^{3-}$ anionic unit, in addition to one nitromethane molecule. **2** also contains two crystallographically distinct Fe(II) centers, Fe1 and Fe2. As discussed above, the coordination geometry parameters ϕ and Σ were found to be $\phi = 176.1°$ and $\Sigma = 87°$ for Fe1, while those of Fe2 were found to be $\phi = 156.4°$ and $\Sigma = 158°$. The average Fe1–N and Fe2–N bond lengths are 1.9 Å and 2.2 Å, respectively, at 100 K. In this case, the observed values for Fe1 and Fe2 suggest the existence of both low-spin and high-spin states in the Fe(II) ions, respectively. It can also be noted that Fe1 is located on the center of symmetry and shows no H-bonding interactions between [Fe(dppOH)$_2$] and the POM cluster, while Fe2 shows hydrogen bonding interactions at the pendant hydroxyl groups, analogous to those in compound **1**.

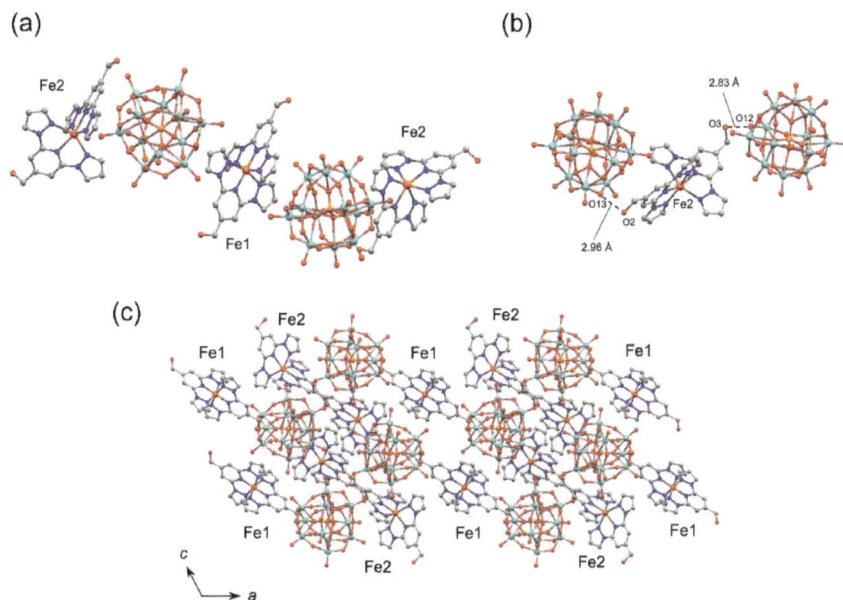

Figure 3. (**a**) X-ray structure of **2**; (**b**) hydrogen bonding interactions between Fe2-complexes and [PMo$_{12}$O$_{40}$]$^{3-}$ and (**c**) the packing structure of **2** viewed along the crystallographic *b*-axis. Color code: C: gray, N: blue, O: red, P: orange, Fe: brown, Mo: light blue (hydrogen atoms and solvent molecules omitted for clarity).

2.3. Temperature Dependence of the Magnetic Susceptibilities

The temperature dependence of the magnetic susceptibility was measured for both **1** and **2** in the range of 1.8–300 K (Figure 4). At 300 K, the $\chi_m T$ value of **1** is 4.51 emu·mol^{-1}·K (g = 2.45). This value confirms that the Fe(II) ion of **1** is in a high spin state (S = 2). At temperatures above 30 K, the $\chi_m T$ value is constant, while below 30 K it decreases sharply due to intermolecular interactions and/or zero-field splitting effects associated with the Fe(II) ions. The $\chi_m T$ value of **2** at 300 K was found to be slightly low for three isolated high spin Fe(II) ions at 11.05 emu·mol^{-1}·K, suggesting that **2** may not be in a fully high spin state at 300 K (i.e., the spin state of **2** may be more accurately described as $(2 + x)$ HS Fe^{II} + (1 − x) LS Fe^{II}). Furthermore, upon cooling, the $\chi_m T$ value of **2** gradually decreases to 9.11 emu·mol^{-1}·K at 100 K, and this magnetic behavior can be ascribed to partial thermal spin crossover behavior. From this data, it can be suggested that **2** contains one low-spin Fe(II) ion and two high-spin Fe(II) ions (g = 2.46) at 100 K, whilst at higher temperatures (300 K), a fraction of low spin Fe(II) ions may undergo thermally-driven spin crossover to the high spin state.

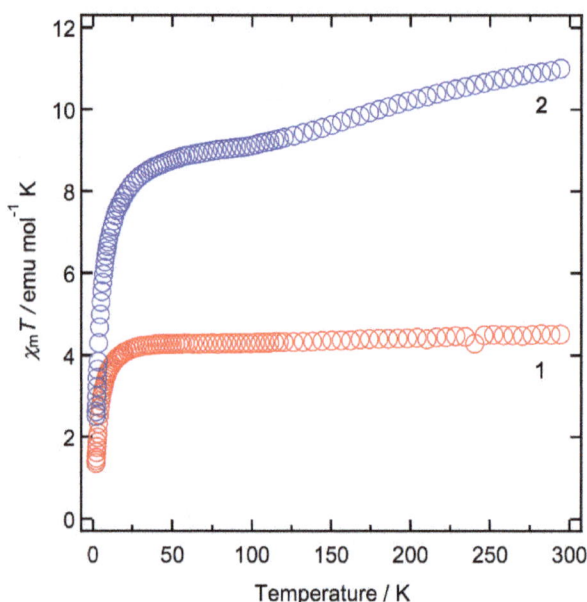

Figure 4. Temperature dependence of the $\chi_m T$ product for **1** (red) and **2** (blue).

3. Discussion

It was previously reported that the analogous complexes [Fe(dppOH)$_2$](BF$_4$)$_2$ and [Fe(dppOH)$_2$](HClO$_4$)$_2$ show spin crossover behavior with $T_{1/2}$ = 271 K and 284 K, respectively [10], whilst the Fe(II) ion in **1** is "locked" in the high-spin state and shows no spin crossover behavior. It follows that the high spin state of Fe(II) is therefore stabilized in the crystal structure of **1**, and we have considered the following explanations. Firstly, the influence of the packing structure formed by the [Fe(dppOH)$_2$]$^{2+}$ cations and [PMo$_{12}$O$_{40}$]$^{3-}$ anions is crucial. From our results, it appears clear that close-packing and H-bonding interactions between the POM clusters and the hydroxyl groups on the pendant (dppOH) ligands lead to significant distortion in the coordination environment of the Fe(II) ions (ϕ = 158.1°, Σ = 151°). This highly-distorted coordination geometry around the Fe(II) ion stabilizes a high-spin state at all temperatures. In the case of compound **1**, the uniform 1:1 dense packing arrangement of the [Fe(dppOH)$_2$] and POM units (seen clearly in Figure 2b) facilitated by these close-packing interactions is supported by an additional layer of charge-balancing TBA cations which can be found in a distinct intermediate space between the groups of closely-associated POM and [Fe(dppOH)$_2$]$^{2+}$ moieties.

On the other hand, **2** is shown to exhibit partial spin crossover behavior. In the crystal structure of **2** at 100 K, two high-spin Fe(II) moieties (Fe2) and one low-spin Fe(II) center (Fe1) can be identified. Where the crystal structure of **2** is similar to that of **1** (in which an observably similar 1:1 dense packing arrangement can be found between the Fe2-complexes and the POM anions), a highly distorted coordination geometry around the Fe2 ions is favored which stabilizes a high spin state in a manner analogous to that in compound **1** (ϕ = 156.4° and Σ = 158°). Conversely, the Fe1 complexes show no significant intermolecular interactions with the POM counter-anions (with a minimum distance d_{O-O} = 3.52 Å between the POM and the pendant hydroxyl group on the dppOH ligands) and are located in a larger crystallographic space (analogous to the TBA cations in compound **1**), with markedly less dense packing than the Fe2 complexes (see Figure 3). As a result, the Fe1 ions experience minimal distortion (ϕ = 176.1° and Σ = 87°) and the structural changes associated with the spin crossover of the

Fe2 complex are unhindered. As a result, the Fe2 complexes show moderate spin crossover behavior with an estimated $T_{1/2}$ value $\approx$ 230 K. We attempted to explore the details of these structural changes by single crystal structural analysis, however it failed as the result of **2** having lost crystallinity at 300 K. In addition, Mössbauer spectroscopy could not be employed in this instance to aid elucidation of the electronic state of the Fe ions because the Mo-containing POM clusters strongly absorb γ-rays.

4. Materials and Methods

4.1. Materials

The dppOH ligand and Fe(II) SCO complex, $[Fe(dppOH)_2](BF_4)_2$ were synthesized according to the literature methods [13]. $(TBA)_3[PMo_{12}O_{40}]$ was synthesized by performing a simple cation exchange by adding excess TBABr to an aqueous solution of $H_3PMo_{12}O_{40} \cdot nH_2O$ and collecting the resulting yellow precipitate by vacuum filtration. All other chemicals and solvents are commercially available and were used as received without further purification.

4.2. Synthesis of [Fe(dppOH)₂][POM] Composites

$(TBA)[Fe(dppOH)_2][PMo_{12}O_{40}] \cdot 2CH_3NO_2$ (**1**): A solution of $(TBA)_3[PMo_{12}O_{40}]$ (255 mg, 0.10 mmol) in 10 mL nitromethane was prepared and then separated into 10 equal batches in glass tubes (ϕ = 8 mm). A mixture of nitromethane and acetonitrile (1:1 *v/v*, 0.5 mL) was then layered on top of the POM solution as a buffer layer to facilitate slow diffusion. Finally, a pre-prepared solution of $[Fe(dppOH)_2](BF_4)_2$ (95 mg, 0.13 mmol) in 10 mL acetone/acetonitrile (2:1 *v/v*) was separated into 10 equal batches and carefully layered on top of the buffer layer. The vials were then sealed and left undisturbed, and after approximately one week, yellow crystals of **1** suitable for X-ray analysis were obtained. The crystalline sample was collected via pipette and dried under vacuum prior to analysis (yield = 55 mg, 15%). Anal. Calcd. for $1 \cdot 2CH_3NO_2$, $C_{42}H_{64}N_{13}FeMo_{12}O_{46}P$; C, 18.51; H, 2.37; N, 6.68. Found: C, 19.34; H, 2.11; N, 6.32.

$[Fe(dppOH)_2]_3[PMo_{12}O_{40}]_2 \cdot 15H_2O$ (**2**): A solution of $H_3PMo_{12}O_{40} \cdot nH_2O$ (190 mg) in 10 mL nitromethane was separated into 10 equal batches and placed in glass tubes (ϕ = 8 mm). A mixture of nitromethane and acetonitrile (1:1 *v/v*, 0.5 mL) was layered on top of the POM solution as a buffer layer to facilitate slow diffusion. Finally, a solution of $[Fe(dppOH)_2](BF_4)_2$ (95 mg, 0.13 mmol) in 10 mL acetone/acetonitrile (2:1 *v/v*) was separated into 10 equal batches and carefully layered on top of the buffer layer. After approximately 1 week, yellow crystals of **2** suitable for X-ray analysis were obtained. The crystalline sample was collected via pipette and dried under vacuum prior to analysis (yield = 33 mg, 4%). Anal. Calcd. for $2 \cdot 15H_2O$, $C_{72}H_{96}N_{30}Fe_3Mo_{24}O_{101}P_2$; C, 15.64; H, 1.75; N, 7.60. Found: C, 15.77; H, 1.36; N, 7.15. We note that the elemental analysis was measured after the samples had been dried under vacuum and that, whilst care was taken in all cases to measure the dry material as promptly as possible, re-exposure to ambient conditions might help to explain the anomalous hydration of the sample (as is best fitted to the elemental analysis data) when compared to the single crystal data, where two nitromethane molecules are found per formula unit.

4.3. X-ray Crystallography

Crystals were mounted in oil on a micromount, and data were collected at 100 K on a SMART APEXII diffractometer (Bruker, Billerica, MA, USA) coupled with a CCD area detector and with graphite monochromated Mo Kα (λ = 0.71073 Å) radiation. The structure was solved using direct methods and expanded using Fourier techniques within the *SHELXTL* program [23]. Empirical absorption corrections were calculated using SADABS. In the structure analyses, non-hydrogen atoms were refined with anisotropic thermal parameters. Hydrogen atoms were included in calculated positions and refined with isotropic thermal parameters riding on those of the parent atoms.

Full details of the crystallographic analysis and accompanying cif files may be obtained free of charge from the Cambridge Crystallographic Data Centre (CCDC numbers 1559207 and 1559208) via http://www.ccdc.cam.ac.uk/conts/retrieving.html (or from the CCDC, 12 Union Road, Cambridge CB2 1EZ, UK; Fax: +44 1223 336033; E-mail: deposit@ccdc.cam.ac.uk).

4.4. Magnetic Measurements

Variable-temperature magnetic susceptibility measurements were carried out on polycrystalline samples under an applied field of 20,000 Oe using an MPMS-XL SQUID magnetometer (Quantum Design, San Diego, CA, USA). Diamagnetic corrections for the sample holder and $[PMo_{12}O_{40}]^{3-}$ were collected experimentally and the contribution of the TBA cations and the (dppOH) ligands were calculated using Pascal's constants. Data collections were conducted at a rate of 1 K/min.

5. Conclusions

New composite crystals **1** and **2**, incorporating spin crossover Fe(II) complexes and POM anions were synthesized and their crystal structures have been obtained at 100 K. While [Fe(dppOH)$_2$](BF$_4$)$_2$ shows high-spin to low-spin SCO behavior with $T_{1/2}$ = 271 K, **1** is trapped in the high-spin state below 300 K and shows no spin crossover behavior. **2** contains three [Fe(dppOH)$_2$] cations per formula unit, two of which are similarly locked in the high-spin state configuration whilst the remaining Fe(II) ion is found to exist in a low-spin state at 100 K and shows partial spin crossover behavior. The coordination geometry around the Fe(II) ions is strongly affected by interactions between the pendant hydroxyl groups of the dppOH ligand and the POM anions, which has a direct result on the spin state of the composite compound. From this result, we show that combining spin crossover complexes with functional anions can be used to modify spin crossover properties, and future work will demonstrate how switchable behaviors can be obtained by facilitating proton-transfer interactions between the ligand and the POM.

Supplementary Materials: The following are available online at www.mdpi.com/2304-6740/5/3/48/s1. Cif and cif-checked files.

Acknowledgments: This work was supported by JSPS KAKENHI Grant Number JP16H06523 (Coordination Asymmetry). This work was also supported by Grant-in-Aid for JSPS Research Fellow Grant Number J02555. (Satoshi Kuramochi) and a JSPS Postdoctoral Fellowship for Foreign Researchers (Jamie M. Cameron).

Author Contributions: Hiroki Oshio conceived and supervised the project. Satoshi Kuramochi and Takuya Shiga conducted and interpreted the synthetic and analytical experiments. Jamie M. Cameron and Graham N. Newton provided assistance in the direction of the project. Satoshi Kuramochi prepared the manuscript with assistance from Takuya Shiga and Jamie M. Cameron.

Conflicts of Interest: The authors declare no conflict of interest.

References

1. Real, J.; Gaspar, A.; Muñoz, C.M. Thermal, Pressure and Light Switchable Spin-Crossover Materials. *Dalton Trans.* **2005**, 2062–2079. [CrossRef] [PubMed]
2. Bousseksou, A.; Molnár, G.; Salmon, L.; Nicolazzi, W. Molecular Spin Crossover Phenomenon: Recent Achievements and Prospects. *Chem. Soc. Rev.* **2011**, *40*, 3313–3335. [CrossRef] [PubMed]
3. Halcrow, M.A. The Spin-States and Spin-Transitions of Mononuclear Iron(II) Complexes of Nitrogen-Donor Ligands. *Polyhedron* **2007**, *26*, 3523–3576. [CrossRef]
4. Nihei, M.; Shiga, T.; Maeda, Y.; Oshio, H. Spin Crossover Iron(III) Complexes. *Coord. Chem. Rev.* **2007**, *251*, 2606–2621. [CrossRef]
5. Krivokapic, I.; Zerara, M.; Daku, M.; Vargas, A.; Enachescu, C.; Ambrus, C.; Tregenna-Piggott, P.; Amstutz, N.; Krausz, E.; Hauser, A. Spin-Crossover in Cobalt(II) Imine Complexes. *Coord. Chem. Rev.* **2007**, *251*, 364–378. [CrossRef]

6. Holland, J.M.; McAllister, J.A.; Lu, Z.; Kilner, C.A.; Thornton-Pett, M.; Halcrow, M.A. An Unusual Abrupt Thermal Spin-State Transition in [FeL₂][BF₄]₂ [L = 2,6-Di(Pyrazol-1-yl)Pyridine]. *Chem. Commun.* **2001**, 577–578. [CrossRef]

7. Holland, J.M.; McAllister, J.A.; Kilner, C.A.; Thornton-Pett, M.; Bridgeman, A.J.; Halcrow, M.A. Stereochemical Effects on the Spin-State Transition Shown by Salts of [FeL₂]²⁺ [L = 2,6-Di(Pyrazol-1-yl)Pyridine]. *J. Chem. Soc. Dalton Trans.* **2002**, 548–554. [CrossRef]

8. Money, V.A.; Elhaïk, J.; Halcrow, M.A.; Howard, J.A.K. The Thermal and Light Induced Spin Transition in [FeL₂](BF₄)₂ (L = 2,6-Dipyrazol-1-yl-4-Hydroxymethylpyridine). *Dalton Trans.* **2004**, 1516–1518. [CrossRef] [PubMed]

9. Halcrow, M.A. The Synthesis and Coordination Chemistry of 2,6-Bis(Pyrazolyl)Pyridines and Related Ligands—Versatile Terpyridine Analogues. *Coord. Chem. Rev.* **2005**, *249*, 2880–2908. [CrossRef]

10. Carbonera, C.; Costa, J.; Money, V.A.; Elhaïk, J.; Howard, J.A.K.; Halcrow, M.A.; Létard, J.-F. Photomagnetic Properties of Iron(II) Spin Crossover Complexes of 2,6-Dipyrazolylpyridine and 2,6-Dipyrazolylpyrazine Ligands. *Dalton Trans.* **2006**, 3058–3066. [CrossRef] [PubMed]

11. Halcrow, M.A. Structure: Function Relationships in Molecular Spin-Crossover Complexes. *Chem. Soc. Rev.* **2011**, *40*, 4119–4142. [CrossRef] [PubMed]

12. Cook, L.J.; Mohammed, R.; Sherborne, G.; Roberts, T.D.; Alvarez, S.; Halcrow, M.A. Spin State Behavior of Iron(II)/Dipyrazolylpyridine Complexes. New Insights from Crystallographic and Solution Measurements. *Coord. Chem. Rev.* **2015**, *289*, 2–12. [CrossRef]

13. Nihei, M.; Maeshima, T.; Kose, Y.; Oshio, H. Synthesis, Structure and Magnetic Properties of an Iron(II) Complex with Nitronyl Nitroxides. *Polyhedron* **2007**, *26*, 1993–1996. [CrossRef]

14. Pope, M.T.; Müller, A. Polyoxometalate Chemistry: An Old Field with New Dimensions in Several Disciplines. *Angew. Chem. Int. Ed.* **1991**, *30*, 34–48. [CrossRef]

15. Tzirakis, M.D.; Lykakis, I.N.; Orfanopoulos, M. Decatungstate as an Efficient Photocatalyst in Organic Chemistry. *Chem. Soc. Rev.* **2009**, *38*, 2609–2621. [CrossRef] [PubMed]

16. Long, D.L.; Tsunashima, R.; Cronin, L. Polyoxometalates: Building Blocks for Functional Nanoscale Systems. *Angew. Chem. Int. Ed.* **2010**, *49*, 1736–1758. [CrossRef] [PubMed]

17. Sadakane, M.; Steckhan, E. Electrochemical Properties of Polyoxometalates as Electrocatalysts. *Chem. Rev.* **1998**, *98*, 219–238. [CrossRef] [PubMed]

18. Yamase, T. Photoredox Chemistry of Polyoxometalates as a Photocatalyst. *Catal. Surv. Asia* **2003**, *7*, 203–217. [CrossRef]

19. Streb, C. New Trends in Polyoxometalate Photoredox Chemistry: From Photosensitisation to Water Oxidation Catalysis. *Dalton Trans.* **2012**, *41*, 1651–1659. [CrossRef] [PubMed]

20. Nakamura, O.; Kodama, T.; Ogino, I.; Miyake, Y.; Nakamura, O.; Kodama, T.; Ogino, I.; Miyake, Y. High-Conductivity Solid Proton Conductors: Dodecamolybdophosphoric Acid and Dodecatungstophosphoric Acid Crystals. *Chem. Lett.* **1979**. [CrossRef]

21. Miras, H.N.; Yan, J.; Long, D.-L.L.; Cronin, L. Engineering Polyoxometalates with Emergent Properties. *Chem. Soc. Rev.* **2012**, *41*, 7403–7430. [CrossRef] [PubMed]

22. Sawada, Y.; Kosaka, W.; Hayashi, Y.; Miyasaka, H. Coulombic Aggregations of Mn^{III} salen-Type Complexes and Keggin-Type Polyoxometalates: Isolation of Mn₂ Single-Molecule Magnets. *Inorg. Chem.* **2012**, *51*, 4824–4832. [CrossRef] [PubMed]

23. Sheldrick, G. A short history of *SHELX. Acta Crystallogr. Sect. A* **2008**, *64*, 112–122. [CrossRef] [PubMed]

![inorganics logo] *inorganics*

MDPI

Article

High-Temperature Wide Thermal Hysteresis of an Iron(II) Dinuclear Double Helicate

Shiori Hora [1] and Hiroaki Hagiwara [2,*]

[1]   Graduate School of Education, Gifu University, Yanagido 1-1, Gifu 501-1193, Japan; v1131032@edu.gifu-u.ac.jp

[2]   Department of Chemistry, Faculty of Education, Gifu University, Yanagido 1-1, Gifu 501-1193, Japan

*   Correspondence: hagiwara@gifu-u.ac.jp; Tel.: +81-58-293-2253

Received: 28 June 2017; Accepted: 25 July 2017; Published: 28 July 2017

Abstract: Two new dinuclear iron(II) complexes ($1 \cdot PF_6$ and $1 \cdot AsF_6$) of the general formula $[Fe^{II}_2(L2^{C3})_2](X)_4 \cdot nH_2O \cdot mMeCN$ (X = PF_6, n = m = 1.5 for $1 \cdot PF_6$ and X = AsF_6, n = 3, m = 1 for $1 \cdot AsF_6$) have been prepared and structurally characterized, where $L2^{C3}$ is a bis-1,2,3-triazolimine type Schiff-base ligand, 1,1′-[propane-1,3-diylbis(1H-1,2,3-triazole-1,4-diyl)]bis{N-[2-(pyridin-2-yl)ethyl]methanimine}. Single crystal X-ray structure analyses revealed that $1 \cdot PF_6$ and $1 \cdot AsF_6$ are isostructural. The complex-cation $[Fe^{II}_2(L2^{C3})_2]^{4+}$ of both has the same dinuclear double helicate architecture, in which each iron(II) center has an N6 octahedral coordination environment. Neighboring helicates are connected by intermolecular π–π interactions to give a chiral one-dimensional (1D) structure, and cationic 1D chains with the opposite chirality exist in the crystal lattice to give a heterochiral crystal. Magnetic and differential scanning calorimetry (DSC) studies were performed only for $1 \cdot AsF_6$, since the thermal stability in a high-temperature spin crossover (SCO) region of $1 \cdot PF_6$ is poorer than that of $1 \cdot AsF_6$. $1 \cdot AsF_6$ shows an unsymmetrical hysteretic SCO between the low-spin–low-spin (LS–LS) and high-spin–high-spin (HS–HS) states at above room temperature. The critical temperatures of warming ($T_c\uparrow$) and cooling ($T_c\downarrow$) modes in the abrupt spin transition area are 485 and 401 K, respectively, indicating the occurrence of 84 K-wide thermal hysteresis in the first thermal cycle.

Keywords: double helicate; hysteresis; spin crossover; iron; 1,2,3-triazolimine; π–π interaction

1. Introduction

The interconversion between a high-spin (HS) and a low-spin (LS) state, the so-called "spin crossover" (SCO), is one of the most attractive phenomena in the field of molecular bistability [1–3]. This phenomenon is observed in $3d^4$–$3d^7$ transition metal complexes with various coordination geometries (e.g., octahedral, square-pyramidal, trigonal-bipyramidal, tetrahedral, etc.) [1,4–7], and is triggered by an external stimulus such as temperature, pressure, light irradiation, or a magnetic field [1–3]. While SCO originates from an individual metal center, a spin transition profile, for example, abruptness and multi-steps with or without hysteresis, is provided from the cooperative effect between SCO metal sites through intra- or intermolecular pathways [1–3]. There is a need for SCO materials which are capable of abrupt and complete HS–LS interconversion at around room temperature (RT) with wide thermal hysteresis width (ΔT) for potential applications in molecular memories, switches, and display devices. As such, various cooperative SCO compounds have been produced with supramolecular assembly via intermolecular hydrogen bonding [8–10] and π–π stacking [11–13], or coordination polymeric architecture by using bridging ligands [14–18]. Multinuclear clusters have also been widely studied in the hope of developing multistable compounds with more accessible spin states towards denser information storage, or hybrid SCO materials with multifunctional properties (e.g., SCO with charge transfer, magnetic coupling, luminescence, etc.) arising from the combination of the SCO metal center and other metal centers [2,19–24]. In particular, dinuclear species have received much attention since they are a simple model for revealing intra- and intermolecular interactions for enhancing cooperativity [25–34]. From a more practical point of view, the importance of studying the reproducible nature and scan rate dependence of the hysteresis loop and high-temperature SCO with sufficient thermal stability has recently been pointed out in order to define the limiting characteristics of SCO materials [33,35–39].

Recently, we have focused on SCO molecules with 1-R-1H-1,2,3-triazole-containing Schiff-base (1-R-1H-1,2,3-triazolimine; R = Me and Ph) ligands since the ligand system can easily modify its structure with the choice of a substituent R of a precursor, 1-R-1H-1,2,3-triazole-4-carbaldehyde, and another amino precursor by simple click reaction [40–42], Schiff-base reaction [43] and replacement of R using the method reported by L'abbé and coworkers [44,45]. The iron(II) complexes with multidentate 1-R-1H-1,2,3-triazolimine ligand show a variety of SCO properties such as gradual, abrupt, and two-step transition with (or without) hysteresis in a wide temperature range (around RT, below RT and above RT over ca. 100 °C) [34,46,47]. In one of these studies, we reported the SCO iron(II) dinuclear double helicate for the first time [34]. This compound, with the formula [Fe^{II}$_2$(L2^{C2})$_2$](PF$_6$)$_4$·5H$_2$O·MeCN, has two bis-tridentate type ligand strands (L2^{C2}) in which two 1,2,3-triazolimine moieties of the ligand strand are bridged by an ethylene chain (Scheme 1), and shows an anomalous two-step SCO with 11 K-wide hysteresis in a second step centered at 432 K, which is the highest hysteretic transition temperature in the dinuclear SCO system reported so far [26–29,33,48].

From the fascinating structure and function of the [Fe^{II}$_2$(L2^{C2})$_2$]$^{4+}$ system, our own interest in the system has increasingly concerned the effect of the bridging alkyl chain length. We report here the synthesis, thermal stability, and structure of the corresponding propylene-bridged iron(II) dinuclear double helicate complexes [Fe^{II}$_2$(L2^{C3})$_2$](PF$_6$)$_4$·1.5H$_2$O·1.5MeCN (**1·PF$_6$**) and [Fe^{II}$_2$(L2^{C3})$_2$](AsF$_6$)$_4$·3H$_2$O·MeCN (**1·AsF$_6$**), where L2^{C3} = 1,1'-[propane-1,3-diylbis(1H-1,2,3-triazole-1,4-diyl)]bis{N-[2-(pyridin-2-yl)ethyl]methanimine} (Scheme 1), and the high-temperature hysteretic SCO of **1·AsF$_6$**.

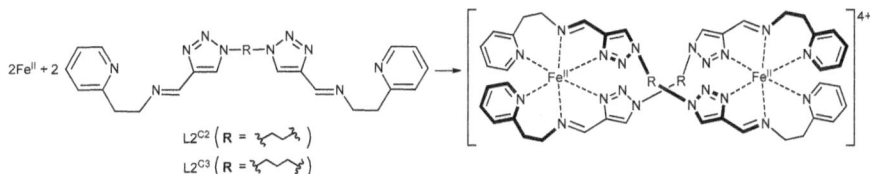

Scheme 1. Synthesis of the dinuclear double helicate complex-cation [Fe^{II}$_2$(L2^R)$_2$]$^{4+}$ (R = C2 = ethylene [34], R = C3 = propylene (this work)).

2. Results and Discussion

*2.1. Synthesis and Characterization of [Fe^{II}$_2$(L2^{C3})$_2$](PF$_6$)$_4$·1.5H$_2$O·1.5MeCN (**1·PF$_6$**) and [Fe^{II}$_2$(L2^{C3})$_2$](AsF$_6$)$_4$·3H$_2$O·MeCN (**1·AsF$_6$**)*

The ligand L2^{C3} was prepared stoichiometric by the 1:2 condensation reaction of 1,1'-(propane-1,3-diyl)bis(1H-1,2,3-triazole-4-carbaldehyde) (**4**) and 2-(2-aminoethyl)pyridine. **4** was synthesized starting from 1-phenyl-1H-1,2,3-triazole-4-carbaldehyde and 1,3-diaminopropane (Scheme 2) based on the previously reported procedure of a related ethylene-bridged compound [34]. The dinuclear **1·PF$_6$** and **1·AsF$_6$** were prepared by mixing the ligand L2^{C3}, Fe^{II}Cl$_2$·4H$_2$O, and KX (X = PF$_6$ and AsF$_6$ for **1·PF$_6$** and **1·AsF$_6$**, respectively) with a 1:1:2 molar ratio in MeOH/H$_2$O mixed solution, which was then recrystallized from the MeCN/MeOH mixed solution. All the synthetic procedures were performed in air and both complexes were obtained as dark-red block crystals. The infrared spectra of both showed a characteristic band at ca. 1609 cm^{-1}, corresponding to the C=N stretching vibration of the Schiff-base ligand [34,49]. In addition, characteristic bands at ca. 844 and 701 cm^{-1} were observed, corresponding to the counter anions, PF_6^- and AsF_6^- for **1·PF$_6$** and **1·AsF$_6$**, respectively [34,49,50]. The formula of both complexes was confirmed by elemental analyses and thermogravimetric analyses. Thermogravimetry and differential thermal analysis (TG/DTA) curves of both are shown in Figure 1. As shown in Figure 1a, when the powdered sample of **1·PF$_6$** was heated from 30 °C (303 K) at a rate of 10 °C·min^{-1}, the sample weight decreased gradually and a weight loss of 4.3% at 224 °C (497 K) was observed, which corresponds to the calculated weight percentages of $0.5H_2O$ and 1.5MeCN molecules per $[Fe^{II}_2(L2^{C3})_2](PF_6)_4 \cdot 1.5H_2O \cdot 1.5MeCN$ (**1·PF$_6$**) (4.2%). Above this temperature, the weight loss became abrupt, and then the weight loss of the remaining one H_2O molecule (1.1%) was observed at 242 °C (515 K). In this region, a sharp thermal anomaly in DTA appeared (T_{max} = 232 °C (505 K)). Finally, above 242 °C (515 K), the weight loss became more and more abrupt, indicating the decomposition of **1·PF$_6$**. In the same way, as shown in Figure 1b, when the powdered sample of **1·AsF$_6$** was heated from 30 °C (303 K) at a rate of 10 °C·min^{-1}, the sample weight decreased gradually and the total weight loss was 5.0% at 215 °C (488 K) in agreement with the calculated weight percentage of $3H_2O$ and one MeCN molecule per $[Fe^{II}_2(L2^{C3})_2](AsF_6)_4 \cdot 3H_2O \cdot MeCN$ (**1·AsF$_6$**) (5.1%). On further increasing the temperature from 215 °C (488 K), the desolvated sample of **1·AsF$_6$** was stable up to ca. 263 °C (536 K), and began to decompose above this temperature. Although a sharp thermal anomaly in DTA was also observed at T_{max} = 212 °C (485 K) in **1·AsF$_6$**, this peak was far from the decomposition temperature (ca. 263 °C), which is obviously different from that of **1·PF$_6$**. To access the reason for exhibiting a sharp thermal anomaly in DTA around 485–505 K, thermochromic properties of both compounds were also investigated (Figure 2). Although the orange-brown color of powdered samples of **1·PF$_6$** becomes slightly lighter from 300 K to ca. 483 K, possibly due to the partial SCO, further elevation of the temperature to 493 K induces rapid black coloration, suggesting the decomposition of the compound (Figure 2a). This thermochromic tendency and TG/DTA results suggest that the decomposition temperature overlaps the spin transition temperature in **1·PF$_6$**. On the other hand, powdered samples of **1·AsF$_6$** showed a clearly distinguishable thermochromism from orange-brown at 300 K to orange-yellow at ca. 498 K, suggesting the occurrence of SCO (Figure 2b). Then, further increasing the temperature to 503 K induces slight yellowish coloration, and ca. 5 min. later, a black coloration related to decomposition proceeds. This color change of **1·AsF$_6$** in a step-by-step-manner is comparable to the clear distinction between sharp thermal anomaly at T_{max} = 485 K and decomposition temperature observed in TG/DTA curves. Consequently, we decided to investigate SCO properties only for **1·AsF$_6$** which is more stable than **1·PF$_6$** in a high-temperature SCO region.

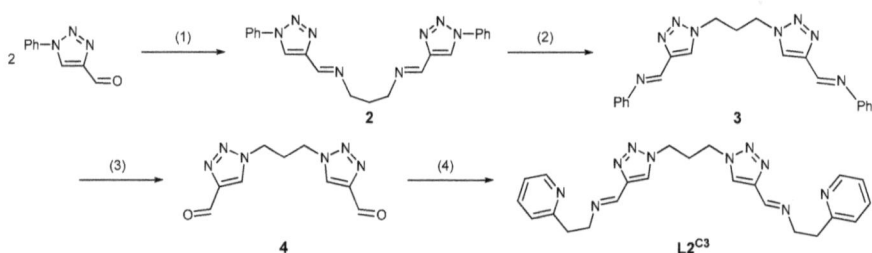

Scheme 2. Synthetic scheme of the ligand L2^{C2}. Reagents and conditions: (1) 1,3-diaminopropane, MeOH, 1 h, 60 °C; 78%; (2) 1-PrOH, 16 h, 80 °C; 78%; (3) HCl (aq), 2 h, room temperature (RT); 86%; (4) 2-(2-aminoethyl)pyridine (2 equiv), MeOH, 1 h, RT.

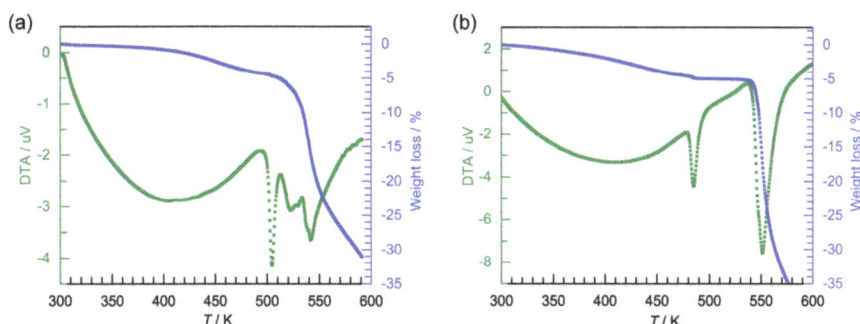

Figure 1. Thermogravimetry and differential thermal analysis (TG/DTA) curves of **1·PF₆** (a) and **1·AsF₆** (b).

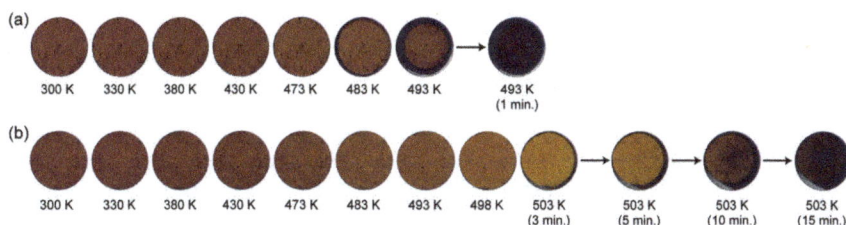

Figure 2. Temperature-dependent optical microscope images of grinding samples of **1·PF₆** (a) and **1·AsF₆** (b) deposited on a heating stage. Unless otherwise noted, after increasing the sample temperature at each target point, the sample was held for 5 min, and then an image was taken at each temperature.

2.2. Crystal Structure of **1·PF₆** and **1·AsF₆**

The crystal structure of **1·PF₆** and **1·AsF₆** was determined at 120 K (low-spin–low-spin (LS–LS) state). As shown in Table 1, both complexes crystallize in the same monoclinic space group type C2/c (number 15) with Z = 4 and are isostructural. The structure consists of a $[Fe^{II}_2(L2^{C3})_2]^{4+}$ complex-cation, two disordered monovalent counterions (PF_6^- and AsF_6^- for **1·PF₆** and **1·AsF₆**, respectively), and H_2O and MeCN molecules as lattice solvents. In the crystal lattice, both solvent molecules are in the one solvent site in a severely disordered manner in both compounds. Thus, these regions, including solvent molecules, are treated with the *PLATON* SQUEEZE program [51] (see Experimental Section 3.3 including the details about analyses of disordered counter anions and solvent molecules).

Figure 3 shows the molecular structure of the complex-cation $[Fe^{II}_2(L2^{C3})_2]^{4+}$ of **1·AsF₆** at 120 K. As shown in Table 2, the structural feature of the complex-cation of **1·PF₆** is similar to that of **1·AsF₆**. Thus, we only discuss the structure of **1·AsF₆** below. The complex-cation has a dinuclear double-helical architecture, in which each ligand L2^{C3} bound as a bis-tridentate ligand to two different iron centers, and conversely, the Fe^{II} centers are N6-coordinated pseudo-octahedral coordination environments bound to two different ligand strands. The intra-helical metal–metal distance is 10.894(3) Å, which is much longer than that of the related ethylene-bridged helicate $[Fe^{II}_2(L2^{C2})_2]^{4+}$ (7.855(3) Å) [34] due to the elongation of the alkyl chain from two to three C atoms. Each metal site is chiral, with either a Δ or a Λ configuration, and the complex-cation depicted in Figure 3 has a homochiral Λ–Λ pair. In the crystal, complex-cations with the Δ–Δ and Λ–Λ pairs coexist to form a racemic crystal since the complex crystallizes in a centrosymmetric space group, C2/c. The Fe–N coordination bond distances (1.954(3)–2.007(3) and 1.945(3)–2.007(3) Å for Fe1–N and Fe2–N, respectively) indicate that both centers are in the LS Fe^{II} state. While the average Fe–N bond distance and octahedral volume of

LS–LS $[Fe^{II}_2(L2^{C3})_2]^{4+}$ are similar to those of LS–LS $[Fe^{II}_2(L2^{C2})_2]^{4+}$, the Θ [52] and continuous shape measures (CShMs) relative to the regular octahedron with the center as the reference shape [53] of LS–LS $[Fe^{II}_2(L2^{C3})_2]^{4+}$ are lower than those of LS–LS $[Fe^{II}_2(L2^{C2})_2]^{4+}$, indicating that the Fe^{II} centers of LS–LS $[Fe^{II}_2(L2^{C3})_2]^{4+}$ have a more regular octahedral geometry than those of LS–LS $[Fe^{II}_2(L2^{C2})_2]^{4+}$ with the one minor exception of the Σ [54] of LS–LS $[Fe^{II}_2(L2^{C2})_2]^{4+}$, which is slightly lower than that of LS–LS $[Fe^{II}_2(L2^{C3})_2]^{4+}$ (average Fe–N bond distance, Σ, Θ, CShMs and octahedral volume of the LS–LS $[Fe^{II}_2(L2^{C2})_2]^{4+}$ for $[Fe^{II}_2(L2^{C2})_2](PF_6)_4 \cdot 5H_2O \cdot MeCN$ [34] are 1.971 Å, 42.5°, 133.7°, 0.696 and 10.096 Å^3, respectively).

Table 1. X-ray crystallographic data for **1·PF$_6$** and **1·AsF$_6$** at 120 K.

	$[Fe^{II}_2L2^{C3}_2](PF_6)_4\ 1.5H_2O\cdot1.5MeCN$ **(1·PF$_6$)** [a]	$[Fe^{II}_2L2^{C3}_2](AsF_6)_4\cdot3H_2O\cdot MeCN$ **(1·AsF$_6$)** [a]
Formula	$C_{49}H_{59.5}N_{21.5}O_{1.5}P_4F_{24}Fe_2$	$C_{48}H_{61}N_{21}O_3As_4F_{24}Fe_2$
Formula weight	1665.26	1847.55
Crystal system	monoclinic	monoclinic
Space group	C2/c (No. 15)	C2/c (No.15)
a/Å	19.049(6)	19.410(3)
b/Å	19.056(5)	19.034(3)
c/Å	20.600(9)	20.772(5)
β/deg.	115.832(3)	115.723(2)
V/Å^3	6731(4)	6914(2)
Z	4	4
$d_{calcd.}$/g·cm^{-3}	1.643	1.775
μ (Mo Kα)/mm^{-1}	0.649	2.441
R_1 [b] ($I > 2$sigma(I))	0.0879	0.0520
wR_2 [c] ($I >$ 2sigma(I))	0.2141	0.1167
R_1 [b] (all data)	0.1054	0.0626
wR_2 [c] (all data)	0.2286	0.1236
u [c]	0.0791	0.0425
v [c]	27.4133	16.3374
S	1.142	1.112
CCDC number	1544125	1544126

[a] The *PLATON* SQUEEZE program [51] was used to treat regions with highly disordered solvent molecules which could not be sensibly modeled in terms of atomic sites; [b] $R_1 = \Sigma||Fo| - |Fc||/\Sigma|Fo|$; [c] $wR_2 = [\Sigma w(|Fo|^2 - |Fc|^2)^2 / \Sigma w|Fo|^2|^2]^{1/2}$, $w = 1/[\sigma^2(|Fo|^2) + (uP)^2 + vP]$ where $P = (|Fo|^2 + 2|Fc|^2)/3$.

The assembly structure of **1·AsF$_6$** is shown in Figure 4. As shown in Figure 4a, a cationic one-dimensional (1D) structure is formed along the *b*-axis by slightly inclined intermolecular π–π interactions between all pyridyl rings of neighboring complex-cations with Cg1···Cg2 [ii] = 4.123(3) Å.

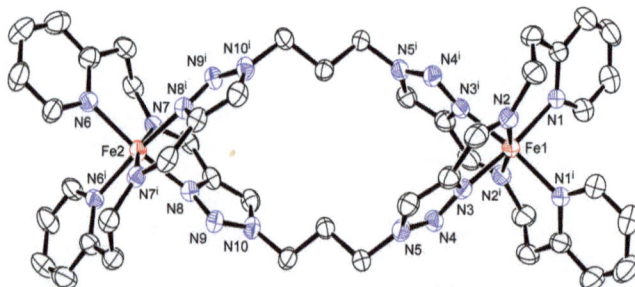

Figure 3. ORTEP drawing of the dinuclear double helicate complex-cation $[Fe^{II}_2(L2^{C3})_2]^{4+}$ of **1·AsF$_6$** at 120 K with the atom numbering scheme except for carbon atoms, where the thermal ellipsoids are drawn with a 50% probability level. H atoms have been omitted for clarity. Symmetry operation: (i) –x, y, 3/2 – z.

(Cg1 = centroid of the N1–C1–C2–C3–C4–C5 ring, Cg2 = centroid of the N6–C12–C13–C14–C15–C16 ring; symmetry operation: (ii) x, −1 + y, z; Table 3). In the light of chiral assembly, homochiral complex-cations are linked together in the π–stacked 1D structure to give a homochiral 1D chain (Δ–Δ···Δ–Δ···Δ–Δ··· or Λ–Λ···Λ–Λ···Λ–Λ···). Figure 4b shows the stacking observed between adjacent cationic 1D chains viewed along the *ac* plane. One-dimensional chains of opposite chirality are alternately arrayed along the *c*-axis to give a heterochiral crystal. Each cationic 1D chain is separated by AsF_6^- ions and solvent molecules occupying the space between the 1D chains in a strongly disordered manner even at 120 K. Although the assembly structure of **1·PF₆** is almost same as that of **1·AsF₆**, these 1D-based structures are remarkably different from the two-dimensional (2D) assembly of the related ethylene-bridged helicate $[Fe^{II}_2(L2^{C2})_2](PF_6)_4 \cdot 5H_2O \cdot MeCN$ by intermolecular π–π and CH/π interactions [34].

Table 2. Coordination bond lengths, angles, and structural parameters for **1·PF₆** and **1·AsF₆** at 120 K. CShMs: continuous shape measures.

Fe1 Site	1·PF₆	1·AsF₆	Fe2 Site	1·PF₆	1·AsF₆
Fe1–N Bond Lengths (Å)			Fe2–N Bond Lengths (Å)		
Fe1–N1	1.998(4)	2.007(3)	Fe2–N6	2.033(4)	2.007(3)
Fe1–N2	1.990(4)	1.970(4)	Fe2–N7	1.911(4)	1.962(3)
Fe1–N3	1.931(4)	1.954(3)	Fe2–N8	1.957(4)	1.945(3)
Fe1–N1 [i]	1.998(4)	2.007(3)	Fe2–N6 [i]	2.034(4)	2.007(3)
Fe1–N2 [i]	1.990(5)	1.970(4)	Fe2–N7 [i]	1.911(4)	1.962(3)
Fe1–N3 [i]	1.931(4)	1.954(3)	Fe2–N8 [i]	1.957(4)	1.945(3)
Average Fe1–N	1.973	1.977	Average Fe2–N	1.967	1.971
N–Fe1–N Bond Angles (°)			N–Fe2–N Bond Angles (°)		
N1–Fe1–N2	89.1(2)	91.35(15)	N6–Fe2–N7	93.38(17)	92.63(12)
N1–Fe1–N3	171.26(19)	170.93(14)	N6–Fe2–N8	172.89(15)	172.40(12)
N1–Fe1–N1 [i]	91.4(2)	92.67(18)	N6–Fe2–N6 [i]	93.9(2)	92.05(17)
N1–Fe1–N2 [i]	98.00(19)	95.70(14)	N6–Fe2–N7 [i]	95.50(16)	96.18(12)
N1–Fe1–N3 [i]	92.58(16)	91.47(12)	N6–Fe2–N8 [i]	89.72(17)	91.84(12)
N2–Fe1–N3	82.65(18)	80.20(13)	N7–Fe2–N8	80.17(17)	80.47(12)
N2–Fe1–N1 [i]	98.00(19)	95.69(15)	N7–Fe2–N6 [i]	95.50(16)	96.17(12)
N2–Fe1–N2 [i]	169.9(3)	169.79(18)	N7–Fe2–N7 [i]	167.0(3)	167.32(16)
N2–Fe1–N3 [i]	89.86(17)	92.27(13)	N7–Fe2–N8 [i]	90.38(16)	90.16(12)
N3–Fe1–N1 [i]	92.58(16)	91.47(12)	N8–Fe2–N6 [i]	89.72(17)	91.84(12)
N3–Fe1–N2 [i]	89.85(17)	92.27(13)	N8–Fe2–N7 [i]	90.38(16)	90.16(12)
N3–Fe1–N3 [i]	84.6(2)	85.61(17)	N8–Fe2–N8 [i]	87.3(3)	85.07(17)
N1 [i]–Fe1–N2 [i]	89.1(2)	91.36(15)	N6 [i]–Fe2–N7 [i]	93.38(17)	92.63(12)
N1 [i]–Fe1–N3 [i]	171.26(19)	170.93(14)	N6 [i]–Fe2–N8 [i]	172.89(15)	172.41(12)
N2 [i]–Fe1–N3 [i]	82.65(18)	80.20(14)	N7 [i]–Fe2–N8 [i]	80.17(17)	80.47(12)
Σ (°) [54]	44.8	48.2		45.3	47.7
Θ (°) [52]	113.9	120.1		106.9	108.1
CShMs [c] [53]	0.594	0.616		0.636	0.590
Octahedral volume (Å³)	10.132	10.183		10.029	10.098

Symmetry operation: (i) −x, y, 3/2 − z; [c] The reference shape is the regular octahedron with center.

Table 3. Intermolecular contacts (Å) of π–π interaction for **1·PF₆** and **1·AsF₆** at 120 K.

	1·PF₆	1·AsF₆
Cg1 [a]···Cg2 [b,ii]	4.139(4)	4.123(3)
C3···C14 [ii]	3.582(13)	3.610(9)
C3···C15 [ii]	3.403(12)	3.423(8)
C4···C14 [ii]	3.432(11)	3.421(7)
C4···C15 [ii]	3.711(11)	3.703(6)

[a] Cg1 = centroid of the N1–C1–C2–C3–C4–C5 ring; [b] Cg2 = centroid of the N6–C12–C13–C14–C15–C16 ring. Symmetry operation: (ii) x, 1 + y, z for **1·PF₆** and x, −1 + y, z for **1·AsF₆**.

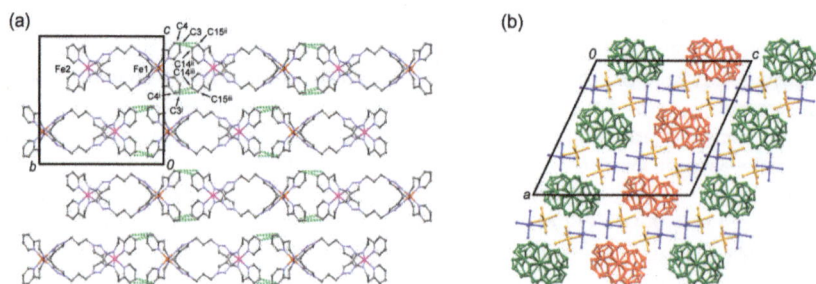

Figure 4. (a) Intermolecular interactions of **1·AsF$_6$** at 120 K. Adjacent complex-cations [Fe^{II}$_2$(L2^{C3})$_2$]$^{4+}$ with the same chirality are connected by intermolecular π–π interactions (green dotted line) to form the homochiral 1D chain structure along the *b*-axis. Symmetry operations: (i) −x, y, 3/2 − z; (ii) x, −1 + y, z; (iii) −x, −1 + y, 3/2 − z; (b) Packing diagram of **1·AsF$_6$** at 120 K viewed along the *ac* plane. Δ–Δ-[Fe^{II}$_2$(L2^{C3})$_2$]$^{4+}$ and Λ–Λ-[Fe^{II}$_2$(L2^{C3})$_2$]$^{4+}$ enantiomers are represented by green and red colors, respectively. Cationic 1D chains with the opposite chirality running along the *b*-axis exist in a crystal lattice to give a heterochiral crystal. Each cationic chain is separated by the space-occupying two disordered AsF$_6$$^−$ ions (blue and orange; only the major component is indicated) and solvent molecules such as 3H$_2$O and one MeCN molecule (not indicated), which could not be sensibly modeled due to the strong disorder and regions including these solvent molecules being treated with the *PLATON* SQUEEZE program. H atoms have been omitted for clarity.

2.3. Magnetic Property of **1·AsF$_6$**

The magnetic susceptibilities for polycrystalline samples of **1·AsF$_6$** were measured upon heating from 300 to 498 K and subsequent cooling to 300 K at a sweep rate of 0.5 K·min^{-1} under an applied magnetic field of 1 T using a superconducting quantum interference device (SQUID) magnetometer with a special heating setup. The χ$_M$T vs. T plots are shown in Figure 5, where χ$_M$ is the molar magnetic susceptibility per Fe and T is the absolute temperature. As shown in Figure 5, **1·AsF$_6$** showed an unsymmetrical hysteretic SCO above room temperature. The initial χ$_M$T value is 0.3 cm^3·K·mol^{-1} at 300 K, which is consistent with the theoretical value for a LS–LS Fe^{II} system. On raising the temperature from 300 K, the χ$_M$T value increases gradually to reach ca. 2.1 cm^3·K·mol^{-1} at 478 K, and then increases abruptly to reach ca. 6.3 cm^3·K·mol^{-1} at 498 K, which is consistent with the theoretical value for a high-spin–high-spin (HS–HS) Fe^{II} system (χ$_M$T = 6.0 cm^3·K·mol^{-1}). On lowering the temperature from 498 K, the χ$_M$T value decreases gradually, reaching ca. 4.3 cm^3·K·mol^{-1} at 409 K, and then decreases abruptly to reach ca. 1.5·cm^3·K mol^{-1} at 379 K. Then, the χ$_M$T value further decreases gradually to reach ca. 0.5 cm^3·K·mol^{-1} at 300 K. The critical temperatures of the warming (T_c↑) and cooling (T_c↓) modes in the abrupt spin transition area are 485 and 401 K, respectively, indicating the occurrence of ΔT = 84 K, which is the widest thermal hysteresis loop in the dinuclear system reported so far [26–29,33,34,48]. It is also noteworthy that the T_c↑ of **1·AsF$_6$** is considerably higher than that of the related ethylene-bridged helicate [Fe^{II}$_2$(L2^{C2})$_2$](PF$_6$)$_4$·5H$_2$O·MeCN (T_c↑ = 437 K) [34] by about 48 K.

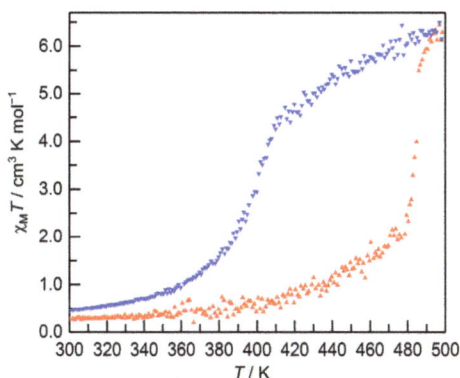

Figure 5. The magnetic behavior of **1·AsF₆** in the form of the $\chi_M \dot{T}$ vs. T plots. **1·AsF₆** was warmed from 300 to 498 K (filled triangles; red), and then cooled from 498 to 300 K (filled inverted triangles; blue) at a sweep rate of 0.5 K·min⁻¹.

2.4. Reproducibility of Hysteretic SCO of **1·AsF₆**

To reveal the desolvation effect and reproducibility of hysteretic SCO of **1·AsF₆**, temperature-dependent magnetic susceptibilities were also measured over three consecutive cycles. As shown in Figure 6a, although **1·AsF₆** shows a more gradual spin transition in the second cycle than in the first cycle, probably due to the in situ solvent liberation in the first heating, an asymmetric wide hysteresis loop is then still observed. Further thermal cycling (third cycle) causes the spin transition pattern to have a much more gradual and partial SCO fashion, which does not occur in the related ethylene-bridged helicate [Fe$^{II}_2$(L2^{C2})₂](PF₆)₄·5H₂O·MeCN [34]. Comparing powder X-ray diffraction (PXRD) patterns of as-synthesized **1·AsF₆** and after first thermal cycle of SQUID measurement confirms that there is no structural phase transition associated with the desolvation, because the peak patterns of the sample are not changed, although the peak positions increase slightly possibly due to the removal of solvent molecules and concomitant lattice contraction (Figure 6b). However, an intensity decrease and peak broadening are observed after three thermal cycles of SQUID measurement, presumably due to a loss of crystallinity and/or partial decomposition.

These properties were also confirmed by differential scanning calorimetry (DSC) experiments of four consecutive thermal cycles at a sweep rate of 10 K·min⁻¹, in the temperature range of 222–502 K (Figure 7). As shown in Figure 7, from the first cycle in heating and following cooling modes, a thermal hysteresis of ca. 85 K was detected ($T_{max}\uparrow$ and $T_{max}\downarrow$ are 498 and 413 K, respectively) and the hysteresis width is consistent with the ΔT value observed by the magnetic measurement. As the thermal cycling is repeated, the shape of DSC peaks becomes broader and smaller, and the position of the peak shifts to a lower temperature in both the heating and cooling modes of the second cycle due to the solvent loss in the first heating ($T_{max}\uparrow$ and $T_{max}\downarrow$ are 487 and 401 K, respectively, in the second cycle). However, further thermal cycling from the second cycle induces the peak shift to the opposite directions, in which the temperature variation in cooling modes is larger than that in heating modes ($T_{max}\uparrow$ and $T_{max}\downarrow$ are 490 and 392 K, respectively, in the third cycle, and 492 and 384 K, respectively, in the fourth cycle). In addition, peak broadening and reduction also continue in the third and fourth cycles. As a whole, thermal hysteresis of **1·AsF₆** is retained for at least four cycles, although the spin transition becomes gradual and incomplete with shifting the transition temperature due to the solvent liberation in the first cycle and a loss of crystallinity and/or partial decomposition upon further thermal cycling.

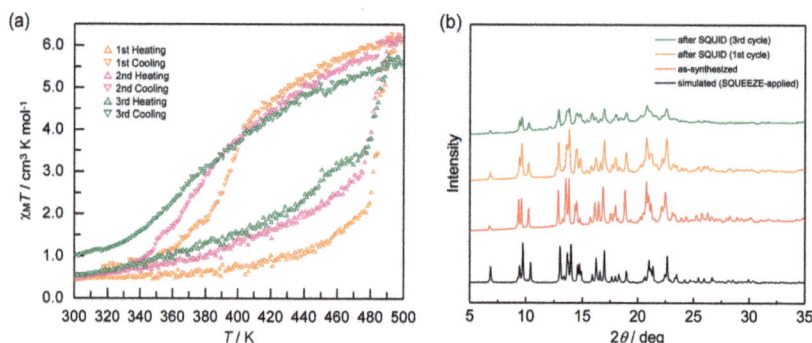

Figure 6. (a) Temperature dependence of the $\chi_M T$ product of **1·AsF₆** in the heating (empty triangles) and cooling (empty inverted triangles) modes over three successive thermal cycles at a sweep rate of $0.5\ \text{K·min}^{-1}$. Orange: first cycle; magenta: second cycle; olive: third cycle; **(b)** Powder X-ray diffraction (PXRD) patterns of **1·AsF₆** at RT in different states. Black: simulated from the SQUEEZE-applied single crystal X-ray data at 120 K; red: as-synthesized **1·AsF₆**; orange: **1·AsF₆** after the first thermal cycle of superconducting quantum interference device (SQUID) measurement; olive: **1·AsF₆** after a third thermal cycle of SQUID measurement.

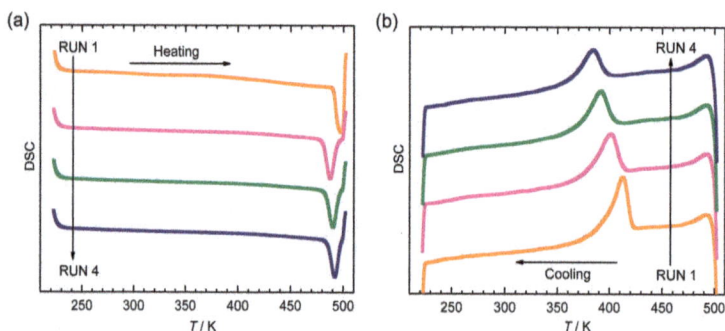

Figure 7. Differential scanning calorimetry (DSC) curves of **1·AsF₆** in the heating mode **(a)** and cooling mode **(b)** recorded over four successive thermal cycles at a sweep rate of $10\ \text{K·min}^{-1}$, in the temperature range of 222–502 K. Orange: first cycle; magenta: second cycle; olive: third cycle; navy blue: fourth cycle.

3. Materials and Methods

3.1. Synthesis of Fe^{II} Complexes

3.1.1. General

All reagents and solvents were purchased from commercial sources and used for the syntheses without further purification. The 1-Phenyl-1*H*-1,2,3-triazole-4-carbaldehyde was prepared according to methods in the literature [47]. The synthetic procedure of 1,1′-(propane-1,3-diyl)bis(1*H*-1,2,3-triazole-4-carbaldehyde) (**4**) was constructed by modifying the reported procedures [34]. All the synthetic procedures were carried out in air.

3.1.2. Preparation of 1,1′-(propane-1,3-diyl)bis(1*H*-1,2,3-triazole-4-carbaldehyde) (**4**)

To a warm pale yellow solution of 1-phenyl-1*H*-1,2,3-triazole-4-carbaldehyde (3.464 g, 20 mmol) in MeOH (30 mL), 1,3-diaminopropane (0.741 g, 10 mmol) in MeOH (1 mL) was added. The resulting solution was stirred for 1 h at 60 °C, and was then cooled in a fridge. The precipitated white powders were collected by suction filtration, washed with Et_2O (10 mL × 3), and dried in vacuo. White solid of *N,N′*-(propane-1,3-diyl)bis[1-(1-phenyl-1*H*-1,2,3-triazol-4-yl)methanimine] (**2**). Yield: 3.005 g (78%). Mp 127–128 °C (from MeOH), ^{1}H NMR (600 MHz; DMSO-d6; TMS): δ 9.24 (s, 2H), 8.55 (s, 2H), 8.00–7.98 (m, 4H), 7.63–7.60 (m, 4H), 7.53–7.51 (m, 2H), 3.72 (t, *J* = 6.5 Hz, 4H), 2.05–2.00 (m, 2H). Anal. Calcd. for **2**·0.2H_2O = $C_{21}H_{20.4}N_8O_{0.2}$: C, 65.00; H, 5.30; N, 28.88%. Found: C, 65.05; H, 5.19; N, 29.01%.

A suspension of **2** (2.898 g, 7.5 mmol) in 1-PrOH (25 mL) was heated at 80 °C for 16 h, turning into a pale yellow solution, and the resulting solution was then cooled in a fridge. The precipitated white powders were collected by suction filtration, washed with Et_2O (10 mL), and dried in vacuo. White solid of 1,1′-[propane-1,3-diylbis(1*H*-1,2,3-triazole-1,4-diyl)]bis(*N*-phenylmethanimine) (**3**). Yield: 2.260 g (78%). m.p. 192–193 °C (from 1-PrOH), ^{1}H NMR (600 MHz; DMSO-d6; TMS): δ 8.78 (s, 2H), 8.67 (s, 2H), 7.43–7.41 (m, 4H), 7.28–7.25 (m, 6H), 4.52 (t, *J* = 6.9 Hz, 4H), 2.59–2.55 (m, 2H). Anal. Calcd. for **3**·0.35H_2O = $C_{21}H_{20.7}N_8O_{0.35}$: C, 64.55; H, 5.34; N, 28.68%. Found: C, 64.15; H, 5.20; N, 29.06%.

Hydrochloric acid (36%) (1.671 g, 16.5 mmol) in H_2O (28 mL) was added to white powders of **3** (2.125 g, 5.5 mmol), and the mixture was stirred for 2 h at room temperature. After turning the mixture into a yellowish-white suspension, the residual white solids of **4** were collected by suction filtration, washed with H_2O (10 mL) and then Et_2O (10 mL), and dried in vacuo. Yield: 1.103 g (86%). m.p. 124–125 °C (from H_2O), ^{1}H NMR (600 MHz; DMSO-d6; TMS): δ 10.03 (s, 2H), 8.89 (s, 2H), 4.52 (t, *J* = 6.9 Hz, 4H), 2.55–2.51 (m, 2H). Anal. Calcd. for $C_9H_{10}N_6O_2$: C, 46.15; H, 4.30; N, 35.88%. Found: C, 45.87; H, 4.19; N, 35.56%.

3.1.3. Synthesis of the Bis-Tridentate Ligand L2^{C3} = 1,1′-[propane-1,3-diylbis(1*H*-1,2,3-triazole-1,4-diyl)]bis{*N*-[2-(pyridin-2-yl)ethyl]methanimine}

The ligand L2^{C3} was prepared by mixing **4** and 2-(2-aminoethyl)pyridine with 1:2 molar ratio in MeOH. The ligand solution thus prepared was used for the synthesis of Fe^{II} complexes without further purification and isolation.

3.1.4. Preparation of [Fe^{II}$_2$(L2^{C3})$_2$](PF$_6$)$_4$·1.5H_2O·1.5MeCN (**1·PF$_6$**)

2-(2-Aminoethyl)pyridine (0.122 g, 1 mmol) in MeOH (1 mL) was added to a solution of **4** (0.118 g, 0.5 mmol) in MeOH (20 mL), and the resulting mixture was stirred at ambient temperature for 1 h. A pale yellow solution of the ligand (0.5 mmol) thus prepared was treated first with a solution of Fe^{II}Cl$_2$·4H_2O (0.099 g, 0.5 mmol) in MeOH (1 mL), turning into a dark orange-red solution, and then with KPF$_6$ (0.184 g, 1 mmol) in 4 mL of a mixed solution of MeOH and H_2O (1/1 by volume). The resulting mixture was stirred at ambient temperature for 1 h, during which time the precipitated orange-brown crude product was collected by suction filtration. The collected precipitate was dissolved in MeCN (4 mL) and then filtered. Dark-red block crystals were obtained by slow diffusion of MeOH (8 mL) into the filtrate (liquid–liquid diffusion) for a day. Yield: 0.154 g (37%). Anal. Calcd. for [Fe^{II}$_2$(L2^{C3})$_2$](PF$_6$)$_4$·1.5MeCN·1.5MeOH = $C_{49}H_{59.5}N_{21.5}O_{1.5}P_4F_{24}Fe_2$: C, 35.34; H, 3.60; N, 18.08%. Found: C, 35.48; H, 3.37; N, 18.06%. IR (KBr): $\nu_{C=N}$ 1609 cm^{-1}, ν_{P-F}(PF$_6$$^-$) 844 cm^{-1}.

3.1.5. Preparation of [Fe^{II}$_2$(L2^{C3})$_2$](AsF$_6$)$_4$·3H_2O·MeCN (**1·AsF$_6$**)

2-(2-Aminoethyl)pyridine (0.122 g, 1 mmol) in MeOH (2 mL) was added to a solution of **4** (0.118 g, 0.5 mmol) in MeOH (20 mL), and the resulting mixture was stirred at ambient temperature for 1 h. A pale yellow solution of the ligand (0.5 mmol) thus prepared was treated first with a solution of Fe^{II}Cl$_2$·4H_2O (0.099 g, 0.5 mmol) in MeOH (2 mL), turning into a dark orange-red solution, and then with KAsF$_6$ (0.228 g, 1 mmol) in 4 mL of a mixed solution of MeOH and H_2O (1/1 by volume). The resulting mixture was stirred at ambient temperature for 30 min, during which time the precipitated red-orange crude product was collected by suction filtration. The collected precipitate was dissolved in MeCN (3 mL) and then filtered. Dark-red block crystals were obtained by slow diffusion of MeOH (7 mL) into the filtrate (liquid–liquid diffusion) for a day. Yield: 0.155 g (33%). Anal. Calcd. for

$[Fe^{II}_2(L2^{C3})_2](AsF_6)_4 \cdot 3H_2O \cdot MeCN = C_{48}H_{61}N_{21}O_3As_4F_{24}Fe_2$: C, 31.21; H, 3.33; N, 15.92%. Found: C, 31.46; H, 3.04; N, 16.25%. IR (KBr): $\nu_{C=N}$ 1609 cm^{-1}, $\nu_{As-F}(AsF_6^-)$ 701 cm^{-1}.

3.2. Physical Measurements

Elemental C, H, and N analyses were performed on a J-Science Lab (Kyoto, Japan) MICRO CORDER JM-10. IR spectra were recorded at room temperature using a JASCO (Tokyo, Japan) FT/IR 460Plus spectrophotometer or a PerkinElmer (Waltham, MA, USA) Spectrum100 FT-IR spectrometer with the samples prepared as KBr disks. ^{1}H NMR spectra were recorded on a JEOL (Tokyo, Japan) ECA-600 spectrometer. The melting point was measured through a Yanaco (Kyoto, Japan) MP-S3 micro melting point meter and was uncorrected. Thermogravimetric data were collected on a TG/DTA6300 (SII Nano Technology Inc., Chiba, Japan) instrument in the temperature range of 30–318 °C (303–591 K) for **1·PF$_6$** and 30–381 °C (303–654 K) for **1·AsF$_6$** at a rate of 10 K·min^{-1} under a nitrogen atmosphere. DSC measurements for **1·AsF$_6$** were performed with a DSC6200 (SII Nano Technology Inc., Chiba, Japan) over the temperature range of 222–502 K, at a sweep rate of 10 K·min^{-1} under a nitrogen atmosphere using aluminum hermetic pans with an empty pan as reference. Magnetic susceptibilities of **1·AsF$_6$** were measured in the temperature range of 300–498 K at a sweep rate of 0.5 K·min^{-1} under an applied magnetic field of 1 T using a Quantum Design (San Diego, CA, USA) MPMS-7 SQUID magnetometer with a special heating setup of a sample space oven option. The sample was wrapped in an aluminum foil and was then inserted into a quartz glass tube with a small amount of glass wool filler. Corrections for diamagnetism of the sample were made using Pascal's constants [55] and a background correction for the sample holder was applied. PXRD patterns were recorded at room temperature on a portion of polycrystalline powders placed on a non-reflecting silicon plate, using a Rigaku (Tokyo, Japan) MiniFlex600 diffractometer with Cu Kα radiation (λ = 1.5418 Å) operated at 0.4 kW power (40 kV, 10 mA). Images of grinding samples were recorded under an optical microscope SMZ800N (Nikon, Tokyo, Japan). The sample temperature was controlled by using a Linkam (Tadworth, UK) THMS 600 heating and freezing stage.

3.3. Crystallographic Data Collection and Structure Analyses

X-ray diffraction data of **1·PF$_6$** and **1·AsF$_6$** were collected by a Rigaku (Tokyo, Japan) AFC7R Mercury CCD diffractometer using graphite monochromated Mo Kα radiation (λ = 0.71075 Å) operated at 5 kW power (50 kV, 100 mA). A single crystal was mounted on a MiTeGen (Ithaca, NY, USA) MicroMount (200 μm) with liquid paraffin, then rapidly frozen in a nitrogen-gas stream at 120 K to avoid the loss of crystal solvents and the diffraction data were collected at 120 K. The temperature of the crystal was maintained by means of a Rigaku cooling device to within an accuracy of ±2 K. The data were corrected for Lorentz, polarization, and absorption effects. The structures were solved by the direct method [56] and refined on F^2 data using the full-matrix least-squares algorithm using *SHELXL*-2014 [57]. Non-hydrogen atoms were refined anisotropically. The structure of **1·PF$_6$** contains two disordered PF$_6^-$ ions. Both PF$_6^-$ ions were disordered over two positions and the occupancy factors for the possible two positions of F atoms (F1A–F6A:F1B–F6B = 0.518(15):0.482(15) for P(1)F$_6^-$ and F7A–F12A:F7B–F12B = 0.635(16):0.365(16) for P(2)F$_6^-$) were refined using the tools available from the *SHELXL*-2014 program package. In the refinement of these disordered PF$_6^-$ ions, the SADI command for all P–F bonds and F···F distances and the RIGU command for all P and F atoms were also applied. In addition, diffraction quality of the crystal of **1·PF$_6$** was not sufficient to model highly disordered solvent molecules. The *PLATON* SQUEEZE program (290617, A.L. Spek, Utrecht, NL-UT, The Netherlands) [51] was used to treat regions with disordered solvent molecules which could not be sensibly modelled in terms of atomic sites. Available void volume was 892 Å^3. Then, 196 electrons per unit cell were located and these were assigned to 1.5 H$_2$O and 1.5 MeCN molecules per complex (196/4 = 49 e per complex; 1.5 H$_2$O (15) + 1.5 MeCN (33) = 48 electrons). In the same way, the structure of **1·AsF$_6$** contains two disordered AsF$_6^-$ ions and severely disordered solvent molecules. In **1·AsF$_6$**, both AsF$_6^-$ ions were disordered over three positions and the occupancy factors for the possible three positions of F atoms (F1A–F6A:F1B–F6B:F1C–F6C = 0.434(3):0.291(3):0.275(3) for As(1)F$_6$$^-$ and F7A–F12A:F7B–F12B:F7C–F12C = 0.513(3):0.219(3):0.268(3) for As(2)F$_6$$^-$) were refined using the SUMP command to keep the sum of three occupancy factors of each ion as 1.0. In the refinement of these disordered AsF$_6$$^-$ ions, the SADI command for all As–F bonds and F···F distances, the RIGU command for all As and F atoms, and the EADP command for F5A, F5B, F5C and F6A, F6B, F6C were also applied. The *PLATON* SQUEEZE program was also used to treat regions with disordered solvent molecules in the structure of **1·AsF$_6$**. Available void volume was 926 Å^3, and 220 electrons per unit cell were located. These were assigned to 3 H$_2$O and 1 MeCN molecules per complex (220/4 = 55 e per complex; 3 H$_2$O (30) + 1 MeCN (22) = 52 electrons). All H atoms were placed in geometrically calculated positions, with the distances of C–H = 0.95 (aromatic) and 0.99 (CH$_2$) Å, and refined as riding atoms, with *U*iso(H) = 1.2 *U*eq(C). All calculations were performed by using the Yadokari-XG software package [58]. The CShMs of the Fe^{II} centers relative to the ideal octahedron were calculated by SHAPE 2.1 [53]. The octahedral volumes of the Fe^{II} centers were calculated by *OLEX2* [59]. CCDC 1544125–1544126 contains the supplementary crystallographic data for this paper. These data can be obtained free of charge via http://www.ccdc.cam.ac.uk/conts/retrieving.html or from the CCDC (12 Union Road, Cambridge CB2 1EZ, UK; Fax: +44 1223 336033; E-mail: deposit@ccdc.cam.ac.uk).

4. Conclusions

In conclusion, the present work reveals the effects of the extension of the bridging alkyl chain length (from C2 to C3) on the structure and SCO properties of the iron(II) dinuclear double helicate [Fe^{II}$_2$(L2^R)$_2$](X)$_4$·solvent, with the concomitant effects of the coexistent counter anions and lattice solvents. While propylene-bridged dinuclear complexes [Fe^{II}$_2$(L2^{C3})$_2$](PF$_6$)$_4$·1.5H$_2$O·1.5MeCN (**1·PF$_6$**) and [Fe^{II}$_2$(L2^{C3})$_2$](AsF$_6$)$_4$·3H$_2$O·MeCN (**1·AsF$_6$**) are isostructural, the thermal stability of **1·PF$_6$** is lower than that of **1·AsF$_6$** in a high-temperature SCO region. **1·AsF$_6$** shows an unsymmetrical hysteretic SCO between the LS–LS and HS–HS states at above room temperature with Δ*T* = 84 K (*T*$_c$↑ = 485 K and *T*$_c$↓ = 401 K in the abrupt spin transition area) in the first thermal cycle, which is the widest hysteresis loop in the dinuclear system reported so far. In addition, the *T*$_c$↑ of **1·AsF$_6$** is considerably higher than that of the related ethylene-bridged compound, by about 48 K. This high-temperature wide thermal hysteresis may be related to the 1D assembly structure composed of inter-helicate π–π interactions, which is obviously different from the 2D assembly of the related ethylene-bridged compound. Although the desolvation of **1·AsF$_6$** does not directly affect the assembly structure, consecutive thermal cycles cause a loss of crystallinity and/or partial decomposition, and subsequent modification of the hysteretic SCO loop to a more gradual and partial fashion. As a consequence, the extension of the bridging alkyl chain length (from C2 to C3) and subsequent elongation of the dinuclear cation makes the 2D supramolecular assembly one-dimensional, and enhances the width of the hysteresis loop. Simultaneous raising of the *T*$_c$↑ may also cause the overlapping between spin transition temperature and decomposition temperature, which is related to the thermal durability upon SCO.

Supplementary Materials: The following are available online at www.mdpi.com/2304-6740/5/3/49/s1. Cif and cif-checked files.

Acknowledgments: This work was partly funded by the Gifu University for the promotion of the research of young scientists. Part of this work was conducted at the Institute for Molecular Science, supported by the Nanotechnology Platform Program (Molecule and Material Synthesis) of the Ministry of Education, Culture, Sports, Science and Technology (MEXT), Japan. The authors would like to thank Professor O. Sakurada (Gifu University, Japan) for his assistance in collecting PXRD data.

Author Contributions: The experimental work was performed mainly by Shiori Hora with assistance from Hiroaki Hagiwara. Hiroaki Hagiwara supervised the experiments. Both authors analyzed the data and contributed to the preparation of the manuscript.

Conflicts of Interest: The authors declare no conflict of interest.

References

1. Gütlich, P.; Goodwin, H.A. (Eds.) *Spin Crossover in Transition Metal Compounds I-III*; Topics in Current Chemistry; Springer: Berlin, Germany, 2004; Volume 233–235.
2. Halcrow, M.A. (Ed.) *Spin-Crossover Materials–Properties and Applications*; John Wiley & Sons: Chichester, UK, 2013.
3. Gütlich, P.; Gaspar, A.B.; Garcia, Y. Spin state switching in iron coordination compounds. *Beilstein J. Org. Chem.* **2013**, *9*, 342–391. [CrossRef] [PubMed]
4. Jenkins, D.M.; Peters, J.C. Solution and Solid-State Spin-Crossover Behavior in a Pseudotetrahedral d^7 Ion. *J. Am. Chem. Soc.* **2003**, *125*, 11162–11163. [CrossRef] [PubMed]
5. Li, J.; Lord, R.L.; Noll, B.C.; Baik, M.-H.; Schulz, C.E.; Scheidt, W.R. Cyanide: A Strong-Field Ligand for Ferrohemes and Hemoproteins? *Angew. Chem. Int. Ed.* **2008**, *47*, 10144–10146.
6. Scepaniak, J.J.; Harris, T.D.; Vogel, C.S.; Sutter, J.; Meyer, K.; Smith, J.M. Spin Crossover in a Four-Coordinate Iron(II) Complex. *J. Am. Chem. Soc.* **2011**, *133*, 3824–3827. [PubMed]
7. Mossin, S.; Tran, B.L.; Adhikari, D.; Pink, M.; Heinemann, F.W.; Sutter, J.; Szilagyi, R.K.; Meyer, K.; Mindiola, D.J. A Mononuclear Fe(III) Single Molecule Magnet with a 3/2↔5/2 Spin Crossover. *J. Am. Chem. Soc.* **2012**, *134*, 13651–13661. [PubMed]
8. Weber, B.; Bauer, W.; Obel, J. An Iron(II) Spin-Crossover Complex with a 70 K Wide Thermal Hysteresis Loop. *Angew. Chem. Int. Ed.* **2008**, *47*, 10098–10101. [CrossRef] [PubMed]
9. Weber, B.; Bauer, W.; Pfaffender, T.; Dîrtu, M.M.; Naik, A.D.; Rotaru, A.; Garcia, Y. Influence of Hydrogen Bonding on the Hysteresis Width in Iron(II) Spin-Crossover Complexes. *Eur. J. Inorg. Chem.* **2011**, *2011*, 3193–3206.
10. Zhao, X.-H.; Zhang, S.-L.; Shao, D.; Wang, X.-Y. Spin Crossover in [Fe(2-Picolylamine)$_3$]$^{2+}$ Adjusted by Organosulfonate Anions. *Inorg. Chem.* **2015**, *54*, 7857–7867. [CrossRef] [PubMed]
11. Hayami, S.; Gu, Z.-Z.; Yoshiki, H.; Fujishima, A.; Sato, O. Iron (III) spin-crossover compounds with a wide apparent thermal hysteresis around room temperature. *J. Am. Chem. Soc.* **2001**, *123*, 11644–11650. [PubMed]
12. Schäfer, B.; Rajnák, C.; Šalitroš, I.; Fuhr, O.; Klar, D.; Schmitz-Antoniak, C.; Weschke, E.; Wende, H.; Ruben, M. Room temperature switching of a neutral molecular iron(II) complex. *Chem. Commun.* **2013**, *49*, 10986–10988. [CrossRef] [PubMed]
13. Iasco, O.; Rivière, E.; Guillot, R.; Cointe, M.B.-L.; Meunier, J.-F.; Bousseksou, A.; Boillot, M.-L. Fe^{II}(pap-5NO$_2$)$_2$ and Fe^{II}(qsal-5NO$_2$)$_2$ Schiff-Base Spin-Crossover Complexes: A Rare Example with Photomagnetism and Room-Temperature Bistability. *Inorg. Chem.* **2015**, *54*, 1791–1799. [CrossRef] [PubMed]
14. Kröber, J.; Codjovi, E.; Kahn, O.; Grolière, F.; Jay, C. A Spin Transition System with a Thermal Hysteresis at Room Temperature. *J. Am. Chem. Soc.* **1993**, *115*, 9810–9811. [CrossRef]
15. Kahn, O.; Martinez, C.J. Spin-Transition Polymers: From Molecular Materials toward Memory Devices. *Science* **1998**, *279*, 44–48. [CrossRef]
16. Muñoz-Lara, F.J.; Gaspar, A.B.; Aravena, D.; Ruiz, D.; Muñoz, M.C.; Ohba, M.; Ohtani, R.; Kitagawa, S.; Real, J.A. Enhanced bistability by guest inclusion in Fe(II) spin crossover porous coordination polymers. *Chem. Commun.* **2012**, *48*, 4686–4688. [CrossRef] [PubMed]
17. Lochenie, C.; Bauer, W.; Railliet, A.P.; Schlamp, S.; Garcia, Y.; Weber, B. Large Thermal Hysteresis for Iron(II) Spin Crossover Complexes with N-(Pyrid-4-yl)isonicotinamide. *Inorg. Chem.* **2014**, *53*, 11563–11572. [CrossRef] [PubMed]
18. Dîrtu, M.M.; Naik, A.D.; Rotaru, A.; Spinu, L.; Poelman, D.; Garcia, Y. Fe^{II} Spin Transition Materials Including an Amino–Ester 1,2,4-Triazole Derivative, Operating at, below, and above Room Temperature. *Inorg. Chem.* **2016**, *55*, 4278–4295. [CrossRef] [PubMed]
19. Murray, K.S. Advances in Polynuclear Iron(II), Iron(III) and Cobalt(II) Spin-Crossover Compounds. *Eur. J. Inorg. Chem.* **2008**, 3101–3121. [CrossRef]
20. Nihei, M.; Sekine, Y.; Suganami, N.; Nakazawa, K.; Nakao, A.; Nakao, H.; Murakami, Y.; Oshio, H. Controlled Intramolecular Electron Transfers in Cyanide-Bridged Molecular Squares by Chemical Modifications and External Stimuli. *J. Am. Chem. Soc.* **2011**, *133*, 3592–3600. [CrossRef] [PubMed]
21. Hietsoi, O.; Dunk, P.W.; Stout, H.D.; Arroyave, A.; Kovnir, K.; Irons, R.E.; Kassenova, N.; Erkasov, R.; Achim, C.; Shatruk, M. Spin Crossover in Tetranuclear Fe(II) Complexes, {[(tpma)Fe(µ-CN)]$_4$}X$_4$ (X = ClO$_4^-$, BF$_4^-$). *Inorg. Chem.* **2014**, *53*, 13070–13077. [CrossRef] [PubMed]

22. Matsumoto, T.; Newton, G.N.; Shiga, T.; Hayami, S.; Matsui, Y.; Okamoto, H.; Kumai, R.; Murakami, Y.; Oshio, H. Programmable spin-state switching in a mixed-valence spin-crossover iron grid. *Nat. Commun.* **2014**, *5*, 3865. [CrossRef] [PubMed]

23. Wu, S.-Q.; Wang, Y.-T.; Cui, A.-L.; Kou, H.-Z. Toward Higher Nuclearity: Tetranuclear Cobalt(II) Metallogrid Exhibiting Spin Crossover. *Inorg. Chem.* **2014**, *53*, 2613–2618. [CrossRef] [PubMed]

24. Steinert, M.; Schneider, B.; Dechert, S.; Demeshko, S.; Meyer, F. Spin-State Versatility in a Series of Fe$_4$ [2 × 2] Grid Complexes: Effects of Counteranions, Lattice Solvent, and Intramolecular Cooperativity. *Inorg. Chem.* **2016**, *55*, 2363–2373. [CrossRef] [PubMed]

25. Charbonnière, L.J.; Williams, A.F.; Piguet, C.; Bernardinelli, G.; Rivara-Minten, E. Structural, Magnetic, and Electrochemical Properties of Dinuclear Triple Helices: Comparison with Their Mononuclear Analogues. *Chem. Eur. J.* **1998**, *4*, 485–493. [CrossRef]

26. Ksenofontov, V.; Gaspar, A.B.; Niel, V.; Reiman, S.; Real, J.A.; Gütlich, P. On the Nature of the Plateau in Two-Step Dinuclear Spin-Crossover Complexes. *Chem. Eur. J.* **2004**, *10*, 1291–1298. [CrossRef] [PubMed]

27. Weber, B.; Kaps, E.S.; Obel, J.; Achterhold, K.; Parak, F.G. Synthesis and Characterization of a Dinuclear Iron(II) Spin Crossover Complex with Wide Hysteresis. *Inorg. Chem.* **2008**, *47*, 10779–10787. [CrossRef] [PubMed]

28. Min, K.S.; Swierczek, K.; DiPasquale, A.G.; Rheingold, A.L.; Reiff, W.M.; Arif, A.M.; Miller, J.S. A dinuclear iron(II) complex, [(TPyA)Fe^{II}(THBQ^{2-})Fe^{II}(TPyA)](BF$_4$)$_2$ [TPyA = tris(2-pyridylmethyl)amine; THBQ^{2-} = 2,3,5,6-tetrahydroxy-1,4-benzoquinonate] exhibiting both spin crossover with hysteresis and ferromagnetic exchange. *Chem. Commun.* **2008**, 317–319. [CrossRef]

29. Sunatsuki, Y.; Kawamoto, R.; Fujita, K.; Maruyama, H.; Suzuki, T.; Ishida, H.; Kojima, M.; Iijima, S.; Matsumoto, N. Structures and Spin States of Bis(tridentate)-Type Mononuclear and Triple Helicate Dinuclear Iron(II) Complexes of Imidazole-4-carbaldehyde azine. *Inorg. Chem.* **2009**, *48*, 8784–8795. [CrossRef] [PubMed]

30. Sunatsuki, Y.; Kawamoto, R.; Fujita, K.; Maruyama, H.; Suzuki, T.; Ishida, H.; Kojima, M.; Iijima, S.; Matsumoto, N. Structures and spin states of mono- and dinuclear iron(II) complexes of imidazole-4-carbaldehyde azine and its derivatives. *Coord. Chem. Rev.* **2010**, *254*, 1871–1881. [CrossRef]

31. Archer, R.J.; Hawes, C.S.; Jameson, G.N.L.; Mckee, V.; Moubaraki, B.; Chilton, N.F.; Murray, K.S.; Schmitt, W.; Kruger, P.E. Partial spin crossover behaviour in a dinuclear iron(II) triple helicate. *Dalton Trans.* **2011**, *40*, 12368–12373. [CrossRef] [PubMed]

32. Schneider, C.J.; Cashion, J.D.; Chilton, N.F.; Etrillard, C.; Fuentealba, M.; Howard, J.A.K.; Létard, J.-F.; Milsmann, C.; Moubaraki, B.; Sparkes, H.A.; et al. Spin Crossover in a 3,5-Bis(2-pyridyl)-1,2,4-triazolate-Bridged Dinuclear Iron(II) Complex [{Fe(NCBH3)(py)}$_2$-(μ-L^1)$_2$]—Powder versus Single Crystal Study. *Eur. J. Inorg. Chem.* **2013**, *2013*, 850–864. [CrossRef]

33. Kulmaczewski, R.; Olguín, J.; Kitchen, J.A.; Feltham, H.L.C.; Jameson, G.N.L.; Tallon, J.L.; Brooker, S. Remarkable Scan Rate Dependence for a Highly Constrained Dinuclear Iron(II) Spin Crossover Complex with a Wide Thermal Hysteresis Loop. *J. Am. Chem. Soc.* **2014**, *136*, 878–881. [CrossRef] [PubMed]

34. Hagiwara, H.; Tanaka, T.; Hora, S. Synthesis, structure, and spin crossover above room temperature of a mononuclear and related dinuclear double helicate iron(II) complexes. *Dalton Trans.* **2016**, *45*, 17132–17140. [CrossRef] [PubMed]

35. Halcrow, M.A. Spin-crossover Compounds with Wide Thermal Hysteresis. *Chem. Lett.* **2014**, *43*, 1178–1188. [CrossRef]

36. Brooker, S. Spin crossover with thermal hysteresis: Practicalities and lessons learnt. *Chem. Soc. Rev.* **2015**, *44*, 2880–2892. [CrossRef] [PubMed]

37. Fujinami, T.; Nishi, K.; Hamada, D.; Murakami, K.; Matsumoto, N.; Iijima, S.; Kojima, M.; Sunatsuki, Y. Scan Rate Dependent Spin Crossover Iron(II) Complex with Two Different Relaxations and Thermal Hysteresis *fac*-[Fe^{II}(HIL$^{n\text{-Pr}}$)$_3$]Cl·PF$_6$ (HIL$^{n\text{-Pr}}$ = 2 Methylimidazol-4-yl-methylideneamino-*n*-propyl). *Inorg. Chem.* **2015**, *54*, 7291–7300. [CrossRef] [PubMed]

38. Bao, X.; Guo, P.-H.; Liu, W.; Tucek, J.; Zhang, W.-X.; Leng, J.-D.; Chen, X.-M.; Gural'skiy, I.; Salmon, L.; Bousseksou, A.; et al. Remarkably high-temperature spin transition exhibited by new 2D metal–organic frameworks. *Chem. Sci.* **2012**, *3*, 1629–1633. [CrossRef]

39. Liu, W.; Bao, X.; Li, J.-Y.; Qin, Y.-L.; Chen, Y.-C.; Ni, Z.-P.; Tong, M.-L. High-Temperature Spin Crossover in Two Solvent-Free Coordination Polymers with Unusual High Thermal Stability. *Inorg. Chem.* **2015**, *54*, 3006–3011. [CrossRef] [PubMed]

Inorganics **2017**, *5*, 49

40. Rostovtsev, V.V.; Green, L.G.; Fokin, V.V.; Sharpless, K.B. A Stepwise Huisgen Cycloaddition Process: Copper(I)-Catalyzed Regioselective "Ligation" of Azides and Terminal Alkynes. *Angew. Chem. Int. Ed.* **2002**, *41*, 2596–2599. [CrossRef]

41. Tornøe, C.W.; Christensen, C.; Meldal, M. Peptidotriazoles on Solid Phase: [1,2,3]-Triazoles by Regiospecific Copper(I)-Catalyzed 1,3-Dipolar Cycloadditions of Terminal Alkynes to Azides. *J. Org. Chem.* **2002**, *67*, 3057–3064. [CrossRef] [PubMed]

42. Pathigoolla, A.; Pola, R.P.; Sureshan, K.M. A versatile solvent-free azide–alkyne click reaction catalyzed by in situ generated copper nanoparticles. *Appl. Catal. A: Gen.* **2013**, *453*, 151–158. [CrossRef]

43. Schiff, H. Mittheilungen aus dem Universitätslaboratorium in Pisa: Eine neue Reihe organischer Basen. *Justus Liebigs Ann. Chem.* **1864**, *131*, 118–119. [CrossRef]

44. L'abbé, G.; Bruynseels, M. Replacement of aryl by alkyl in 1-substituted 1H-1,2,3-triazole-4-carbaldehydes. *J. Chem. Soc. Perkin Trans. 1* **1990**, 1492–1493. [CrossRef]

45. L'abbé, G.; Bruynseels, M.; Delbeke, P.; Toppet, S. Molecular rearrangements of 4-iminomethyl-1,2,3-triazoles. Replacement of 1-aryl substituents in 1H-1,2,3-triazole-4-carbaldehydes. *J. Heterocycl. Chem.* **1990**, *27*, 2021–2027. [CrossRef]

46. Hagiwara, H.; Minoura, R.; Okada, S.; Sunatsuki, Y. Synthesis, Structure, and Magnetic Property of a New Mononuclear Iron (II) Spin Crossover Complex with a Tripodal Ligand Containing Three 1, 2, 3-Triazole Groups. *Chem. Lett.* **2014**, *43*, 950–952. [CrossRef]

47. Hagiwara, H.; Okada, S. A polymorphism-dependent $T_{1/2}$ shift of 100 K in a hysteretic spin-crossover complex related to differences in intermolecular weak CH···X hydrogen bonds (X = S vs. S and N). *Chem. Commun.* **2016**, *52*, 815–818. [CrossRef] [PubMed]

48. Schneider, C.J.; Moubaraki, B.; Cashion, J.D.; Turner, D.R.; Leita, B.A.; Batten, S.R.; Murray, K.S. Spin crossover in di-, tri- and tetranuclear, mixed-ligand tris(pyrazolyl)methane iron(II) complexes. *Dalton Trans.* **2011**, *40*, 6939–6951. [CrossRef] [PubMed]

49. Nakamoto, K. *Infrared and Raman Spectra of Inorganic and Coordination Compounds*, 6th ed.; John Wiley & Sons: Hoboken, NJ, USA, 2009.

50. Yamada, M.; Hagiwara, H.; Torigoe, H.; Matsumoto, N.; Kojima, M.; Dahan, F.; Tuchagues, J.-P.; Re, N.; Iijima, S. A Variety of Spin-Crossover Behaviors Depending on the Counter Anion: Two-Dimensional Complexes Constructed by NH···Cl⁻ Hydrogen Bonds, [Fe^{II}H$_3$L^{Me}]Cl·X (X = PF$_6^-$, AsF$_6^-$, SbF$_6^-$, CF$_3$SO$_3^-$; H$_3$L^{Me} = Tris[2-{[(2-methylimidazol-4-yl)methylidene]amino}ethyl]amine). *Chem. Eur. J.* **2006**, *12*, 4536–4549. [CrossRef] [PubMed]

51. Spek, A.L. *PLATON SQUEEZE*: A tool for the calculation of the disordered solvent contribution to the calculated structure factors. *Acta Crystallogr. Sect. C Struct. Chem.* **2015**, *71*, 9–18. [CrossRef] [PubMed]

52. Marchivie, M.; Guionneau, P.; Létard, J.-F.; Chasseau, D. Photo-induced spin-transition: The role of the iron(II) environment distortion. *Acta Crystallogr. Sect. B* **2005**, *61*, 25–28. [CrossRef] [PubMed]

53. Llunell, M.; Casanova, D.; Cirera, J.; Alemany, P.; Alvarez, S. *SHAPE2.1. Program for Calculating Continuous Shape Measures of Polyhedral Structures*; Universitat de Barcelona: Barcelona, Spain, 2013.

54. Guionneau, P.; Marchivie, M.; Bravic, G.; Létard, J.-F.; Chasseau, D. Structural aspects of spin crossover. Examples of the [Fe^{II}L$_n$(NCS)$_2$] complexes. *Top. Curr. Chem.* **2004**, *234*, 97–128.

55. Kahn, O. *Molecular Magnetism*; VCH: Weinheim, Germany, 1993.

56. Burla, M.C.; Caliandro, R.; Carrozzini, B.; Cascarano, G.L.; Cuocci, C.; Giacovazzo, C.; Mallamo, M.; Mazzone, A.; Polidori, G. Crystal structure determination and refinement via *SIR2014*. *J. Appl. Cryst.* **2015**, *48*, 306–309. [CrossRef]

57. Sheldrick, G.M. Crystal structure refinement with *SHELXL*. *Acta Cryst.* **2015**, *C71*, 3–8.

58. Kabuto, C.; Akine, S.; Nemoto, T.; Kwon, E. Release of Software (Yadokari-XG 2009) for Crystal Structure Analyses. *J. Cryst. Soc. Jpn.* **2009**, *51*, 218–224. [CrossRef]

59. Dolomanov, O.V.; Bourhis, L.J.; Gildea, R.J.; Howard, J.A.K.; Puschmann, H. *OLEX2*: A complete structure solution, refinement and analysis program. *J. Appl. Cryst.* **2009**, *42*, 339–341. [CrossRef]

![inorganics logo] *inorganics*

MDPI

Article

Heteroleptic and Homoleptic Iron(III) Spin-Crossover Complexes; Effects of Ligand Substituents and Intermolecular Interactions between Co-Cation/Anion and the Complex

Wasinee Phonsri, Luke C. Darveniza, Stuart R. Batten and Keith S. Murray *

School of Chemistry, Building 23, 17 Rainforest Walk, Monash University, Clayton, VIC 3800, Australia; wasinee.phonsri@monash.edu (W.P.); lcdar2@student.monash.edu (L.C.D.); stuart.batten@monash.edu (S.R.B.)
* Correspondence: keith.murray@monash.edu; Tel.: +61-399-054-512; Fax: +61-399-054-597

Received: 19 June 2017; Accepted: 28 July 2017; Published: 1 August 2017

Abstract: The structural and magnetic properties of a range of new iron(III) bis-tridentate Schiff base complexes are described with emphasis on how intermolecular structural interactions influence spin states and spin crossover (SCO) in these d^5 materials. Three pairs of complexes were investigated. The first pair are the neutral, heteroleptic complexes [Fe(3-OMe-SalEen)(thsa)] **1** and [Fe(3-MeOSalEen)(3-EtOthsa)] **2**, where 3-R-HSalEen = (E)-2-(((2-(ethylamino)ethyl)imino)methyl)-6-R-phenol and 3-R-H_2thsa = thiosemicarbazone-3-R-salicylaldimine. They display spin transitions above room temperature. However, **2** shows incomplete and gradual change, while SCO in **1** is complete and more abrupt. Lower cooperativity in **2** is ascribed to the lack of π–π interactions, compared to **1**. The second pair, cationic species [Fe(3-EtOSalEen)$_2$]NO$_3$ **3** and [Fe(3-EtOSalEen)$_2$]Cl **4** differ only in the counter-anion. They show partial SCO above room temperature with **3** displaying a sharp transition at 343 K. Weak hydrogen bonds from cation to Cl$^-$ probably lead to weaker cooperativity in **4**. The last pair, CsH$_2$O[Fe(3-MeO-thsa)$_2$] **5** and Cs(H$_2$O)$_2$[Fe(5-NO$_2$-thsa)$_2$] **6**, are anionic homoleptic chelates that have different substituents on the salicylaldiminate rings of thsa^{2-}. The Cs cations bond to O atoms of water and the ligands, in unusual ways thus forming attractive 1D and 3D networks in **5** and **6**, respectively, and **5** remains HS (high spin) at all temperatures while **6** remains LS (low spin). Comparisons are made to other literature examples of Cs salts of [Fe(5-R-thsa)$_2$]$^-$ (R = H and Br).

Keywords: iron(III) complexes; spin crossover; heteroleptic; homoleptic; magnetism; structure; intermolecular interactions; cation–anion interactions

---

1. Introduction

In recent reviews and monographs on spin-crossover (SCO) materials [1–3], there is a dearth of examples of heteroleptic iron(III) complexes, apart from those of type [Fe(O$_2$N$_2$-tetradentate)(L)$_2$]$^+$ where L = pyridine, imidazole etc.; tetradentate = Schiff base [4]. We recently described the first example of an heteroleptic iron(III) spin-crossover complex [Fe(3-OMe-SalEen)(thsa)] **1** containing two *mer*-tridentate Schiff base ligands where 3-OMe-HSalEen = (E)-2-(((2-(ethylamino)ethyl)imino)methyl)-6-methoxyphenol and H$_2$thsa = thiosemicarbazone-salicylaldimine [5]. The donor set in **1** was N$_3$O$_2$S. Notably, the complex showed a gradual, complete spin transition above room temperature at 344 K, with enhanced spin-crossover properties compared to homoleptic cationic [Fe(3-OMe-SalEen)$_2$]$^+$ and anionic [Fe(thsa)$_2$]$^-$ analogues. π–π intermolecular interactions between each ligand type were the key reasons for the high $T_{1/2}$ value.

Here, we have extended this work by making reasonably systematic changes to both the tridentate ligands particularly by changing the substituent groups on the salicylaldehyde rings (Scheme 1).

We describe the neutral, heteroleptic complex [Fe(3-MeO-SalEen)(5-NO$_2$thsa)] **2** and make comparisons to the thsa parent **1** [5]. A pair of cationic homoleptic complexes [Fe(3-EtOSalEen)$_2$]NO$_3$ **3** and [Fe(3-EtOSalEen)$_2$]Cl **4** are explored to identify any anion effects. In a similar manner, a pair of anionic complexes, CsH$_2$O[Fe(3-MeOthsa)$_2$] **5** and Cs(H$_2$O)$_2$[Fe(5-NO$_2$-thsa)$_2$] **6**, are described that allow us to probe how both ligand substituent and cation–anion effects influence spin states and SCO. Crystallography is used extensively to probe intermolecular interactions in these new materials.

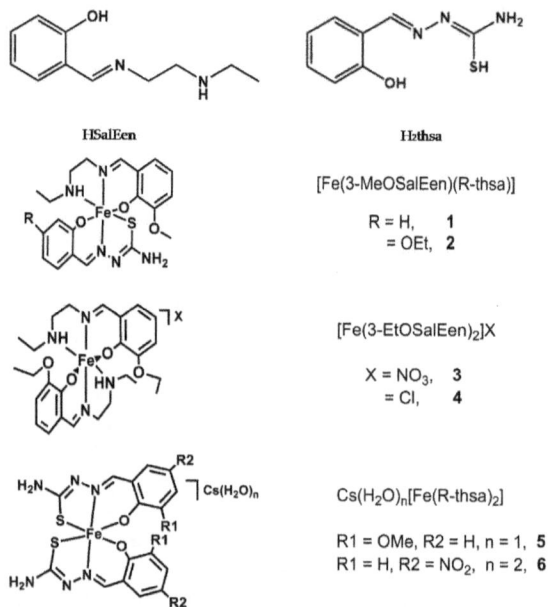

Scheme 1. Molecular structures of HSalEen and H$_2$thsa ligands together with Fe(III) compounds **1–6**. Note that the actual coordination of the Fe(III) compounds will be discussed *vide infra*, the pictures shown here clarify the molecular components only.

2. Results and Discussion

2.1. Preparation of Iron(III) Complexes

The complexes were all formed using layered diffusion techniques. Different combinations of solvents in a reaction depend on the solubility of reactants in each reaction and the layers were constructed according to the density of the solvents. The heteroleptic compound, [Fe(3-MeOSalEen)(3-EtOthsa)] **2**, was formed using an aqueous layer of the 3-EtO-H$_2$thsa ligand with CsOH in H$_2$O at the bottom, and a layer of FeCl$_3$ in *n*-butanol was in the middle. Then a layer of another ligand, 3-MeO-HSalEen ligand with triethylamine, in methanol, was on the top. Notably, the solubility of the R-H$_2$thsa ligands is very poor in water but it can be improved by adding a base to the mixture. The 3-EtOSalEen homoleptic complexes, [Fe(3-EtOSalEen)$_2$]NO$_3$ **3** and [Fe(3-EtOSalEen)$_2$]Cl **4**, were formed using a layer of ligand in dichloromethane. The bulk sample of **3** required solvated MeOH and water to fit the micro-analytical data. The substituted thsa homoleptic complexes, CsH$_2$O[Fe(3-MeOthsa)$_2$] **5** and Cs(H$_2$O)$_2$[Fe(5-NO$_2$-thsa)$_2$] **6**, were formed using an aqueous layer of ligand and CsOH with a layer of Fe(III) salts, in methanol, on the top. Phase purity of samples used for magnetic study was confirmed by comparison of powder X-ray diffractograms (PXRD) to simulated diffractograms (see Supplementary Figure S1). The PXRD for **6(bulk)** fitted well to the crystallographic refinement model (see Sections 2.5 and 3.1); however, the closest fit for the microanalytical data was a formula $Cs(H_2O)_2[Fe(5-NO_2-thsa)_2]\cdot CsOH$ (i.e., $6\cdot CsOH$), suggesting the presence of some amorphous CsOH contaminant in the sample. Thus, as seen above, we have labelled the bulk sample **6(bulk)** and the crystals **6**.

2.2. Magnetism

The variable-temperature magnetic susceptibility data for the complexes were obtained within the 5–400 K range without any protective coating applied to the sample. All the experiments were conducted under a DC field of 0.5 T and at a heating/cooling rate of $10~K\cdot min^{-1}$ in the settle mode, apart from compound **1** that was examined under various sweep rates with the results being reported previously [5]. For compounds **2–4**, the plots in Figure 1 all show incomplete SCO up to 360 K. At low temperature, the magnetic susceptibilities are about 0.4–0.5 $cm^3\cdot mol^{-1}\cdot K$, indicative of the compounds being in the LS (low spin) state. Upon warming, the spin transitions begin to take place at around 250 K (for **2**) and 340 K (for **3** and **4**), and show 48%, 36% and 16% degree of spin crossover up to 360 K, respectively. The $\chi_M T$ values of the compounds tend to keep increasing at higher temperature but, unfortunately, do not reach fully the HS (high spin) value at 400 K. It is noted that compound **3** exhibits an abrupt, reproducible spin transition with a 3 K hysteresis width ($T\uparrow = 343$ K), while **2** and **4** show gradual changes. The small change in $\chi_M T$ at the abrupt transition in **3** suggests that only some fraction of the material is showing this transition. On the other hand, compounds **5** and **6(bulk)** show invariant HS and LS behaviour, respectively (Figure S2). All the magnetic results agree well with the single-crystal structure data of **1–6**.

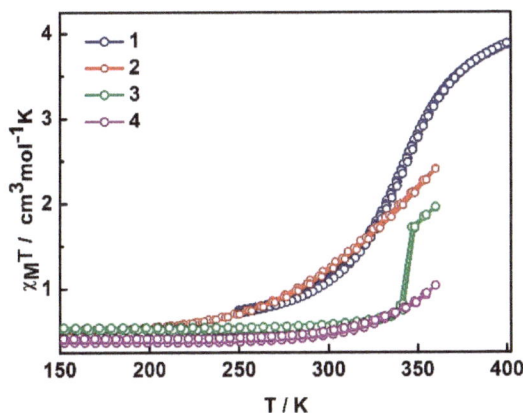

Figure 1. Variable-temperature magnetic susceptibility ($\chi_M T$) measurements for compounds **1–4**.

2.3. Structural Study of Heteroleptic Fe(III) Compounds 1 and 2

The crystal structures of [Fe(3-MeOSalEen)(thsa)] **1** and [Fe(3-MeOSalEen)(3-EtOthsa)] **2** have been examined at 100 K. The data revealed that the structure of **2** crystallizes in the monoclinic, $P2_1/c$ space group (Table 1). Details of the crystal structure of **1** have been presented [5] and are given in Tables 1 and 2. In the asymmetric unit, for both **1** and **2**, there is a neutral molecule of the Fe(III) compound where the metal centres coordinate to N_3O_2S donors belonging to 3-MeOSalEen$^-$ and R-thsa^{2-} ligands where R = H **1**, 3-EtO **2** (Figure S3). For compound **2**, the Fe–L bond lengths and the octahedral distortion parameters are shown in Table 2. Fe–O, Fe–N and Fe–S bond lengths are 1.88–1.93, 1.91–1.94 (for 3° amine) 2.05 (for 2° amine), and 2.25 Å, respectively. These bond lengths agree with the low values of the octahedral distortion parameters [6,7] ($\Sigma = 51°$ and $\Theta = 98°$) and all are indicative of the low spin state (LS) of the Fe(III) centre at 100 K.

Table 1. Crystallographic data and structure refinement for 1–6.

Complex	1 [5]	2	3	4	5	6
Temp.	100 K	100 K	100 K	100 K	123 K	123 K
Molecular weight/g·mol^{-1}	470.35	514.40	588.45	561.90	653.28	701.25
Crystal system	Triclinic	Monoclinic	Triclinic	Triclinic	Orthorhombic	Tetragonal
Space group	$P\bar{1}$	$P2_1/c$	$P\bar{1}$	$P\bar{1}$	$Pnma$	$P\bar{4}n2$
a/Å	8.3940 (17)	14.122 (3)	9.646 (2)	9.698 (2)	19.0649 (7)	20.6177 (3)
b/Å	9.3500 (19)	13.889 (3)	10.632 (2)	10.633 (2)	9.1084 (4)	20.6177 (3)
c/Å	13.675 (3)	13.470 (3)	14.242 (3)	13.515 (3)	13.1902 (5)	11.7467 (3)
α/°	82.26 (3)	90	100.25 (3)	96.15 (3)	90	90
β/°	73.44 (3)	118.40 (3)	105.21 (3)	103.05 (3)	90	90
γ/°	82.14 (3)	90	102.93 (3)	103.49 (3)	90	90
Cell volume/Å^3	1013.9 (4)	2324.1 (10)	1329.5 (5)	1301.3 (4)	2290.5 (2)	4993.4 (2)
Z	2	4	2	2	4	8
Absorption coefficient/mm^{-1}	0.880	0.778	0.622	0.722	2.451	2.266
Reflections collected	27,497	23,856	24,637	21,725	42,177	188,772
Independent Reflections, R_{int}	5929, 0.0501	6297, 0.0369	6224, 0.0792	5703, 0.1664	3544, 0.0582	10084, 0.0651
Max., min. transmission	0.9913, 0.9741	0.9923, 0.977	0.9969, 0.9695	0.9964, 0.9857	0.9301, 0.7493	0.7727, 0.6600
Restraints/parameters	1/277	0/301	0/364	0/329	0/156	12/331
Final R indices [$I > 2\sigma(I)$]: R_1, wR_2	0.0401, 0.1123	0.0408, 0.1118	0.0419, 0.1111	0.0573, 0.1518	0.0298, 0.0648	0.0747, 0.2314
CCDC No.	1420398	1552518	1552515	1552514	1552517	1552516

Data for **2–6**, this work. CCDC number from Cambridge Crystallographic Data Base deposition number.

Table 2. Selected bond lengths and octahedral distortion parameters for **1–6**.

Complex	1 [5]	2		3	4		5		6
Temp.	100 K	100 K		100 K	100 K		123 K		123 K
Fe1–O1/Å	1.8875 (14)	1.882 (1)	Fe1–O2/Å	1.877 (1)	1.874 (2)	Fe1–O1/Å	1.987 (2)	Fe1–O1/Å	1.917 (6)
Fe1–O3/Å	1.8937 (13)	1.9315 (1)	Fe1–O4/Å	1.889 (2)	1.895 (2)	Fe1–O1 ⁱⁱⁱ/Å	1.987 (2)	Fe1–O2/Å	1.922 (5)
Fe1–N1/Å	2.2590 (8)	2.050 (2)	Fe1–N1/Å	1.926 (2)	1.923 (3)	Fe1–N1/Å	2.179 (2)	Fe1–N1/Å	1.931 (6)
Fe1–N2/Å	2.0588 (17)	1.915 (2)	Fe1–N2/Å	2.037 (2)	2.041 (3)	Fe1–N1 ⁱⁱⁱ/Å	2.179 (2)	Fe1–N4/Å	1.911 (6)
Fe1–N3/Å	1.9198 (14)	1.942 (2)	Fe1–N3/Å	2.060 (2)	2.046 (3)	Fe1–S1/Å	2.416 (1)	Fe1–S1/Å	2.214 (2)
Fe1–S1/Å	1.9424 (14)	2.248 (1)	Fe1–N4/Å	1.923 (2)	1.922 (3)	Fe1–S1 ⁱⁱⁱ/Å	2.416 (1)	Fe1–S2/Å	2.237 (2)
Σ/°	44	51	Σ/°	47	48	Σ/°	129	Σ/°	39
Θ/°	80	98	Θ/°	70	72	Θ/°	404	Θ/°	67

Symmetry codes: (iii) x, $-y+1/2$, $-z+1/2$. Data for **2–6**, this work.

Although both **1** and **2** are in the LS state at low temperatures, compound **2**, with its extra ethoxy group substituted on the 3 position of the thsa^{2-} ligand, has a higher degree of distortion around the metal centre. It is obvious that the octahedral distortion parameters are higher in the case of **2** than in **1**, especially in the case of the Θ value that differs by 18°. Accordingly, the angle between the planes of thsa^{2-} ligand in **2** are larger than that angle is in **1** by about 10° (Table S1). On the other hand, for the 3-OMeSalEen$^-$ ligand, the angle between the ligand planes in **1** and **2** shows smaller differences by about 3° (Figure S4). The superimposition of the Fe(III) molecules of **1** and **2** is illustrated in Figure 2. Unexpectedly, the orientation of the ethane moiety in the 3-MeOSalEen$^-$ ligand for **1** and **2** bends out of the plane in the opposite direction. Therefore, the 3-EtO substituent on the thsa^{2-} ligand affects the arrangement of both thsa^{2-} itself as well as the 3-MeOSalEen$^-$ ligand.

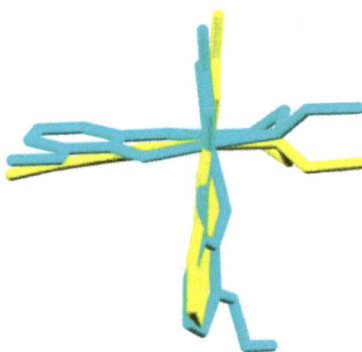

Figure 2. Superimposition of the Fe(III) molecules of 1 (yellow) and 2 (cyan) in a "wireframe" model.

Crystal packing in solid-state spin-crossover complexes is important in regard to explaining cooperativity and other spin-crossover nuances. According to the packing of **1** and **2**, the 1D chain motif along the *b* axis is similar and involves interactions via C–H···S and N–H···N. This is the only one direction that the 3-EtO group in **2** does not affect the crystal packing and shows similar packing as **1** (Figure 3). As the 3-EtO substituent on the thsa^{2-} ligand occupies space above the 3-OMeSalEen$^-$ ligand (Figure S3b), it prevents another ligand from coming close and forming π–π interactions. Therefore, there are no π–π interactions present in **2**. In **1**, however, strong π–π interactions were claimed to be a key reason for the high $T_{1/2}$ value observed [5]. The Fe(III) moieties in **2** form a chain through C–H···C/π interactions instead of connecting through two sets of strong π–π interactions as was found in **1** (Figure S5). Likewise, in the last dimension along the *a* axis, a continuous chain from two sets of π–π interactions links a plane of **1** into a high dimensional structure with a high degree of packing order (Figure S5a). Whilst there are weak C–H···C/π interactions found in the case of **2**, it is suggested that adding an extra 3-EtO on the thsa^{2-} ligand in **2** reduces the structural order in the structure, not only in the isolated Fe(III) molecule, but also in the overall packing of the structure (Figure S6b). Consequently, the molecules are unable to form π–π interactions. In summary, this is suggested to lead to a decrease in the cooperativity in **2** and result in the incomplete SCO (see above).

(a)

(b)

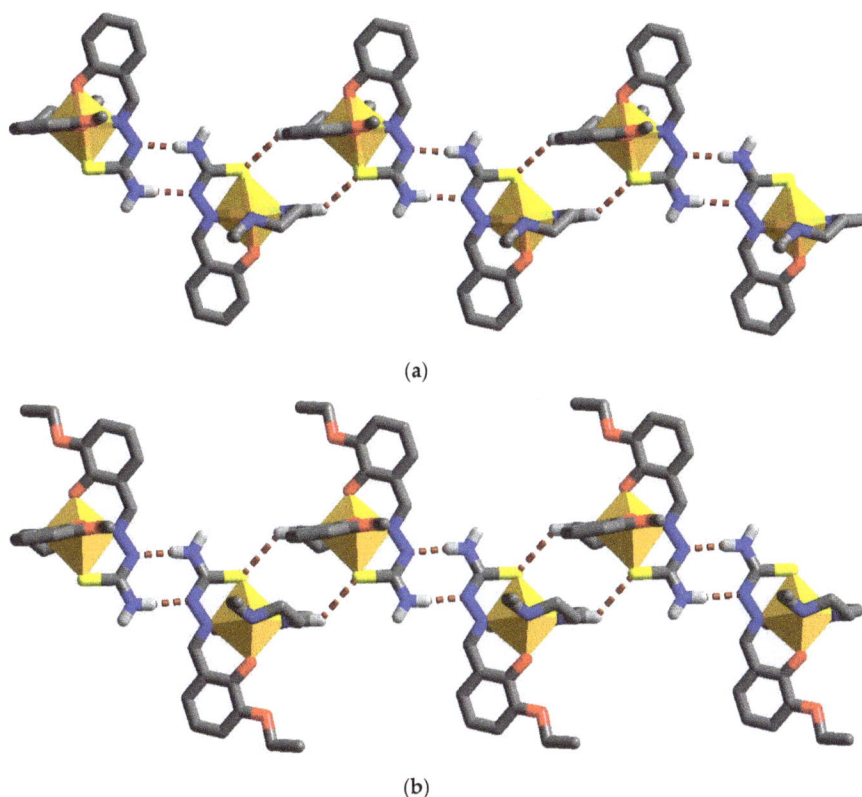

Figure 3. C–H···S and N–H···N interactions that show similarly connecting Fe(III) molecules in (**a**) **1** and (**b**) **2**. Colour coding: O, red; N, blue; S, yellow; C, grey; H, white; yellow octahedra show coordination environment around Fe.

2.4. Structural Study of Homoleptic Cationic Fe(III) Compounds **3** and **4**

The crystal structures of [Fe(3-EtOSalEen)$_2$]NO$_3$ **3** and [Fe(3-EtOSalEen)$_2$]Cl **4** have also been investigated at 100 K. The data revealed that both structures are crystallized in the triclinic, $P\bar{1}$ space group (Table 1). With a slightly bigger anion, the cell volume of **3** is rather larger than that of **4** by about 28 Å^3. It is noted that the c parameters of the unit cell in **3** are significantly larger than that in **4** compared to other dimensions. This is because the NO$_3^-$ and Cl$^-$ anions occupy space between the Fe(III) sheets along the c axis (Figure S7). The size of the anions then influences the size of the unit cell along the c axis more than the other axes. For the asymmetric unit of **3** and **4**, a cationic molecule of [Fe(3-EtOSalEen)$_2$]$^+$ co-exists with the counter anions (Figure 4). A Fe(III) centre coordinates to the N$_4$O$_2$ donors from two of 3-EtOSalEen$^-$ ligands disposed in the meridional fashion. The Fe–L bond lengths and octahedral distortion parameters shown in Table 2 are in the same ranges as in **1** and **2**, which again suggests that the LS state exists at the Fe(III) centres of **3** and **4** at low temperature.

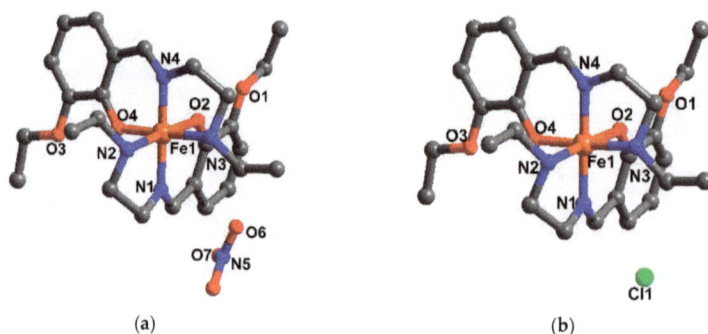

Figure 4. Representation of the asymmetric unit components for (**a**) **3** and (**b**) **4**. Hydrogen atoms are omitted for clarity.

Compounds **3** and **4** are isostructural. Both the Fe (III) moieties and the counter anions are located on the same position (Figure S8). Along the *a* axis, [Fe(3-EtOSalEen)$_2$]$^+$ interacts with the neighbouring molecules mainly through the parallel fourfold aryl embrace (P4AE) [8] and C–H···π interactions (Figure 5). All details of the interactions are shown in Tables S3 and S4. It is noted that these P4AE interactions do not show any relation to the magnetic behaviour of the compounds, as can be seen in Table S5. This is in agreement with our previous work that suggested a strong correlation between the π–π interactions (not P4AE) and SCO properties in this system [5]. In other dimension, the molecules connect through weak hydrogen bonds mainly involving the anion groups (Figure S9).

As mentioned, **3** and **4** are isostructural and only the anions are different between these two compounds. It is interesting to note that **3** shows an abrupt 50% SCO with a small thermal hysteresis width, while **4** exhibits a gradual incomplete spin change (see magnetism section). To try to understand these differences, the crystal structures of **3** and **4** are thoroughly investigated particularly in regard to intermolecular interactions involving the anions. It is found that some of the interactions relating to NO$_3$$^-$ in **3**,where there are linking Fe(III) moieties forming an extended structure, utilise moderate hydrogen bonds, while, those for Cl$^-$ in **4** are all of the weak hydrogen bond type [9] (Table S4). Thus, it is suggested to give rise to poorer cooperativity in **4** and lead to a gradual incomplete SCO appearing in this compound. It is possible that electronic effects provided by NO$_3$$^-$ and Cl$^-$ are partly responsible for the magnetic properties of the compounds but theoretical calculation would be required to prove this.

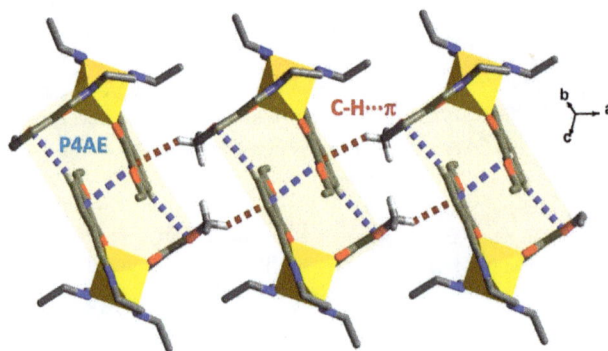

Figure 5. The 1D chain packing motif in **3** and **4** involving P4AE [8] and C–H···π interactions. Colour coding: O, red; N, blue; C, grey; H, white; yellow octahedra show coordination environment around Fe.

2.5. Structural Study of Homoleptic Anionic Fe(III) Compounds **5** and **6**

Six-coordinate Fe(III) compounds containing two substituted di-anionic ligands, R-H$_2$thsa with Cs$^+$ counter cations, viz. CsH$_2$O[Fe(3-MeOthsa)$_2$] **5** and Cs(H$_2$O)$_2$ [Fe(5-NO$_2$-thsa)$_2$] **6**, have been studied. Single-crystal data for the compounds are given in Table 1. The data show that the structure of **5** is crystallized in the orthorhombic, *Pnna* space group, while the crystal structure of **6** is in the tetragonal, *P$\bar{4}$n2* system. The contents of the asymmetric units of the compounds are illustrated in Figure 6 and show a half molecule of [Fe(3-MeOthsa)] with a half occupancy of Cs cation in that of **5**, while for **6** a molecule of [Fe(5-NO$_2$-thsa)$_2$]$^-$ exists together with two Cs cations with a half occupancy at each site of Cs$^+$. Selected Fe–L and Cs–O bond lengths are shown in Table S2 and Table S6. Notably, Cs cations in both **5** and **6** form bonds with O atoms of the ligands and water molecules, which is different from previous reports of Fe(III)-thsa with Cs cations that also showed Cs–N and Cs–S bonds with R-thsa^{2-} ligands, where R = H [10] and Br [11].

The Fe(III) centres in **5** and **6** possess an octahedral geometry coordinating via N$_2$O$_2$S$_2$ donors in a meridional fashion provided by two R-thsa$_2^{2-}$ ligands. According to the Fe–L bond lengths in Table 2, the data for **6** are similar to the Fe–L bond lengths of Fe(III) to 3-EtOthsa^{2-} ligand in **2**, which also suggests the LS state at the metal centre. In the case of **5**, the Fe–L lengths are larger than those in **6** by about 0.07, 025 and 0.2 Å for Fe–O, Fe–N and Fe–S, respectively. Moreover, the high values of octahedral distortion parameters, particularly the Θ ca. 400°, indicate the high degree of distortions around the Fe(III) centre. All the data are indicative to the HS state being populated in **5** at low temperature.

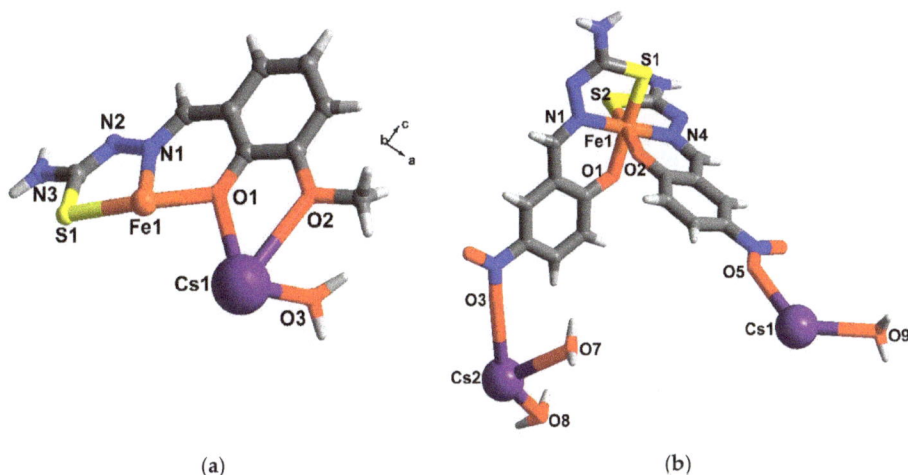

Figure 6. Representation of the asymmetric unit of (**a**) **5** and (**b**) **6**. Notably, some atoms contain 0.5 occupancy in the asymmetric unit i.e., Fe1, Cs1 and O3 in **5**, and Cs1, Cs2, O8 and O9 in **6**.

In an extended structure of **5**, two Cs$^+$ atoms form coordination bonds with [Fe(3-MeOthsa)$_2$]$^-$ through mono-dentate O2, and μ$_4$-O1 that acts as a bridging atom between two Cs cations (Figure 7a). This building unit links to the other neighbouring units through Cs–O3(H$_2$O) and forms a polymer chain along the *b* axis (Figure 7b). In higher dimension, N–H···π/S interactions play a role in connecting polymer chains into a pseudo 3D structure (Figure S11). As mentioned, the distortion parameters (Σ and Θ) for **5** are surprisingly high compared to other HS structures of Fe(III) compounds [3]. According to Figure 7a, one Cs$^+$ bonds to two O1 from two different 3MeO-thsa^{2-} ligands, and one O1 atom also bridges two Cs atoms. As O1 is a direct donor to the Fe(III) centre, these Cs–O1 coordination bonds thus give rise to an unexpectedly high degree of the distortion surrounding the Fe(III) centre

(Figure S10). Moreover, Cs–O1 bonds are suggested to reduce the electron density around the Fe(III) ion. Consequently, a small energy gap (Δ_{oct}) results in the invariant HS behaviour being observed in **5** (see magnetism section).

In the case of the crystal packing in **6**, each Cs atom connects to the same Cs type in different fashions as shown in Figure 8, to give Cs_2 (for Cs1) and Cs_4 (for Cs2, ignoring the partial occupancy (see Section 3.1)) clusters. The $[Fe(5\text{-}NO_2\text{-}thsa)_2]^+$ cation in turn coordinates to one Cs_4 and one Cs_2 cluster through $Cs\text{-}O(NO_2\text{-}thsa^{2-})$ interactions (Figure S12b), and a second Cs_2 cluster through interactions between Cs1 and one of the phenol oxygens (O1). All the Cs–O bond lengths in **5** and **6** are shown in Table S6. For Cs1, Cs1–O bonds are formed with four different Fe(III) molecules (Figure 8a), yielding a chain along the *c* axis. The Cs1 chain bridges to the adjacent Cs1 chain via $\mu_2\text{-}O9(H_2O)$, giving rise to a double polymeric chain of Cs1 and $[Fe(5\text{-}NO_2\text{-}thsa)_2]$ (Figure 8c). On the other hand, Cs2 atoms form a four-member cluster with two $\mu_2\text{-}O(H_2O)$ bridges (Figure 8b). Overall, each dimer of Cs1 atoms is thus connected to eight Fe(III) complexes (four through each Cs1 atom), while the tetramer of Cs2 atoms is connected to four Fe(III) complexes (one per Cs2 atom). In turn, the Fe complex is coordinated to three Cs clusters, to give a complicated 3,4,8-c 3D net with $(4.8^2)_4(8^6)(4^{12}.8^8.10^8)$ topology (Figure 8d).

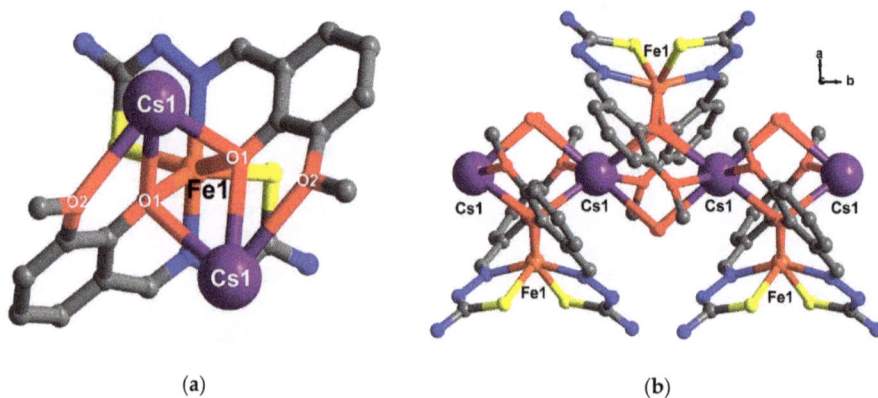

(a)   (b)

Figure 7. Representation of (a) the coordination bonds between Cs cations and adjacent Fe(III) molecules and (b) a 1D chain polymeric motif in **5**.

For homoleptic Fe(III)-thsa compounds containing Cs cations, in addition to **5** and **6**, there are $Cs[Fe(thsa)_2]$ [10] and $Cs[Fe(5\text{-}Br\text{-}thsa)_2]$ [11] that have been reported to show HS and LS behaviour, respectively, at 293 K. It is interesting to note that the Cs cation always form a Cs–O bond with the oxygen that directly bonds to the Fe(III) centre and this can be expected to affect the electronic structure. Unfortunately, in **5** and **6** it gives rise to an inappropriate ligand field energy gap for SCO to take place and, subsequently, leads to invariant HS or LS behaviour in these Fe(III)-thsa-Cs systems within the experimental ranges measured. Interestingly, in the related Fe(III)-thsa compound, $K[Fe(5\text{-}Br\text{-}thsa)_2]$ [12] that has a similar structure to $Cs[Fe(5\text{-}Br\text{-}thsa)_2]$ [11], the compound was reported to show SCO with thermal hysteresis above 350 K (for $T_{1/2}\uparrow = 358$ K) [12]. Thus, the s-block cation plays a role but one that is hard to predict.

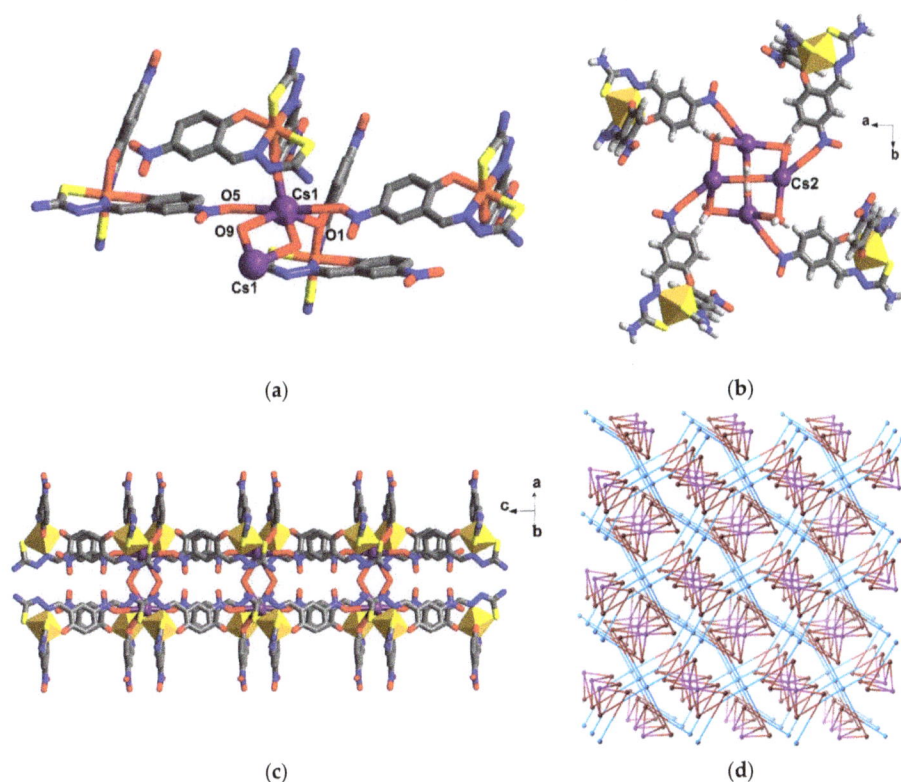

(a)

(b)

(c)

(d)

Figure 8. Representations of **6** show (**a**) the coordinating atoms surrounding Cs1 (**b**) the cluster of Cs2 cations, (**c**) a 1D polymer of [Fe(5-NO$_2$-thsa)$_2$] and Cs1 with water bridges and (**d**) the underlying 3,4,8-c 3D network with $(4.8^2)_4(8^6)(4^{12}.8^8.10^8)$ topology (brown spheres represent [Fe(5-NO$_2$-thsa)$_2$] moieties, blue spheres represent Cs$_4$ clusters, and purple spheres represent Cs$_2$ clusters).

3. Materials and Methods

3.1. General

All reagents and solvents were purchased from Sigma–Aldrich Australia (Castle Hill, NSW, 1765, Australia) and used as received. Infrared spectra were measured with a Bruker Equinox 55 FTIR spectrometer fitted with a 71Judson MCT detector and Specac Golden Gate diamond ATR. Microanalyses were performed by Campbell Microanalytical Laboratory, Department of Chemistry, University of Otago, Dunedin, New Zealand. Variable-temperature magnetic susceptibility data were collected with either a Quantum Design MPMS 5 superconducting quantum interference device (SQUID) magnetometer or a MPMS XL-7 SQUID magnetometer, with a scan speed of 10 K·min^{-1} followed by a one-minute wait after each temperature change.

X-ray crystallographic measurements for compounds **2**, **3** and **4** were collected at the Australian Synchrotron operating at approximately 16 keV ($\lambda = 0.71073$ Å). Single crystals were mounted on a glass fibre using oil. The collection temperature, 100 K, was maintained at specified temperatures using an open-flow N$_2$ cryostream. Data were collected using Blue Ice software [13]. Initial data processing was carried out using the XDS package [14]. X-ray data for **1** have been published [5].

X-ray crystallographic measurements on **5** and **6** were collected at 123 K using a Bruker Smart Apex X8 diffractometer with Mo-$K\alpha$ radiation ($\lambda = 0.71073$ Å). Single crystals were mounted on a glass fibre using oil. The data collection and integration were performed within SMART and SAINT+ software programs and corrected for absorption using the Bruker SADABS program [15]. For **6**, one of the Cs atoms (Cs2) was refined at half occupancy; higher occupancies led to unreasonably high anisotropic displacement parameters. Cambridge Crystallographic Data Base CCDC numbers are 1552514–1552518 for compounds **2–6**. The CCDC number for compound **1** is 1420398 [5].

X-ray powder diffraction patterns on **4** and **6(bulk)** were recorded using a Bruker D8 Advance powder diffractometer operating at Cu Kα wavelength (1.5418 Å), with samples mounted on a zero-background silicon single-crystal stage. Scans were performed at room temperature in the 2θ range 5°–55°.

3.2. Synthesis of Ligands

R-HSalEen, where R = 3-MeO and 3-EtO and R-H$_2$thsa, where R = 5-NO$_2$, 3-MeO and 3-EtO were synthesized according to the literature methods given in references [16,17], respectively.

3.3. Synthesis of Iron(III) Complexes

3.3.1. Synthesis of [Fe(3-MeOSalEen)(3-EtOthsa)] 2

A solution of 3-MeO-HSalEen (0.4 mmol) in MeOH (3 mL) was on the top of the layered diffusion method in which NEt$_3$ (56 µL, 0.4 mmol) had been added as a base. The middle layer is the metal solution. FeCl$_3$ (36 mg, 0.2 mmol) was dissolved in n-BuOH (2 mL). The solution was stirred for 5 min and then layered onto the solution of 3-EtO-H$_2$thsa (50 mg, 0.2 mmol) in H$_2$O (3 mL), in which CsOH (72 mg, 0.4 mmol) had been added as a base. After 3 weeks, black hexagonal plate-crystals formed together with colourless plate-crystals of by-product salts. The crystals were filtered and dried under ambient conditions. After a day, the black crystals were separated manually under the microscope; yield 19 mg (18%). $\tilde{\upsilon}_{max}$/cm^{-1} 3235 (υ_{NH2}), 3046 (υ_{Ar-H}), 1595 ($\upsilon_{C=N}$), 1297 (υ_{C-O}), 731 (υ_{CS}) cm^{-1}. m/z (ESI) 515.2 [Fe(3-MeOSalEen)(3-EtOthsa)]. Calcd. for [Fe(3-MeOSalEen)(3-EtOthsa)] (found %) C$_{22}$H$_{28}$FeN$_5$O$_4$S: C, 51.37 (51.25); H, 5.49 (5.59); N, 13.61 (13.48).

3.3.2. Synthesis of [Fe(3-EtOSalEen)$_2$]NO$_3$ 3

Fe(NO$_3$)$_3$·9H$_2$O (51 mg, 0.2 mmol) was dissolved in MeOH (5 mL). The solution was stirred for 5 min and then layered onto a solution of 3-EtO-HSalEen (95 mg, 0.4 mmol) in CH$_2$Cl$_2$ (2 mL), which was in the bottom layer to which NEt$_3$ (56 µL, 0.4 mmol) had been added as a base. After 7 days, the homogenous black solution was allowed to slowly evaporate in air. After a few days, black crystals formed, which were washed with acetone (2 × 1 mL) and then air-dried; yield 22 mg (18%). $\tilde{\upsilon}_{max}$/cm^{-1} 3170 (υ_{NH}), 2933 (υ_{Ar-H}), 1596 ($\upsilon_{C=N}$), 1246 (υ_{C-N}), 1217 (υ_{C-O}) cm^{-1}. m/z (ESI) 526.2 [Fe(3-EtOSalEen)$_2$]$^+$. Calcd. for [Fe(3-EtOSalEen)$_2$]NO$_3$·MeOH·0.5H$_2$O (found %) C$_{27}$H$_{43}$FeN$_5$O$_{8.5}$: C, 51.51 (51.16); H, 6.88 (7.13); N, 11.12 (11.74).

3.3.3. Synthesis of [Fe(3-EtOSalEen)$_2$]Cl 4

The same synthesis procedure for compound **3** was used to prepare compound **4** as well. FeCl$_3$ (34 mg, 0.2 mmol) has been used instead of Fe(NO$_3$)$_3$·9H$_2$O; yield 26 mg (23%). $\tilde{\upsilon}_{max}$/cm^{-1} 3055 (υ_{NH}), 2922 (υ_{Ar-H}), 1596 ($\upsilon_{C=N}$), 1247 (υ_{C-N}), 1218 (υ_{C-O}) cm^{-1}. m/z (ESI) 526.3 [Fe(3-EtOSalEen)$_2$]$^+$. Calcd. for [Fe(3-EtO-SalEen)$_2$]Cl (found %) C$_{26}$H$_{38}$ClFeN$_4$O$_4$: C, 55.58 (57.16); H, 6.82 (7.50); N, 9.97 (9.35). This analysis is on the sample giving the PXRD in Figure S1.

3.3.4. Synthesis of CsH$_2$O[Fe(3-MeOthsa)$_2$] 5

FeCl$_3$ (39 mg, 0.2 mmol) was dissolved in MeOH (3 mL). The solution was stirred for 5 min and then layered on the solution of 3-MeO-H$_2$thsa (92 mg, 0.4 mmol) in H$_2$O (2 mL), which was in the bottom, in which CsOH (134 mg, 0.8 mmol) had been added as a base. After 7 days, black crystals formed, these were washed with acetone (2 × 1 mL) and then air-dried; yield 73 mg (56%).

$\tilde{\nu}_{max}/cm^{-1}$ 3291 (ν_{NH2}), 2959 (ν_{Ar-H}), 1594 ($\nu_{C=N}$), 1237 (ν_{C-N}), 1215 (ν_{C-O}) 728 (ν_{C-S}) cm^{-1}. m/z (ESI) 501.9 [Fe(3-MeOthsa)$_2$]$^-$. Calcd. for CsH$_2$O[Fe(3-MeOthsa)$_2$] (found %) C$_{18}$H$_{20}$CsFeN$_6$O$_5$S$_2$: C, 33.09 (33.27); H, 3.08 (2.94); N, 12.86 (12.75).

3.3.5. Synthesis of Cs(H$_2$O)$_2$[Fe(5-NO$_2$-thsa)$_2$] 6(bulk)

FeCl$_3$ (39 mg, 0.2 mmol) was dissolved in MeOH (3 mL). The solution was stirred for 5 min and then layered on the solution of 5-NO$_2$-H$_2$thsa (97 mg, 0.4 mmol) in H$_2$O (3 mL), which was in the bottom, in which CsOH (146 mg, 0.8 mmol) had been added as a base. After 7 days, black crystals formed, which were washed with acetone (2 × 1 mL) and then air-dried; yield 34 mg (24%). $\tilde{\nu}_{max}/cm^{-1}$ 3255 (ν_{NH2}), 3013 (ν_{Ar-H}), 1587 ($\nu_{C=N}$), 1277 (ν_{C-N}) 737 (ν_{C-S}) cm^{-1}. m/z (ESI) 532.0 [Fe(5-NO$_2$-thsa)$_2$]$^-$. Calcd. for Cs(H$_2$O)$_2$ [Fe(5-NO$_2$-thsa)$_2$].CsOH (found %) C$_{16}$H$_{17}$Cs$_2$FeN$_8$O$_9$S$_2$: C, 22.50 (22.73); H, 1.90 (1.90); N, 13.18 (11.87). This analysis is on the sample giving the PXRD in Figure S1.

4. Conclusions

Three groups of Fe(III) compounds with various tridentate Schiff base ligands, i.e., heteroleptic: [Fe(3-OMeSalEen)(thsa)] **1** and [Fe(3-MeOSalEen)(3-EtOthsa)] **2**, cationic homoleptic: [Fe(3-EtOSal Een)$_2$]NO$_3$ **3** and [Fe(3-EtOSalEen)$_2$]Cl **4** and anionic homoleptic: CsH$_2$O[Fe(3-MeOthsa)$_2$] **5** and Cs(H$_2$O)$_2$[Fe(5-NO$_2$-thsa)$_2$] **6** have been investigated. Complexes **2–6** were newly synthesized and studied for the first time. Magnetic studies of the compounds revealed incomplete SCO for **2–4**, while **5** and **6(bulk)** showed HS and LS behaviour up to 360 K, respectively. It is interesting to note that the spin transition in **2–4** starts to take place at high temperature especially in the cases of **3** and **4**, above 340 K.

In comparison among each group, for **1** and **2**, an ethoxy substitution on the thsa^{2-} ligand results in less order of the molecular packing and, thus, a lower ability to form a potential intermolecular interaction. Consequently, it has poorer cooperativity than the analogous unsubstituted compound **1**, thus exhibiting more gradual and incomplete SCO up to 360 K. In the case of the homoleptic cationic compounds **3** and **4**, which are isostructural, stronger hydrogen bonds between the anions and the Fe(III) moieties are suggested to be responsible for an abrupt spin change observed in **3**, at 343 K, accompanied by hysteresis, which compares to a small, gradual SCO change in **4**. The anionic homoleptic compounds **5** and **6** are somewhat distinct in behaviour. With a preference for their co-cation Cs to form a bond with oxygen atoms, the resulting Cs–O bonds link the Fe(III) molecules into a coordination polymer forming a 1D chain for **5** and a 3D network for **6**. However, direct bonding of Cs$^+$ to an oxygen donor yields the inappropriate ligand field around the Fe(III) centres in **5** and **6**. Consequently, there is no spin transition taking place in the experimental temperature region for these two compounds.

Supplementary Materials: The following are available online at www.mdpi.com/2304-6740/5/3/51/s1, Tables S1–S7, structural details. Figure S1, Powder X-ray diffractograms, observed and simulated, for **4** and **6(bulk)**. Figure S2, plots of magnetic data for **2** to **6**. Figures S3–S12, structural details and comparisons. CIF files and checkcif reports.

Acknowledgments: This work was supported by an Australian Research Council Discovery grant (to KSM). Access to the Australian Synchrotron is gratefully acknowledged. Boujemaa Moubaraki, Monash University, is thanked for help with magnetic studies.

Author Contributions: Luke Darveniza and Wasinee Phonsri synthesized and characterized all compounds, Keith Murray and Wasinee Phonsri planned the research. Wasinee Phonsri collected the X-ray data, refined the crystal structures, with Stuart Batten's involvement, and measured the magnetic properties. Wasinee Phonsri wrote the first draft, Keith Murray and Stuart Batten proofread and prepared the final version of the manuscript.

Conflicts of Interest: The authors declare no conflict of interest.

Inorganics **2017**, *5*, 51

References

1. Murray, K.S. The development of spin-crossover research. In *Spin-Crossover Materials: Properties and Applications*; Halcrow, M.A., Ed.; John Wiley & Sons Ltd.: Chichester, UK, 2013; pp. 1–54.
2. Gütlich, P.; Goodwin, H.A. *Spin Crossover in Transition Metal Compounds I–III*; Springer: Berlin/Heidelberg, Germany, 2004; pp. 232–235.
3. Harding, D.J.; Harding, P.; Phonsri, W. Spin crossover in iron(III) complexes. *Coord. Chem. Rev.* **2016**, *313*, 38–61. [CrossRef]
4. Kennedy, B.J.; McGrath, A.C.; Murray, K.S.; Skelton, B.W.; White, A.H. Variable-temperature magnetic, spectral, and X-ray crystallographic studies of "spin-crossover" iron(III) schiff-base-lewis-base adducts. Influence of noncoordinated anions on spin-state interconversion dynamics in [Fe(salen)(imd)$_2$]Y species (Y = ClO$_4^-$, BF$_4^-$, PF$_6^-$, BPh$_4^-$; imd = imidazole). *Inorg. Chem.* **1987**, *26*, 483–495.
5. Phonsri, W.; Martinez, V.; Davies, C.G.; Jameson, G.N.L.; Moubaraki, B.; Murray, K.S. Ligand effects in a heteroleptic bis-tridentate iron(III) spin crossover complex showing a very high $T_{1/2}$ value. *Chem. Commun.* **2016**, *52*, 1443–1446. [CrossRef] [PubMed]
6. McCusker, J.K.; Rheingold, A.L.; Hendrickson, D.N. Variable-temperature studies of laser-initiated $^5T_2 \rightarrow {}^1A_1$ intersystem crossing in spin-crossover complexes: Empirical correlations between activation parameters and ligand structure in a series of polypyridyl ferrous complexes. *Inorg. Chem.* **1996**, *35*, 2100–2112. [CrossRef]
7. Marchivie, M.; Guionneau, P.; Letard, J.-F.; Chasseau, D. Photo-induced spin-transition: The role of the iron(II) environment distortion. *Acta Crystallogr. Sect. B Struct. Sci.* **2005**, *61*, 25–28. [CrossRef] [PubMed]
8. Russell, V.; Scudder, M.; Dance, I. The crystal supramolecularity of metal phenanthroline complexes. *J. Chem. Soc. Dalton Trans.* **2001**, *6*, 789–799. [CrossRef]
9. Steiner, T. The hydrogen bond in the solid state. *Angew. Chem. Int. Ed.* **2002**, *41*, 48–76. [CrossRef]
10. Ryabova, N.A.; Ponomarev, V.I.; Zelentsov, V.V.; Atovmyan, L.O. Kristallografiya. *Russ. Crystallogr. Rep.* **1981**, *26*, 101.
11. Powell, R.E.; Schwalbe, C.H.; Tizzard, G.J.; Koningsbruggen, P.J.V. Caesium bis(5-bromosalicylaldehyde thiosemicarbazonato-κ^3 O,N,S)ferrate(III): Supramolecular arrangement of low-spin Fe^{III} complex anions mediated by Cs$^+$ cations. *Acta Cryst.* **2015**, *C71*, 169–174.
12. Kang, S.; Shiota, Y.; Kariyazaki, A.; Kanegawa, S.; Yoshizawa, K.; Sato, O. Heterometallic Fe^{III}/K coordination polymer with a wide thermal hysteretic spin transition at room temperature. *Chem. Eur. J.* **2016**, *22*, 532–538. [CrossRef] [PubMed]
13. McPhillips, T.M.; McPhillips, S.E.; Chiu, H.-J.; Cohen, A.E.; Deacon, A.M.; Ellis, P.J.; Garman, E.; Gonzalez, A.; Sauter, N.K.; Phizackerley, R.P.; et al. Blu-ice and the distributed control system: Software for data acquisition and instrument control at macromolecular crystallography beamlines. *J. Synchrotron Radiat.* **2002**, *9*, 401–406. [CrossRef] [PubMed]
14. Kabsch, W. Automatic processing of rotation diffraction data from crystals of initially unknown symmetry and cell constants. *J. Appl. Crystallogr.* **1993**, *26*, 795–800. [CrossRef]
15. Sheldrick, G.M. *SADABS, Program for Area Detector Adsorption Correction*; Institute for Inorganic Chemistry, University of Göttingen: Göttingen, Germany, 1996.
16. Hui, R.-H.; Zhou, P.; You, Z.-L. Syntheses and crystal structures of two azide-bridged dinuclear zinc(II) complexes with schiff bases and halides. *Synth. React. Inorg. Met. Org. Nano Met. Chem.* **2012**, *42*, 135–139. [CrossRef]
17. Mehta, B.H.; Shaikh, J.A. Synthesis, characterisation, X-ray diffraction and antimicrobial studies of Pd(II), Rh(III) and Ru(III) complexes of thiosemicarbazones. *J. Ind. Counc. Chem.* **2009**, *26*, 1–6.

inorganics

MDPI

Article

Pybox-Iron(II) Spin-Crossover Complexes with Substituent Effects from the 4-Position of the Pyridine Ring (Pybox = 2,6-Bis(oxazolin-2-yl)pyridine)

Akifumi Kimura and Takayuki Ishida *

Department of Engineering Science, The University of Electro-Communications, Chofu, Tokyo 182-8585, Japan; kimura@ttf.pc.uec.ac.jp
* Correspondence: takayuki.ishida@uec.ac.jp; Tel.: +81-42-443-5490; Fax: +81-42-443-5501

Received: 30 June 2017; Accepted: 31 July 2017; Published: 8 August 2017

Abstract: Spin-crossover (SCO) behavior of a series of [Fe(X-pybox)$_2$](ClO$_4$)$_2$ was investigated, where X-pybox stands for 4-X-substituted 2,6-bis(oxazolin-2-yl)pyridine with X = H, Cl, Ph, CH$_3$O, and CH$_3$S. We confirmed that the mother compound [Fe(H-pybox)$_2$](ClO$_4$)$_2$ underwent SCO above room temperature. After X was introduced, the SCO temperatures ($T_{1/2}$) were modulated as 310, 230, and 330 K for X = Cl, Ph, and CH$_3$S, respectively. The CH$_3$O derivative possessed the high-spin state down to 2 K. Crystallographic analysis for X = H, Cl, CH$_3$O, and CH$_3$S was successful, being consistent with the results of the magnetic study. Distorted coordination structures stabilize the HS (high-spin) state, and the highest degree of the coordination structure distortion is found in the CH$_3$O derivative. A plot of $T_{1/2}$ against the Hammett substituent constant σ_p showed a positive relation. Solution susceptometry was also performed to remove intermolecular interaction and rigid crystal lattice effects, and the $T_{1/2}$'s were determined as 260, 270, 240, 170, and 210 K for X = H, Cl, Ph, CH$_3$O, and CH$_3$S, respectively, in acetone. The substituent effect on $T_{1/2}$ became very distinct, and it is clarified that electron-donating groups stabilize the HS state.

Keywords: spin crossover; spin transition; iron(II) ion; crystal structure

---

1. Introduction

Spin-crossover (SCO) is a reversible transition between low-spin (LS) and high-spin (HS) states by external stimuli like heat, light, pressure, or magnetic field [1–7]. Iron(II) (3d^6) coordination compounds attract a great deal of attention in various SCO complexes, because SCO takes place between dia- and paramagnetic states to show drastic change in magnetic and chromic properties. The six-nitrogen donor structures (i.e., Fe^{II}N$_6$) have been studied extensively, and in particular diimines like 2-pyridylmethyleneimine and bipyridyl derivatives [8–13] and triimines like 2,6-bis(azaaryl)pyridine derivatives and tripodal tris(azaaryl) compounds [14–19] are the most popular ligands for this purpose. The structural similarity found between the meridional SCO ligands 1-bpp [14] and 3-bpp [15] (2,6-bis(pyrazol-1-yl)- and 2,6-bis(pyrazol-3-yl)pyridine, respectively) affords us a clue for the development of robust SCO ligands with respect to the ligand-field engineering.

In this study, the synthesis and physical properties of Fe^{2+} complexes were investigated (Scheme 1), in which pybox (2,6-bis(oxazolin-2-yl)pyridine) ligands were involved as a triimine ligand. The pybox coordination compounds have been intensively studied for various reaction catalysts often carrying a chiral center in the oxazoline rings [20,21] and luminescent lanthanoid complexes as relatively recent topics [22–24]. More recently, Gao et al. explored the SCO application of pybox-iron(II) systems; the mother X = H compound [Fe(pybx)$_2$](ClO$_4$)$_2$ [25] and a tetramethylated compound [Fe(L)$_2$](ClO$_4$)$_2$ (L = 2,6-bis(4,4-dimethyloxazolin-2-yl)pyridine) [26] underwent SCO around $T_{1/2}$ = 345 and 162–176 K, respectively, with attention concentrated on the solvent and counter anion dependence. Furthermore,

Halcrow et al. reported that chiral (*S,S*)- or (*R,R*)-4,4′-dimethylated derivatives were subjected to the SCO study [27]. On the other hand, we are now focusing on intramolecular substituent effects, because the introduction of substituents would bring about drastic SCO tuning through the covalent bonds, rather than through van der Waals interaction. Furthermore, the physical properties should be designed and controlled in a non-serendipitous way.

We planned to modulate the SCO temperature with the aid of electronic and steric effects from a substituent group which is bound at the 4-position of the pyridine ring. The electronic substituent effects will be mainly discussed in this work, since the steric effects are hardly parameterized. After the difference of crystal packing motif is ignored, the Hammett substituent constants were applied to the relationship analysis. A role of the coordination structure distortion will be clarified. At the next stage, solution susceptometry was performed to remove intermolecular interaction and rigid crystal lattice effects. The intrinsic substituent dependence on the SCO behavior will be revealed, as discussed in connection with the known parallel results on the SCO complexes having a related meridional ligand.

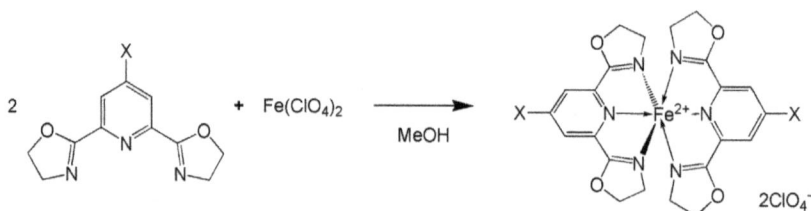

Scheme 1. Synthesis of [Fe(X-pybox)$_2$](ClO$_4$)$_2$ (X = H, Cl, Ph, CH$_3$O, CH$_3$S). Pybox: (2,6-bis(oxazolin-2-yl)pyridine.

2. Results

2.1. Preparation

New compounds [Fe(X-pybox)$_2$](ClO$_4$)$_2$ (X = Cl, Ph, CH$_3$O, CH$_3$S) were prepared according to the conventional method [23,28–30]. All the ligands were known, except for X = CH$_3$S. The complex formation using Fe(ClO$_4$)$_2$·6H$_2$O was conducted in methanol. The resulting dark red polycrystalline compound was isolated on a filter, and the product was purified by recrystallization from methanol. The elemental and spectroscopic analyses supported the formula of [Fe(X-pybox)$_2$](ClO$_4$)$_2$. Solvated molecules were found in the X = Ph and CH$_3$S case.

2.2. Crystal Structures

The X-ray crystallographic analysis on [Fe(X-pybox)$_2$](ClO$_4$)$_2$ (X = H, Cl, CH$_3$O, CH$_3$S) was successful at 100 and/or 400 K (Table 1 and Figure 1). The crystal structure of the mother compound [Fe(H-pybox)$_2$](ClO$_4$)$_2$ at 173 K has already been reported [25]; the crystal system is monoclinic *P*2$_1$/*n* with *a* = 15.386(3), *b* = 10.700(2), *c* = 16.934(3) Å, β = 103.31(3)°, and *V* = 2713.2(10) Å^3. We determined the crystal structure of this compound at 400 K as well as 100 K. The space group was kept at 400 K, and the cell volume was enlarged by 8.4% (Table 1). The Fe–N distances are listed in Table 2, and suggest that the population is almost shifted to the HS state at 400 K.

The space group of the [Fe(Cl-pybox)$_2$](ClO$_4$)$_2$ crystal is *P*2$_1$/*c*, and there is a unique crystallographically-independent molecule in a unit cell. The cell parameters (Table 1) were converted to a space group *P*2$_1$/*n* to give *a* = 19.314(3), *b* = 11.170(2), *c* = 14.787(2) Å, and β = 103.31(3)° at 100 K. The molecular arrangement motif is different from that of [Fe(H-pybox)$_2$](ClO$_4$)$_2$. The Fe–N bond lengths (Table 2) guarantee the LS state at 100 K. The crystal structure at 400 K was also determined. The space group was maintained, and the cell volume was somewhat enlarged by 7.5% (Table 1). The

Fe–N bond lengths were elongated by 10% (Table 2). This finding suggests that most of the population occupies the HS state at 400 K.

Table 1. Selected crystallographic parameters of [Fe(X-pybox)$_2$](ClO$_4$)$_2$ (X = H, Cl, CH$_3$O, CH$_3$S).

X	H	Cl	Cl	CH$_3$O	CH$_3$S
T/K	400	100	400	100	100
Formula weight	689.20	758.09	758.09	749.25	845.46
Crystal system	monoclinic	monoclinic	monoclinic	monoclinic	orthorhombic
Space group	$P2_1/n$	$P2_1/c$	$P2_1/c$	Cc	$P2_12_12_1$
a/Å	16.140(3)	14.787(2)	15.857(4)	12.084(3)	12.656(3)
b/Å	10.9603(17)	11.170(2)	10.588(3)	12.822(3)	15.676(4)
c/Å	17.049(3)	18.164(3)	19.084(4)	19.328(5)	16.897(4)
β/°	102.736(9)	109.071(8)	107.947(11)	101.716(12)	90.
V/Å^3	2941.8(9)	2835.4(8)	3048.1(13)	2932.3(12)	3352.1(13)
Z	4	4	4	4	4
$d_{calcd.}$/g·cm^{-3}	1.556	1.776	1.652	1.697	1.675
μ (MoKα)/mm^{-1}	0.763	0.983	0.914	0.778	0.811
No. of unique reflections	5760	6481	6966	6488	7660
R (F) ($I > 2\sigma$ (I)) [a]	0.0830	0.0698	0.0830	0.0361	0.0503
wR (F^2) (all reflections) [b]	0.2959	0.1978	0.2863	0.0908	0.1193
Goodness-of-fit parameter	0.981	1.047	1.009	1.052	1.068
Flack parameter	-	-	-	−0.008(4)	0.002(9)

[a] $R = \Sigma[|F_o| - |F_c|]/\Sigma|F_o|$; [b] $wR = [\Sigma w(F_o^2 - F_c^2)/\Sigma w F_o^4]^{1/2}$.

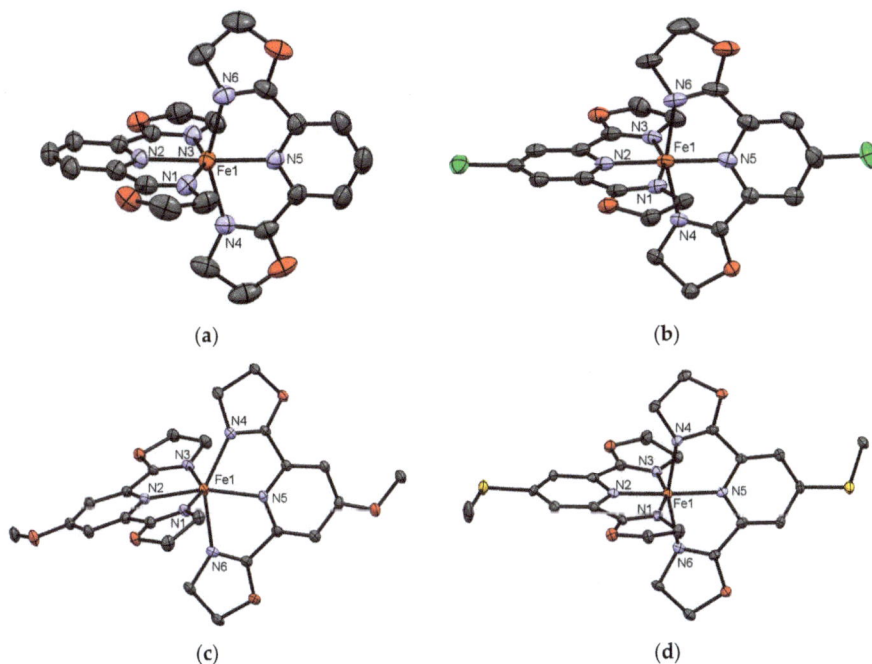

Figure 1. X-ray crystal structures of [Fe(X-pybox)$_2$](ClO$_4$)$_2$ for X = (a) H (400 K); (b) Cl (100 K); (c) CH$_3$O (100 K); and (d) CH$_3$S (100 K). The thermal ellipsoids are drawn at the 20%, 50%, 50%, and 50% probability levels, respectively. Hydrogen atoms and counter anions are omitted for clarity.

The space group of the crystal of [Fe(CH$_3$O-pybox)$_2$](ClO$_4$)$_2$ is monoclinic Cc, and the Flack parameter was satisfactorily reduced. There is a unique crystallographically-independent molecule in a unit cell. The molecular packing motif is completely different from those of the H and Cl derivatives.

From the Fe–N bond lengths (Table 2), the state is suggested to be HS. Table 3 summarizes the N–Fe–N bond angles. As a common feature, the N–Fe–N bite angles are largely deviated from the right angle owing to the five-membered chelate rings. However, the molecular structure of the CH_3O derivative clearly exhibits a bent N(py)–Fe–N(py) axis (the angle of 156.74(12)°) (Figure 1c). In contrast, the 400 K structures of the X = H and Cl derivatives show practically linear N(py)–Fe–N(py) axes (177.6(2) and 176.50(15)°, respectively).

Compound [Fe(CH_3S-pybox)$_2$](ClO$_4$)$_2$ crystallizes in a space group orthorhombic $P2_12_12_1$. The Flack parameter was satisfactorily reduced. There is a unique crystallographically-independent molecule. The crystal involves two molar methanol molecules; namely, the composition formula is [Fe(CH_3S-pybox)$_2$](ClO$_4$)$_2$(CH$_3$OH)$_2$, being consistent with the results of the elemental analysis and magnetic measurements (see below). From the Fe–N bond lengths (Table 2), the state is suggested to be LS at 100 K. The molecular structure (Figure 1d) displays a relatively linear N(py)–Fe–N(py) axis (176.62(19)°), as usually found in the LS molecules of this family. At high temperatures (e.g., 400 K), the crystals were broken down and the crystal structure analysis was unsuccessful. The elemental analysis suggests that some methanol molecules would be liberated from the crystal lattice.

Table 2. Fe–N bond distances (*d*) in Å for [Fe(X-pybox)$_2$](ClO$_4$)$_2$ (X = H, Cl, CH_3O, CH_3S).

X	H (400 K)	Cl (100 K)	Cl (400 K)	CH_3O (100 K)	CH_3S (100 K)
d (Fe1–N1)	2.134(6)	1.956(4)	2.153(4)	2.209(3)	1.991(4)
d (Fe1–N2)	2.060(5)	1.896(4)	2.098(4)	2.115(3)	1.907(4)
d (Fe1–N3)	2.161(5)	1.963(4)	2.159(5)	2.203(3)	1.981(4)
d (Fe1–N4)	2.129(6)	1.960(4)	2.149(4)	2.159(3)	1.956(4)
d (Fe1–N5)	2.067(5)	1.901(4)	2.088(4)	2.136(3)	1.901(4)
d (Fe1–N6)	2.160(6)	1.977(4)	2.194(4)	2.218(3)	1.978(5)
Average	2.12	1.94	2.14	2.17	1.95

Table 3. N–Fe–N bond angles (φ) in ° for [Fe(X-pybox)$_2$](ClO$_4$)$_2$ (X = H, Cl, CH_3O, CH_3S).

X	H (400 K)	Cl (100 K)	Cl (400 K)	CH_3O (100 K)	CH_3S (100 K)
φ (N1–Fe1–N2)	75.2(2)	79.55(17)	74.14(16)	74.73(12)	78.90(18)
φ (N1–Fe1–N3)	150.8(2)	159.31(17)	148.07(18)	147.31(12)	158.41(18)
φ (N1–Fe1–N4)	96.0(2)	92.36(15)	92.48(15)	103.25(12)	93.7(2)
φ (N1–Fe1–N5)	105.6(2)	97.50(16)	102.35(16)	116.23(12)	104.09(18)
φ (N1–Fe1–N6)	91.5(2)	93.41(16)	95.74(16)	88.78(12)	89.80(19)
φ (N2–Fe1–N3)	75.5(2)	79.76(17)	73.93(17)	73.65(12)	79.53(18)
φ (N2–Fe1–N4)	105.9(2)	101.08(16)	105.09(15)	125.05(12)	98.96(19)
φ (N2–Fe1–N5)	177.6(2)	177.03(16)	176.50(15)	156.74(12)	176.62(19)
φ (N2–Fe1–N6)	102.8(2)	99.94(17)	105.62(16)	87.91(12)	102.61(19)
φ (N3–Fe1–N4)	91.17(19)	91.28(15)	95.20(16)	88.33(12)	89.0(2)
φ (N3–Fe1–N5)	103.7(2)	103.19(16)	109.57(18)	96.25(12)	97.45(18)
φ (N3–Fe1–N6)	95.7(2)	90.46(16)	93.30(17)	98.11(12)	95.55(19)
φ (N4–Fe1–N5)	76.4(2)	79.34(16)	74.76(15)	74.30(12)	79.4(2)
φ (N4–Fe1–N6)	158.89(18)	158.89(18)	149.28(16)	146.72(12)	158.42(18)
φ (N5–Fe1–N6)	79.78(18)	79.78(18)	74.58(15)	72.55(12)	79.10(19)

2.3. Mangetic Properties of Polycrystalline Specimens

The magnetic susceptibilities of polycrystalline specimens of [Fe(X-pybox)$_2$](ClO$_4$)$_2$ were measured on a SQUID magnetometer in a temperature range of 10–400 K. As Figure 1 shows, the high temperature limit of the $\chi_m T$ value was 3.5–4.0 cm^3·K·mol^{-1}, being compatible with the S = 2 HS state of the iron(II) ion. On cooling, the $\chi_m T$ value of [Fe(H-pybox)$_2$](ClO$_4$)$_2$ gradually decreased and reached practically null below 270 K. Though the $\chi_m T$ value did not reach the high-temperature limit even at 400 K, the present SCO behavior is supposed to reproduce well the reported data ($T_{1/2}$ = 345 K [25]). Similarly, [Fe(X-pybox)$_2$](ClO$_4$)$_2$ (X = Cl, Ph, CH_3S) showed SCO, and the transition temperatures ($T_{1/2}$) were determined to be 310, 230, and 330 K for X = Cl, Ph, and CH_3S, respectively. The CH_3O derivative possessed the HS state in all the temperature range investigated here.

The CH$_3$S derivative exhibited a $\chi_m T$ jump at 350 K. After the jump, the $\chi_m T$ profile never reproduced the original one because of the loss of the solvent molecules. The crystal structure analysis of the specimen after the thermal cycle was unsuccessful, and the solvent loss was confirmed by the elemental and spectroscopic analyses. We can exclude possibility that the $\chi_m T$ increase would dominantly come from the solvent discharge "reaction". In the present case, from a close look, the $\chi_m T$ reached the half level before the jump, so the $T_{1/2}$ can be correctly defined to be 330 K before the reaction. However, only the solvated form was structurally characterized and will be incorporated to the magneto-structure discussion, implying that the $T_{1/2}$ data of [Fe(CH$_3$S-pybox)$_2$](ClO$_4$)$_2$ involved the effect of the solvation.

Figure 2b depicts the plot of $T_{1/2}$ against the Hammett substituent constant σ_p (-0.27, -0.01, 0, 0.00, 0.23 for the CH$_3$O, Ph, H, CH$_3$S, and Cl groups, respectively) [31]. No SCO appeared in the $\chi_m T(T)$ profile of the CH$_3$O derivative. To incorporate this result into the present discussion, it is enough to suppose that the $G(T)$ level crossing temperature between the HS and LS states must be located much below those of the H, Cl, Ph, and CH$_3$S derivatives (230–350 K). Therefore, we can safely conclude a positive relationship, which indicates that electron-donating groups suppress $T_{1/2}$ while electron-withdrawing groups raise $T_{1/2}$. In Figure 2b, The CH$_3$O data point has tentatively been placed at 0 K. This trend is qualitatively compatible with that of the 1-bpp series [Fe(X-1-bpp)$_2$](BF$_4$)$_2$ (X-1-bpp = 4-X-substituted 1-bpp) [32]. The rigid crystal lattice may play a role of a spin-transition inhibitor. The conclusion seems to have an approximate meaning, because the various space groups and molecular packing motifs were found but neglected in the above discussion.

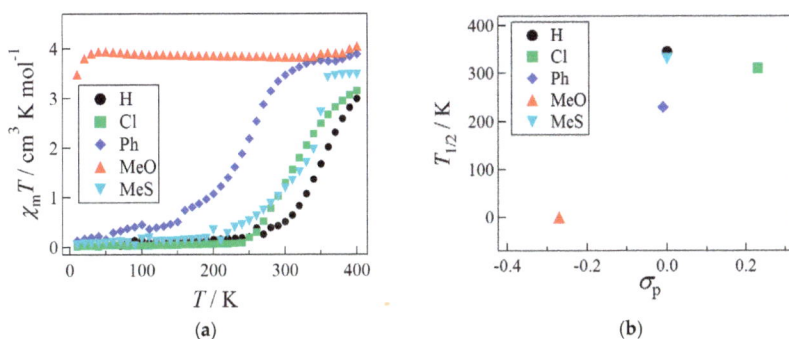

Figure 2. (a) Temperature dependence of $\chi_m T$ for polycrystalline [Fe(X-pybox)$_2$](ClO$_4$)$_2$ measured at 5000 Oe; (b) The spin transition temperature $T_{1/2}$ as a function of the Hammett substituent constant σ_p. The CH$_3$O data point is tentatively placed at 0 K.

2.4. Magnetic Properties of Solution Specimens

In solution, many of the problems from intermolecular interactions in various crystal lattices are removed, so the intrinsic intramolecular substituent effect can be extracted from variable-temperature experiments. We carried out solution susceptometry for the present complexes, inspired by the work on [Fe(X-1-bpp)$_2$](BF$_4$)$_2$ by Halcrow and co-workers [32]. They utilized the Evans method, but we applied the conventional SQUID apparatus. The results of the acetone-solution magnetic susceptibility measurements are summarized in Figure 3a. The data were acquired on cooling until the solution was solidified. We have found no indication of possible ligand-iron ion dissociation reaction. The $T_{1/2}$'s were determined as 260, 270, 240, 170, and 210 K for [Fe(X-pybox)$_2$](ClO$_4$)$_2$ with X = H, Cl, Ph, CH$_3$O, and CH$_3$S, respectively. The CH$_3$O derivatives exhibited a remarkable difference between the solid-state and solution data; namely, the HS state at any temperature in the solid state, whereas $T_{1/2}$ = 170 K was characterized in the acetone solution. The broad transition widths observed are consistent with the complete absence of cooperativity.

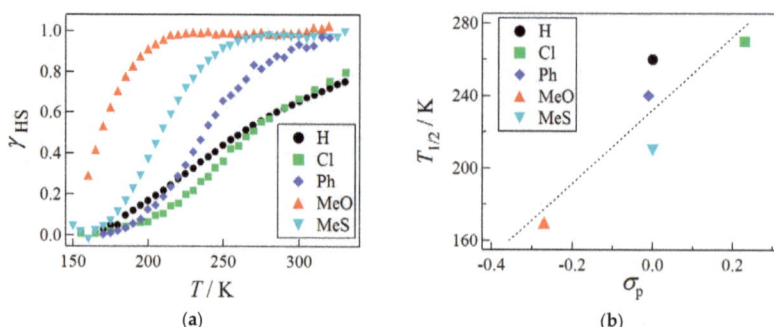

Figure 3. (**a**) Temperature dependence of the HS (high-spin) molar fraction γ_{HS} for acetone solutions of [Fe(X-pybox)$_2$](ClO$_4$)$_2$, measured at 5000 Oe. (**b**) The spin transition temperature $T_{1/2}$ as a function of the Hammett substituent constant σ_p. A dotted line represents the best linear fit.

Similarly to the solid-state experiments, we plotted the solution $T_{1/2}$ against the Hammett substituent constant σ_p [31] (Figure 3b). We can easily find a positive relationship, which indicates that electron-donating groups suppress $T_{1/2}$ while electron-withdrawing groups raise $T_{1/2}$. An empirical relationship equation is described with Equation (1) (a line in Figure 3b). This trend is similar to that of the 1-bpp series [32]. The solution data have the advantage of negligible intermolecular interaction. However, no structural information is afforded from the solution experiment.

$$T_{1/2} = 200(60)\,\text{K} \cdot \sigma_p + 230(10)\,\text{K} \tag{1}$$

3. Discussion

3.1. Substituent Effect

In the present study, the Hammett substituent constants are available to explain the static electronic effect from substituents [33]. There are two major electronic effects from substituents: inductive effect along the σ electron network and mesomeric effect along the π electron network. The 4-position in a pyridine ring is preferable for the mesomeric effect from the pyridine nitrogen site (i.e., the 1-position). Applying the Hammett σ_p constant is reasonable in this system. The π-electron-withdrawing group stabilizes the LS state. This is because the increase of π electron delocalization into the ligand leads to an increase in the ligand-field splitting parameter [34,35]. In other words, the iron(II) t_{2g} orbitals are stabilized owing to the $d\pi$–$p\pi$ orbital interaction, and the ligand-field splitting is enhanced.

3.2. Coordination Structure Effect

The HS states are known to favor distorted coordination geometry in general [34–39], which is related to long Fe–N distances and also accommodation of steric congestion during transition. From the N–Fe–N bond angles summarized in Table 3, we can calculate the Σ value [37] for estimation of distortion degree (Table 4), according to Equation (2). An ideal octahedron (Oh) possesses $\Sigma = 0°$. Other popular parameters are also listed in Table 4. The Θ value represents the deviation of the coordination geometry from an octahedron to a trigonal prism [38]. An ideal trigonal antiprism viewed from the principal axis contributes null, and an ideal Oh leads to $\Theta = 0°$. The α value is the average of the four N–Fe–N bite angles [39]. By using the SHAPE software [40], the continuous shape measures (CShM) are calculated with respect to an Oh. An ideal Oh returns null. The ϕ (N$_{py}$–Fe–N$_{py}$ angle) and θ (dihedral angle between two pybox systems) values are measures of the angular Jahn–Teller distortion, as Halcrow et al. proposed to describe the criterion of SCO [Fe(X-1-bpp)$_2$]$^{2+}$ and related compounds [14]. Very recently, a new empirical rule has been proposed based on the interatomic N–N distance in the chelatable diimine structure [36]. Since there is no structural data of the free ligands, the distances were calculated with the density functional theory on the b3lyp/6-311+G(2d,p) level in Gaussian 03 [41].

$$\Sigma = \sum_{i=1}^{12} |(\angle cis\,\text{N} - \text{Fe} - \text{N})_i - 90°)| \tag{2}$$

Table 4. Distortion parameters for [Fe(X-pybox)$_2$](ClO$_4$)$_2$ (X = H, Cl, CH$_3$O, CH$_3$S).

Parameters [a]	H [b]		Cl		CH$_3$O	CH$_3$S
T/K	173	400 [c]	100	400 [c]	100	100
Σ/°	90.1	130.3	90.8	142.0	158.8	96.6
Θ/°	293	417	310	455	596	326
α/°	79.7	75.5	79.6	74.4	73.8	79.2
CShM (Oh)	2.224	4.327	2.283	5.019	6.919	2.484
ϕ/°	179.14(11)	177.6(2)	177.03(16)	176.50(15)	156.74(12)	176.6(2)
θ/°	92.5	95.12(7)	91.06(4)	93.04(5)	93.41(3)	94.99(5)
$d_{calc.}$ (N–N)/Å [d]		2.855		2.854	2.854	2.852

[a] For the definition, see the text. [b] The 173 K structural data are taken from Ref. [25]. [c] The $\chi_m T$ vs. T plot indicates the spin-crossover is not completed even at 400 K, so the 400 K data imply the "almost" high-spin (HS) data. [d] An averaged value from two distances in a metal-free ligand. CShM: continuous shape measure; Oh: octahedron.

The geometrical parameters of the CH$_3$O derivative could be obtained only for the HS state. Accordingly, the X = CH$_3$O coordination structure at 100 K was compared with the X = H and Cl structures at 400 K, and the highest degree of distortion was found in the CH$_3$O derivative, as indicated with the distortion parameters (Table 4) as well as the determined molecular structure (Figure 1c). As for the LS states, when the X = Cl and CH$_3$S structures at 100 K were compared with the X = H structure at 173 K, their distortions were comparable to each other. These findings are confirmed by the parameters; in particular, Σ and CShM seem to be sensitive and convenient metrics. The ϕ and θ values of the HS [Fe(CH$_3$O-pybox)$_2$](ClO$_4$)$_2$ predict that this compound is not an SCO compound, and in fact it was found to be an HS compound in the whole temperature range (Figure 2a). On the other hand, the calculated N–N distance is insensitive to the substitution. As Figure 4a exemplifies, the CShM vs. σ_p plot has a relation with a negative slope, implying that the electron-donating group would favor a distorted structure. Such distortion would bring about preference for the HS states due to reduction of the ligand-field strength [42–45] and suppress the SCO temperature. Gao and co-workers have already reported the SCO study on a [Fe(H-pybox)$_2$]$^{2+}$ series with counter anion and solvent variation [25] with the substituent X fixed to H, and the magneto-structural relation was clarified to give a partially similar conclusion: highly distorted structures are favorable for the HS state and only intermediately distorted compounds show SCO.

As Figure 2 displays, although the difference of molecular packing motif is ignored, the intramolecular substituent effects seem to be dominant and approximately regulate the overall $T_{1/2}$ trend in the solid state. As for [Fe(CH$_3$O-pybox)$_2$](ClO$_4$)$_2$, the CH$_3$O substituent brings about an excessive effect through the coordination structure distortion, as indicated by the drastic difference from the solution result. The substituents regulate both geometry and ligand-field strength. The plot of $T_{1/2}$ for [Fe(X-1-bpp)$_2$](BF$_4$)$_2$ against $T_{1/2}$ for [Fe(X-pybox)$_2$](ClO$_4$)$_2$ displayed a positive correlation (Figure 4b). The platforms are different, but the common substituent effect is still observable. Consequently, the SCO characteristics can be discussed in connection with the substituent effect. We have to pay attention to an indirect mechanism, where the substituent effect is operative through geometrical modification and coordination structure distortion, together with the direct mechanism from the electronic substituent effect.

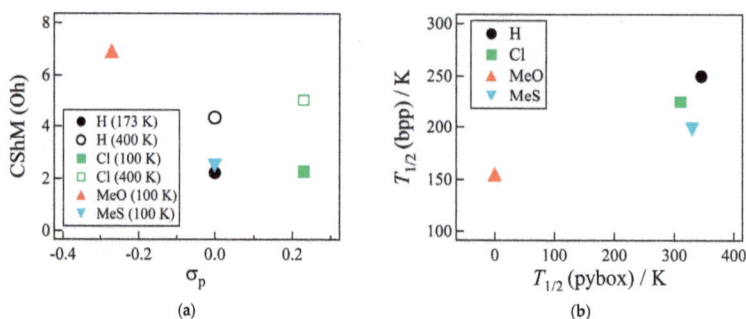

Figure 4. Relations between (**a**) the distortion parameter CShM (continuous shape measures) and Hammett substituent constant σ_p. (**b**) Plot of $T_{1/2}$ for polycrystalline [Fe(X-1-bpp)$_2$](ClO$_4$)$_2$ vs. $T_{1/2}$ for solution [Fe(X-pybox)$_2$](BF$_4$)$_2$. The [Fe(X-1-bpp)$_2$](BF$_4$)$_2$ data are taken from Ref. [32].

3.3. Electronic Substituent Effect

The intrinsic intramolecular substituent effect can be extracted from the solution data. The results on the solution SQUID susceptometry for [Fe(X-pybox)$_2$](ClO$_4$)$_2$ (X = H, Cl, Ph, CH$_3$O, CH$_3$S) are summarized in Figure 3a. The solution $T_{1/2}$ vs. σ_p plot displays a positive relationship (Figure 3b). The solution data have no structural information, and the coordination structures might be modified by the substituent effects. However, we can assume that structural distortion would be minor from thinking of the vast conformational freedom in solution. Thus, the solution results seem to be more reliable than those of the solid-state experiments to evaluate substituent effect. The conclusion is derived here, which is basically the same as the solid-state experiments. Electron-donating groups stabilize the HS state, whereas electron-withdrawing groups stabilize the LS one.

The plot of $T_{1/2}$ of [Fe(X-1-bpp)$_2$](BF$_4$)$_2$ in acetone solutions against the solution $T_{1/2}$ of [Fe(X-pybox)$_2$](ClO$_4$)$_2$ displayed a positive correlation (Figure 5). This relationship is more evident than that of Figure 4b, because both were acquired in solutions. An empirical relationship is formulated as Equation (3) and superposed in Figure 5. In the doubly meridional chelated systems involving 4-X-1-dpp and 4-X-pybox, the structural distortion effects would be similar to each other. Such distortion effects are assumed to be cancelled out in the comparison analysis; the linear relationship observed here is accounted for in terms of the essential electronic substituent effect represented by the Hammett substituent constant σ_p.

$$T_{1/2}(\text{pybox}) = 1.10(4) \cdot T_{1/2}(\text{bpp}) \tag{3}$$

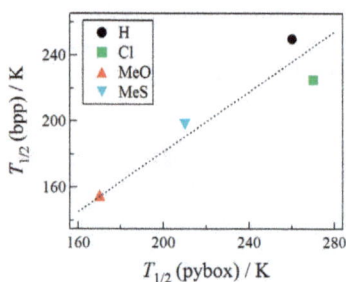

Figure 5. Plot of $T_{1/2}$ for solution [Fe(X-1-bpp)$_2$](ClO$_4$)$_2$ vs. $T_{1/2}$ for solution [Fe(X-pybox)$_2$](BF$_4$)$_2$. The [Fe(X-1-bpp)$_2$](BF$_4$)$_2$ data are taken from Ref. [32]. A dotted line represents the best linear fit.

4. Experimental Section

4.1. Materials

Caution! We have not yet encountered any hazard, but the perchlorate salts should be handled with care.

The ligands X-pybox (X = H [23], Cl [28], Ph [29], and CH$_3$O [30]) were prepared according to the known procedure. A new derivative CH$_3$S-pybox was prepared according to the known procedure for the CH$_3$O-pybox [30] with modification. A mixture of Cl-pybox (0.3707 g; 1.47 mmol) and sodium methylthiolate (0.1212 g; 1.73 mmol) in dry *N,N*-dimethylformamide was stirred at 40 °C for 12 h. After being cooled, the mixture was poured into aqueous NaHCO$_3$, and organic substrates were extracted with ethyl acetate. The organic layer was washed with aqueous NaHCO$_3$, dried over MgSO$_4$, and filtered. Concentration of the filtrate gave CH$_3$S-pybox (0.4266 g; 1.34 mmol) in 91% yield as a colorless solid. m.p. 193–195 °C. ^{1}H NMR (ECA-500, JEOL, Tokyo, Japan) (500 MHz, CDCl$_3$) δ 7.95 (2H, s), 4.53 (4H, t, *J* = 9.6 Hz), 4.11 (4H, t, *J* = 9.6 Hz), 2.18 (3H, s). ^{13}C NMR (126 MHz, CDCl$_3$) δ 163.53, 152.87, 146.27, 121.26, 68.47, 55.09, 13.96. MS (ESI$^+$) (JMS-T100 AccuTOF, JEOL, Tokyo, Japan) *m/z* 286.03 (M + Na$^+$), 264.05 (M + H$^+$). IR (neat, attenuated total reflection (ATR)) (Nicolet 6700, Thermo Scientific, San Jose, CA, USA) 1638, 1572, 1382, 1122, 942, 868, 786, 670, 533 cm^{-1}.

The target complexes were prepared as follows. A mixture of Cl-pybox (0.0559 g; 0.226 mmol), Fe(ClO$_4$)$_2$·6H$_2$O (0.0211 g; 0.113 mmol), and L-ascorbic acid (11 mg) in methanol (18 mL) was allowed to stand in a refrigerator for 24 h. The filtration gave 0.039 g (0.052 mmol) of [Fe(Cl-pybox)$_2$](ClO$_4$)$_2$. Yield 47%. m.p. 204 °C (dec.). The product was purified by recrystallization from methanol prior to the elemental and spectroscopic analyses, crystallographic analysis, and magnetic study. IR (neat, ATR) 1612, 1499, 1274, 1033, 926, 760, 651 cm^{-1}. Anal. Calcd. for C$_{22}$H$_{20}$Cl$_4$FeN$_6$O$_{12}$: C, 34.86%; H, 2.66%; N, 11.09%. Found: C, 35.02%; H, 2.55%; N, 11.37%.

Similarly, compounds [Fe(X-pybox)$_2$](ClO$_4$)$_2$ (X = Ph, CH$_3$O, and CH$_3$S) were prepared, and the yields were 44%, 26%, and 90%, respectively. [Fe(Ph-pybox)$_2$](ClO$_4$)$_2$·H$_2$O: m.p. 209 °C (dec.). IR (neat, ATR) 1584, 1374, 1269, 1068, 913, 757, 620 cm^{-1}. Anal. Calcd. for C$_{34}$H$_{30}$Cl$_2$FeN$_6$O$_{12}$·H$_2$O: C, 47.16%; H, 4.07%; N, 9.43%. Found: C, 47.04%; H, 3.55%; N, 9.48%. [Fe(CH$_3$O-pybox)$_2$](ClO$_4$)$_2$: m.p. 289 °C (dec.). IR (neat, ATR) 1587, 1372, 1239, 1072, 916, 859, 620, 578 cm^{-1}. Anal. Calcd. for C$_{24}$H$_{26}$Cl$_2$FeN$_6$O$_{14}$: C, 38.47%; H, 3.50%; N, 11.22%. Found: C, 38.53; H, 3.29; N, 11.16%. [Fe(CH$_3$S-pybox)$_2$](ClO$_4$)$_2$·2CH$_3$OH: m.p. 208 °C (dec.). IR (neat, ATR) 1574, 1485, 1372, 1270, 1069, 915, 796, 619 cm^{-1}. Anal. Calcd. for C$_{24}$H$_{26}$Cl$_2$FeN$_6$O$_{12}$S$_2$·CH$_4$O·H$_2$O: C, 36.11%; H, 3.88%; N, 10.11%; S, 7.71%. Found: C, 36.14%; H, 3.58%; N, 10.08%; S, 7.70%. After thermal treatment at 400 K in a SQUID magnetometer. Anal. Calcd. for C$_{24}$H$_{26}$Cl$_2$FeN$_6$O$_{12}$S$_2$: C, 36.89%; H, 3.35%; N, 10.76%; S, 8.21%. Found: C, 36.56%; H, 3.18%; N, 10.60%; S, 8.19%.

4.2. Crystallographic Analysis

X-Ray diffraction data of [Fe(X-pybox)$_2$](ClO$_4$)$_2$ (X = H, Cl, CH$_3$O, CH$_3$S) were collected on a Saturn70 CCD diffractometer (Rigaku, Tokyo, Japan) with graphite monochromated Mo Kα radiation (λ = 0.71073 Å). The structures were directly solved by a heavy-atom method and expanded using Fourier techniques in the CRYSTALSTRUCTURE [46]. Numerical absorption correction was used. Hydrogen atoms were located at calculated positions, and their parameters were refined as a riding model. The thermal displacement parameters of non-hydrogen atoms were refined anisotropically. Selected crystallographic data are given in Table 1, and selected bond distances and angles are listed in Tables 2 and 3, respectively. CCDC numbers 1559025, 1559026, 1559027, 1559028, and 1559029 for [Fe(X-pybox)$_2$](ClO$_4$)$_2$ (X = H, Cl (100 K), Cl (400 K), CH$_3$O, CH$_3$S, respectively) include the experimental details and full geometrical parameter tables. These data can be obtained free of charge via http://www.ccdc.cam.ac.uk/conts/retrieving.html.

4.3. Magnetic Study

Magnetic susceptibilities of [Fe(X-pybox)$_2$](ClO$_4$)$_2$ (X = H, Cl, Ph, CH$_3$O, CH$_3$S) were measured on a Quantum Design MPMS-XL7 SQUID magnetometer (San Diego, CA, USA) with a static field of 0.5 T. The magnetic responses were corrected with diamagnetic blank data of the sample holder measured separately. The diamagnetic contribution of the sample itself was estimated from Pascal's constants. The solution magnetic susceptibilities were measured in the same SQUID apparatus. The specimen was dissolved in acetone, and the resultant clear solution was transferred into a 5 mm·ϕ NMR sample tube. After the solution was degassed with argon gas bubbling, the sample tube was sealed and mounted in a SQUID probe. The susceptibility data were acquired on cooling until the acetone solution was solidified (typically around 160 K). In the analysis, the diamagnetic contribution was numerically estimated as a temperature-independent susceptibility term, so that the $\chi_m T$ values of the LS and/or HS regions should be almost constant. Owing to the ambiguity of the sample amount, the S-shaped profile was drawn on the molar fraction basis, $\gamma_{HS} = x(HS)/(x(HS) + x(LS))$ against temperature.

4.4. DFT Calculation Study

Density-functional-theory (DFT) calculation was performed by using the Gaussian 03 package [41]. The geometry was optimized after the b3lyp Hamiltonian and the 6-311+G(2d,p) basis set were chosen. The convergence criterion was below 10^{-8} a.u. in the self-consistent field energy.

5. Conclusions

We synthesized four new iron(II) complexes (Cl, Ph, CH$_3$O, and CH$_3$S). The SCO transition temperature $T_{1/2}$ can be changed by introducing a substituent at the 4-position of the pyridine ring. The $T_{1/2}$ around or slightly above room temperature may be very attractive for future application of the SCO materials. In the solid-state study, there seems to be a correlation between σ_p and Σ or σ_p and CShM, and furthermore between σ_p and $T_{1/2}$. The coordination structure distortion depends on the substituents, and the distortion also indirectly regulates $T_{1/2}$. Distorted structures stabilize the HS state. In the solution study, the substituent dependence on $T_{1/2}$ became very obvious, and electron-donating groups stabilize the HS state. The SCO temperature is regulated by the substituents, being similar to the known parallel work on [Fe(X-1-bpp)$_2$](BF$_4$)$_2$. The platforms are different, but the SCO characteristics can be discussed in connection with the substituent effect in a generalized manner.

Acknowledgments: This work was financially supported from KAKENHI (JSPS/15H03793).

Author Contributions: Akifumi Kimura participated in the preparation, X-ray structural analysis, and magnetic study. Takayuki Ishida designed the study and wrote the manuscript.

Conflicts of Interest: The authors declare no conflict of interest.

References

1. Gütlich, P.; Goodwin, H.A. (Eds.) *Spin Crossover in Transition Metal Compounds I, II, and III*; Springer: Berlin, Germany, 2004.
2. Gütlich, P.; Gaspar, A.B.; Garcia, Y. Spin state switching in iron coordination compounds. *Beilstein J. Org. Chem.* **2013**, *9*, 342–391. [CrossRef] [PubMed]
3. Halcrow, M.A. *Spin-Crossover Materials: Properties and Applications*; John Wiley & Sons, Ltd.: Oxford, UK, 2013.
4. Halcrow, M.A. Spin-Crossover compounds with wide thermal hysteresis. *Chem. Lett.* **2014**, *43*, 1178–1188. [CrossRef]
5. Kahn, O. Chapter 4. In *Molecular Magnetism*; VCH: Weinhein, Germany, 1993.
6. Harding, D.J.; Harding, P.; Phonsri, W. Spin crossover in iron(III) complexes. *Coord. Chem. Rev.* **2016**, *313*, 38–61. [CrossRef]
7. Letard, J.-F.; Guionneau, P.; Nguyen, O.; Costa, J.S.; Marcen, S.; Chastanet, G.; Marchivie, M.; Goux-Capes, L. A guideline to the design of molecular-based materials with long-lived photomagnetic lifetimes. *Chem. Eur. J.* **2005**, *11*, 4582–4589. [CrossRef] [PubMed]

8. Letard, J.-F.; Guionneau, P.; Godjovi, E.; Lavastre, O.; Bravic, G.; Chasseau, D.; Kahn, O. Wide Thermal Hysteresis for the Mononuclear Spin-Crossover Compound *cis*-Bis(thiocyanato)bis[*N*-(2'-pyridylmethylene)-4-(phenylethynyl)anilino]iron(II). *J. Am. Chem. Soc.* **1997**, *119*, 10861–10862. [CrossRef]
9. Yamada, M.; Hagiwara, H.; Torigoe, H.; Matsumoto, N.; Kojima, M.; Dahan, F.; Tuchagues, J.P.; Re, N.; Iijima, S. A variety of spin-crossover behaviors depending on the counter anion: Two-dimensional complexes constructed by NH⋯Cl⁻ hydrogen bonds, [(Fe^{II}H$_3$LMe)]Cl·X (X = PF$_6$⁻, AsF$_6$⁻, SbF$_6$⁻, CF$_3$SO$_3$⁻; H$_3$LMe = tris[2-{[(2-methylimidazol-4-yl)methylidene]amino}ethyl]amine). *Chem. Eur. J.* **2006**, *12*, 4536–4549. [PubMed]
10. Takahashi, K.; Kawakami, T.; Gu, Z.; Einaga, Y.; Fujishima, A.; Sato, O. An abrupt spin transition based on short S...S contacts in a novel Fe(II) complex whose ligand contains a 1,3-dithiole ring. *Chem. Commun.* **2003**, 2374–2375. [CrossRef]
11. Mochida, N.; Kimura, A.; Ishida, T. Spin-Crossover Hysteresis of [Fe^{II}(L$_H$^{iPr})$_2$(NCS)$_2$] (L$_H$^{iPr} = *N*-2-Pyridylmethylene-4-isopropylaniline Accompanied by Isopropyl Conformation Isomerism. *Magnetochemistry* **2015**, *1*, 17–27. [CrossRef]
12. Oso, Y.; Ishida, T. Spin-crossover transition in a mesophase iron(II) thiocyanate complex chelated with 4-hexadecyl-*N*-(2-pyridylmethylene)aniline. *Chem. Lett.* **2009**, *38*, 604–605. [CrossRef]
13. Oso, Y.; Kanatsuki, D.; Saito, S.; Nogami, T.; Ishida, T. Spin-crossover transition coupled with another solid-solid phase transition for iron(II) thiocyanate complexes chelated with alkylated *N*-(di-2-pyridylmethylene)anilines. *Chem. Lett.* **2008**, *37*, 760–761. [CrossRef]
14. Halcrow, M.A. Iron(II) complexes of 2,6-di(pyrazol-1-yl)pyridines—A versatile system for spin-crossover research. *Coord. Chem. Rev.* **2009**, *253*, 2493–2514. [CrossRef]
15. Craig, G.A.; Roubeau, O.; Aromi, G. Spin state switching in 2,6-bis(pyrazol-3-yl)pyridine (3-bpp) based Fe(II) complexes. *Coord. Chem. Rev.* **2014**, *269*, 13–31. [CrossRef]
16. Krober, J.; Codjovi, E.; Kahn, O.; Groliere, F.; Jay, C. A Spin Transition System with a Thermal Hysteresis at Room Temperature. *J. Am. Chem. Soc.* **1993**, *115*, 9810–9811. [CrossRef]
17. Hirosawa, N.; Oso, Y.; Ishida, T. Spin-crossover and light-induced excited spin-state trapping observed for an iron(II) complex chelated with tripodal tetrakis(2-pyridyl)methane. *Chem. Lett.* **2012**, *41*, 716–718. [CrossRef]
18. Yamasaki, M.; Ishida, T. Heating-rate dependence of spin-crossover hysteresis observed in an iron(II) complex having tris(2-pyridyl)methanol. *J. Mater. Chem. C* **2015**, *3*, 7784–7787. [CrossRef]
19. Yamasaki, M.; Ishida, T. Spin-crossover thermal hysteresis and light-induced effect on iron(II) complexes with tripodal tris(2-pyridyl)methanol. *Polyhedron* **2015**, *85*, 795–799. [CrossRef]
20. Johnson, J.S.; Evans, D.A. Chiral Bis(oxazoline) Copper(II) Complexes: Versatile Catalysts for Enantioselective Cycloaddition, Aldol, Michael, and Carbonyl Ene Reactions. *Acc. Chem. Res.* **2000**, *33*, 325–335. [CrossRef] [PubMed]
21. Desimoni, G.; Faita, G.; Quadrelli, P. Pyridine-2,6-bis(oxazolines), Helpful Ligands for Asymmetric Catalysts. *Chem. Rev.* **2003**, *103*, 3119–3154. [CrossRef] [PubMed]
22. Yuasa, J.; Ohno, T.; Miyata, K.; Tsumatori, H.; Hasegawa, Y.; Kawai, T. Noncovalent Ligand-to-Ligand Interactions Alter Sense of Optical Chirality in Luminescent Tris(β-diketonate) Lanthanide(III) Complexes Containing a Chiral Bis(oxazolinyl) Pyridine Ligand. *J. Am. Chem. Soc.* **2011**, *133*, 9892–9902. [CrossRef] [PubMed]
23. De Bettencourt-Dias, A.; Barber, P.S.; Viswanathan, S.; de Lill, D.T.; Rollett, A.; Ling, G.; Altun, S. Para-Derivatized Pybox Ligands As Sensitizers in Highly Luminescent Ln(III) Complexes. *Inorg. Chem.* **2010**, *49*, 8848–8861. [CrossRef] [PubMed]
24. De Bettencourt-Dias, A.; Barber, P.S.; Bauer, S. A Water-Soluble Pybox Derivatives and Its Highly Luminescent Lanthanide Ion Complexes. *J. Am. Chem. Soc.* **2012**, *134*, 6987–6994. [CrossRef] [PubMed]
25. Zhu, Y.-Y.; Li, H.-Q.; Ding, Z.-Y.; Lu, X.-J.; Zhao, L.; Meng, Y.-S.; Liu, T.; Gao, S. Spin transition in a series of [Fe(pybox)$_2$]$^{2+}$ complexes modulated by ligand structures, counter anions, and solvents. *Inorg. Chem. Front.* **2016**, *3*, 1624–1636. [CrossRef]
26. Zhu, Y.-Y.; Liu, C.-W.; Yin, J.; Meng, Z.-S.; Yang, Q.; Wang, J.; Liu, T.; Gao, S. Structural phase transition in a multi-induced mononuclear Fe^{II} spin-crossover complex. *Dalton Trans.* **2015**, *44*, 20906–20912. [CrossRef] [PubMed]

27. Burrows, K.E.; McGrath, S.E.; Kulmaczewski, R.; Cespedes, O.; Barrett, S.A.; Halcrow, M.A. Spin State of Homochiral and Heterochiral Isomers of [Fe(PyBox)$_2$]$^{2+}$ Derivatives. *Chem. Eur. J.* **2017**, *23*, 9067–9075. [CrossRef] [PubMed]

28. De Bettencourt-Dias, A.; Rossini, J.S.K. Ligand Design for Luminescent Lanthanide-Containing Metallopolymers. *Inorg. Chem.* **2016**, *55*, 9954–9963. [CrossRef] [PubMed]

29. Yu, X.; Yang, T.; Wang, S.; Xu, H.; Gong, H. Nickel-Catalyzed Reductive Cross-Coupling of Unactivated Alkyl Halides. *Org. Lett.* **2011**, *13*, 2138–2141. [CrossRef] [PubMed]

30. Vermonden, T.; Branowska, D.; Marcelis, A.T.M.; Sudholter, E.J.R. Synthesis of 4-functionalized terdendate pyridine-based ligads. *Tetrahedron* **2003**, *59*, 5039–5045. [CrossRef]

31. Hansch, C.; Leo, A.; Taft, R.W. A survey of Hammett substituent constants and resonance and field parameters. *Chem. Rev.* **1991**, *91*, 165–195. [CrossRef]

32. Cook, L.J.K.; Rafal, K.; Mohammed, R.; Dudley, S.; Barrett, S.A.; Little, M.A.; Deeth, R.J.; Halcrow, M.A. A Unified Treatment of the Relationship Between Ligand Substituents and Spin State in a Family of Iron(II) Complexes. *Angew. Chem. Int. Ed.* **2016**, *55*, 4327–4331. [CrossRef] [PubMed]

33. Isaccs, N.S. Chapter 4. In *Physical Organic Chemistry*; Wiley: New York, NY, USA, 1987.

34. Tweedle, M.F.; Wilson, L.J. Variable Spin Iron(III) Chelates with Hexadentate Ligands Derived from Triethylenetetramine and Various Salicylaldehydes. Synthesis, Characterization, and Solution State Studies of a New ^{2}T $\leftrightarrow$ ^{6}A Spin Equilibrium System. *J. Am. Chem. Soc.* **1976**, *98*, 4824–4834. [CrossRef]

35. Takahashi, K.; Hasegawa, Y.; Sakamoto, R.; Nishikawa, M.; Kume, S.; Nishibori, E.; Nishihara, H. Solid-State Ligand-Driven Light-Induced Spin Change at Ambient Temperatures in Bis(dipyrazolylstyrylpyridine)iron(II) Complexes. *Inorg. Chem.* **2012**, *51*, 5188–5198. [CrossRef] [PubMed]

36. Phan, H.; Hrudka, J.J.; Igimbayeva, D.; Daku, L.M.L.; Shatruk, M. A Simple Approach for Predicting the Spin State of Homoleptic Fe(II) Tris-diimine Complexes. *J. Am. Chem. Soc.* **2017**, *139*, 6437–6447. [CrossRef] [PubMed]

37. Guionneau, P.; Marchivie, M.; Bravic, G.; Létard, J.-F.; Chasseau, D. Structural Aspects of Spin Crossover. Example of the [Fe^{II}L$_n$(NCS)$_2$] Complexes. *Top. Curr. Chem.* **2004**, *234*, 97–128.

38. Marchivie, M.; Guionneau, P.; Letard, J.F. Photo-induced spin-transition: the role of the iron(II) environment distortion. *Acta Crystallogr. Sect. B Struct. Sci.* **1991**, *15*, 181–190. [CrossRef] [PubMed]

39. Halcrow, M.A. Structure:function relationships in molecular spin-crossover complexes. *Chem. Soc. Rev.* **2011**, *40*, 4119–4142. [CrossRef] [PubMed]

40. Lluncll, M.; Casanova, D.; Circra, J.; Bofill, J.M.; Alcmany, P.; Alvarez, S.; Pinsky, M.; Avnir, D. *SHAPE*; v2.1; University of Barcelona and The Hebrew University of Jerusalem: Barcelona, Spain, 2005.

41. *Gaussian 03*; revision C.02; Gaussian Inc.: Wallingford, CT, USA, 2004.

42. Kroll, N.; Theilacker, K.; Schoknecht, M.; Baabe, D.; Wiedemann, D.; Kaupp, M.; Grohmann, A.; Hörner, G. Controlled ligand distortion and its consequences for structure, symmetry, conformation and spin-state preferences of iron(II) complexes. *Dalton Trans.* **2015**, *44*, 19232–19247. [CrossRef] [PubMed]

43. Matouzenko, G.S.; Jeanneau, E.; Verat, A.Y.; de Gaetano, Y. The Nature of Spin Crossover and Coordination Core Distortion in a Family of Binuclear Iron(II) Complexes with Bipyridyl-Like Bridging Ligands. *Eur. J. Inorg. Chem.* **2012**, 969–977. [CrossRef]

44. Cook, L.J.K.; Thorp-Greenwood, F.L.; Comyn, T.P.; Cespedes, O.; Chastanet, G.; Halcrow, M.A. Unexpected Spin-Crossover and a Low-Pressure Phase Change in an Iron(II) Dipyrazolylpyridine Complex Exhibiting a High-Spin Jahn–Teller Distortion. *Inorg. Chem.* **2015**, *54*, 6319–6330. [CrossRef] [PubMed]

45. Yang, Q.; Cheng, X.; Gao, C.; Wang, B.; Wang, Z.; Gao, S. Structural Distortion Controlled Spin-Crossover Behavior. *Cryst. Growth Des.* **2015**, *15*, 2565–2567. [CrossRef]

46. *CRYSTALSTRUCTURE*, version 4.2.1; Rigaku/MSC: The Woodlands, TX, USA, 2015.

inorganics

MDPI

Communication

Halogen Substituent Effect on the Spin-Transition Temperature in Spin-Crossover Fe(III) Compounds Bearing Salicylaldehyde 2-Pyridyl Hydrazone-Type Ligands and Dicarboxylic Acids

Takumi Nakanishi [1], Atsushi Okazawa [2] and Osamu Sato [1,*]

[1] Institute for Materials Chemistry and Engineering, Kyushu University, 744 Motoka, Nishi-ku, Fukuoka 819-0395, Japan; nakanishi.takumi.316@m.kyushu-u.ac.jp

[2] Department of Basic Science, Graduation School of Arts and Sciences, The University of Tokyo, 3-8-1 Komaba, Meguro-ku, Tokyo 153-8902, Japan; cokazawa@mail.ecc.u-tokyo.ac.jp

* Correspondence: sato@cm.kyushu-u.ac.jp; Tel.: +81-92-802-6208; Fax: +81-92-802-6205

Received: 29 June 2017; Accepted: 11 August 2017; Published: 12 August 2017

Abstract: Four Fe(III) spin-crossover (SCO) compounds, $[Fe(HL1)_2](HCl_4TPA)$ (**1-Cl**), $[Fe(HL1)_2](HBr_4TPA)$ (**1-Br**), $[Fe(HL2)_2](HCl_4TPA)$ (**2-Cl**), and $[Fe(HL2)_2](HBr_4TPA)$ (**2-Br**) (HL1 = 4-chloro-2-nitro-6-(1-(2-(pyridine-2-yl)hydrazono)ethyl)phenolate; HL2 = 4-bromo-2-nitro-6-(1-(2-(pyridine-2-yl)hydrazono)ethyl)phenolate; HCl_4TPA = 2,3,5,6-tetrachloro-4-carboxybenzoate; and HBr_4TPA = 2,3,5,6-tetrabromo-4-carboxybenzoate), were synthesized to investigate the halogen substituent change effect in salicylaldehyde 2-pyridyl hydrazone-type ligands and dicarboxylic acids in SCO complexes to the spin-transition temperature. Crystal structure analyses showed that these compounds were isostructural. In addition, a one-dimensional hydrogen–bonded column was formed by the dicarboxylic acid anion and weak hydrogen bonds between the Fe(III) complexes. From Mössbauer spectroscopy and magnetic property measurements, these compounds were confirmed to exhibit gradual SCO. The spin-transition temperature can be shifted by changing the halogen substituent in the salicylaldehyde 2-pyridyl hydrazone-type ligands and dicarboxylic acids without changing the molecular arrangement in the crystal packing.

Keywords: spin crossover; hydrazone complex; dicarboxylic acid

1. Introduction

Spin-crossover (SCO) complexes have attracted considerable attention as materials for developing memory devices because the spin state can be switched by external stimuli such as temperature, light, and pressure [1–3]. In terms of practical applications, it is preferable for the spin-transition temperature to be near room temperature. Therefore, the effects of replacing anion and solvent molecules and the introduction of substituents into ligands on the spin-transition temperature have been actively studied for many types of SCO complexes [4–16].

As an approach toward shifting the spin-transition temperature, we focused on the effect of halogen substituent changes in the ligands and coexisting molecules, particularly in a carboxylic acid. In our previous study, we developed an Fe(II) SCO compound containing a neutral Fe(II) complex and a halogen-substituted dicarboxylic acid [17]. The halogen substituent change on the dicarboxylic acid molecule was confirmed to affect the spin-transition temperature while maintaining the molecular arrangement of the complex. Fe(III) SCO complexes have a great advantage over Fe(II) SCO complexes in terms of in-air stability; thus, Fe(III) SCO complexes have attracted considerable attention and have been actively developed [18–24]. In this study, we aim to introduce a halogen-substituted dicarboxylic acid into Fe(III) SCO compounds as an anion and confirm the effect of the halogen substituent change in the dicarboxylic acid to the spin-transition temperature. Simultaneously, the effect of the halogen substituent change in the ligand to the spin-transition temperature was also investigated. Salicylaldehyde 2-pyridyl hydrazone-type ligands can form 1+ cationic Fe(III) complexes [25–28]. Those that bear 2-methoxy-6-(pyridine-2-ylhydrazonomethyl)phenol exhibit SCO properties [24]. However, there are no reports on the effect of changing the halogen substituent on the salicylaldehyde 2-pyridyl hydrazone-type ligand to the spin-transition temperature because Fe(III) SCO complexes bearing such ligands are scarcely reported. In this study, we synthesized two ligands, 4-chloro-2-nitro-6-(1-(2-(pyridine-2-ylhydrazonoethyl)phenol)) (H_2L1) and 4-bromo-2-nitro-6-(1-(2-(pyridine-2-ylhydrazonoethyl)phenol)) (H_2L2), and tetrachloroterephthalic acid (H_2Cl_4TPA) and tetrabromoterephthalic acid (H_2Br_4TPA) were used as the halogen-substituted dicarboxylic acids (Scheme 1). In total, four SCO compounds, [Fe(HL1)$_2$](HCl$_4$TPA) (**1-Cl**), [Fe(HL1)$_2$](HBr$_4$TPA) (**1-Br**), [Fe(HL2)$_2$](HCl$_4$TPA) (**2-Cl**), and [Fe(HL2)$_2$](H$_2$Br$_4$TPA) (**2-Br**), were obtained. The crystal structures, Mössbauer spectra, and magnetic properties of these compounds were investigated.

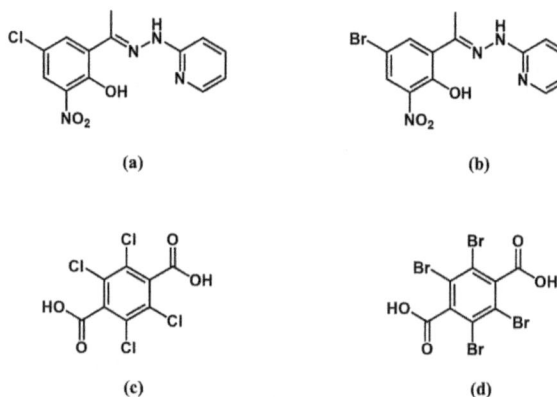

Scheme 1. Structural formula of (**a**) H_2L1; (**b**) H_2L2; (**c**) H_2Cl_4TPA; and (**d**) H_2Br_4TPA.

2. Results and Discussion

2.1. Synthesis and Characterization

The following starting materials were commercially available: 5′-chloro-2′-hydroxy-3′-nitroacetophenone, 5′-bromo-2′-hydroxy-3′-nitroacetophenone, 2-hydrazinopyridine, FeCl$_3$·6H$_2$O, H_2Cl_4TPA, H_2Br_4TPA, and trimethylamine. HL1 and HL2 were obtained from the typical procedure to synthesize salicylaldehyde 2-pyridyl hydrazone-type ligands [29]. Equimolar 5′-chloro-2′-hydroxy-3′-nitroacetophenone or 5′-bromo-2′-hydroxy-3′-nitroacetophenone and hydrazinopyridine were added to ethanol, and the solution was refluxed for 5 h. Subsequently, the solution was left to stand for several days. Yellow needle-shaped crystals were obtained. **1-Cl** and **1-Br** were obtained by the reaction of H_2L1 with FeCl$_3$·6H$_2$O and H_2Cl_4TPA or H_2Br_4TPA with equal parts of trimethylamine in methanol, respectively. The synthesis procedures for **2-Cl** and **2-Br** were identical to those for **1-Cl** and **1-Br**, respectively, except that H_2L2 was used instead of H_2L1. Plate-shaped black-brown crystals were obtained. Each Fe(III) compound was characterized by elemental analysis and single-crystal X-ray measurements.

2.2. Crystal Structure

The crystal structures of **1-Cl** were determined at 123 and 330 K, and those of **1-Br**, **2-Cl**, and **2-Br** were determined at 123 K. The crystallographic data of **1-Cl**, **1-Br**, **2-Cl**, and **2-Br** are summarized in

Table 1. The coordination distances and intermolecular interactions of **1-Cl**, **1-Br**, **2-Cl**, and **2-Br** are summarized in Table 2. The asymmetric unit of **1-Cl** at 123 K shows half of the Fe(III) complex and half of the dicarboxylic acid molecule without any solvent molecules (Figure 1a). The half Fe(III) complex and the half dicarboxylic acid molecule in the asymmetric unit indicate that the two ligands in the Fe(III) complex are crystallographically identical and the two carboxyl groups in the dicarboxylic acid molecule are also crystallographically identical. The molecular arrangement of the Fe(III) complexes and the dicarboxylic acid molecules is illustrated in Figure 1b. The dicarboxylic acid molecules were confirmed to form a one-dimensional (1D) column through hydrogen bonds between each carboxyl group. Because a hydrogen bond is formed between each dicarboxylic acid molecule, it was predicted that (i) one of the two carboxyl groups in each H_2Cl_4TPA molecule was deprotonated and (ii) **1-Cl** is comprised of one HCl_4TPA$^-$ anion and one [Fe^{III}(HL)$_2$]$^+$ cation. The O$^-\cdots$H$^+\cdots$O$^-$-type hydrogen bond found between each HCl_4TPA molecule is generally referred to as a "charge-assisted hydrogen bond", and such types of hydrogen bond tend to be strong and very short [30]. Indeed, the hydrogen bond distance of O5$\cdots$O5A was 2.458 Å. This value can be recognized as a considerably strong and short hydrogen bond. The two ligands in the Fe(III) complexes were linked to a 1D hydrogen–bonded column through the hydrogen bond between N2 in the ligand and O4 in the carboxyl group. The distance of N2$\cdots$O4 was 2.892 Å. This value is classified as a weak hydrogen bond. Overall, each hydrogen–bonded column occupied the area surrounded by Fe(III) complexes and extended along the *c*-axis (Figure 2). The dependence of coordination distances around the Fe(III) atom and temperature in **1-Cl** are also listed in Table 1. At 123 K, the distances between Fe1 and N1, N3, and O1 were 1.975(3), 1.925(3), and 1.857(2) Å, respectively. These values indicated that the Fe(III) complex in **1-Cl** at 123 K was in the low-spin (LS) state. Meanwhile, the same coordination distances at 330 K were 2.098(3), 2.112(3), and 1.882(3) Å, respectively. The increment of coordination distances around Fe1 from 123 to 330 K indicates that the Fe(III) complexes in **1-Cl** exhibited SCO from the LS state (S = 1/2) to the high-spin (HS) state (S = 5/2) with increasing temperature. We also determined the crystal structures of **1-Br**, **2-Cl**, and **2-Br** at 123 K. It was confirmed that the crystal structures of **1-Br**, **2-Cl**, and **2-Br** were isostructural to **1-Cl**. The coordination distances around Fe(III) in **1-Br**, **2-Cl**, and **2-Br** at 123 K were characteristic of the LS Fe(III) complex. The hydrogen bond distances between the dicarboxylic acid and Fe(III) complex and between each dicarboxylic acid in **1-Cl**, **1-Br**, **2-Cl**, and **2-Br** were barely different from each other.

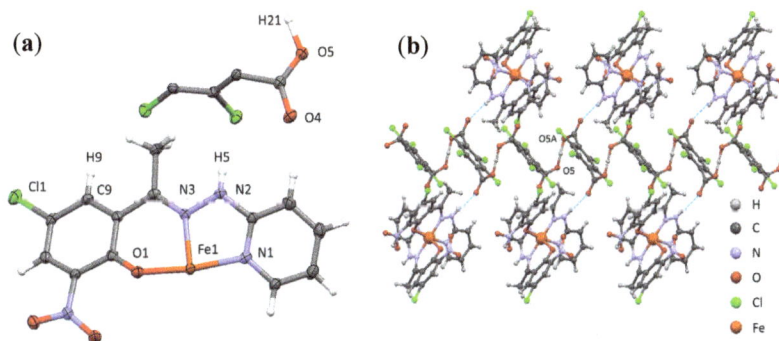

Figure 1. (a) Molecular structure of the asymmetric unit; (b) Molecular arrangement of the [Fe(HL1)$_2$] cation and the HCl_4TPA anion. Blue lines represent the hydrogen bonds (N2–H5$\cdots$O4).

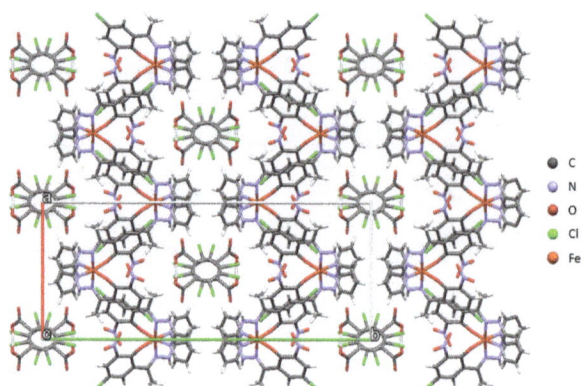

Figure 2. View of the crystal structure of **1-Cl** along the c-axis.

Table 1. Crystallographic data of **1-Cl**, **1-Br**, **2-Cl**, and **2-Br**.

Compound	1-Cl		1-Br	2-Cl	2-Br
			Crystallographic Data		
Formula	$C_{34}H_{21}Cl_6FeN_8O_{10}$		$C_{34}H_{21}Br_4Cl_2FeN_8O_{10}$	$C_{34}H_{21}Br_2Cl_4FeN_8O_{10}$	$C_{34}H_{21}Br_6FeN_8O_{10}$
Formula weight	970.15		1147.96	1059.05	1236.86
Crystal system	Monoclinic		Monoclinic	Monoclinic	Monoclinic
Space group	C2/c		C2/c	C2/c	C2/c
T (K)	123	330	123	123	123
a (Å)	12.479(5)	12.665(3)	12.595(3)	12.627(4)	12.683(7)
b (Å)	30.593(11)	31.287(6)	30.851(7)	30.551(9)	30.984(14)
c (Å)	9.469(3)	9.569(2)	9.607(2)	9.519(3)	9.633(5)
β (°)	96.795(8)	97.326(6)	95.945(5)	96.239(7)	95.330(11)
V (Å³)	3590(2)	3760.4(13)	3713.2(16)	3650.6(19)	3769.(3)
Z	4	4	4	4	4
D_{cal} (g/cm³)	1.795	1.713	2.053	1.927	2.179
F (000)	1956.00	1956.00	2244.00	2100.00	2388.00
Data collected	15,405	16,406	16,626	16,224	17,268
Unique data	4069	4272	4242	4197	4309
R(int)	0.0699	0.0783	0.0699	0.0989	0.1157
GOF on F^2	1.115	1.104	1.130	1.106	1.116
$R_I [I > 2\sigma(I)]$	0.0581	0.0659	0.0518	0.0636	0.0579

Table 2. Coordination and hydrogen bond distances in **1-Cl**, **1-Br**, **2-Cl**, and **2-Br**.

Compound	1-Cl		1-Br	2-Cl	2-Br
			Bond lengths Around Fe(III) (Å)		
Temperature	123 K	330 K	123 K	123 K	123 K
Fe1–N1	1.975(3)	2.098(3)	1.976(4)	1.966(4)	1.980(4)
Fe1–N3	1.925(3)	2.112(3)	1.926(4)	1.923(4)	1.929(4)
Fe1–O1	1.857(2)	1.882(3)	1.857(3)	1.853(3)	1.862(4)
			Hydrogen Bond Distances (Å)		
O5···O5A	2.458(3)	2.464(5)	2.463(4)	2.464(5)	2.458(6)
O5–H21	1.233(5)	1.232(7)	1.236(13)	1.235(9)	1.239(12)
N2···O4	2.892(4)	2.890(5)	2.892(5)	2.886(5)	2.872(6)

2.3. Mössbauer Spectroscopy

The variable-temperature Mössbauer spectra of **1-Cl** were measured to identify the spin states and valence of Fe at 123 and 300 K (Figure 3). The spectrum at 123 K shows a single quadrupole doublet. The isomer shift (δ) and quadrupole splitting (ΔE_Q) were 0.179(1) and 2.178(2) mm·s^{-1}, respectively.

The δ and ΔE_Q at 123 K were assigned to the Fe(III) LS isomer. These results were consistent with the prediction from the crystal structure at 123 K. At 300 K, there was a singlet peak without quadrupole splitting. The isomer shift was 0.281(9) mm·s^{-1}, which was assigned to the Fe(III) HS isomer, and this result was also consistent with the spin-state predictions from the crystal structure at 330 K. Overall, the valence of iron in **1-Cl** was confirmed to be Fe(III) (**1-Br**, **2-Cl**, and **2-Br** were isostructural to **1-Cl** and should therefore also be Fe(III)), and the change of coordination distances around Fe with temperature change in the crystal structures originates from the SCO.

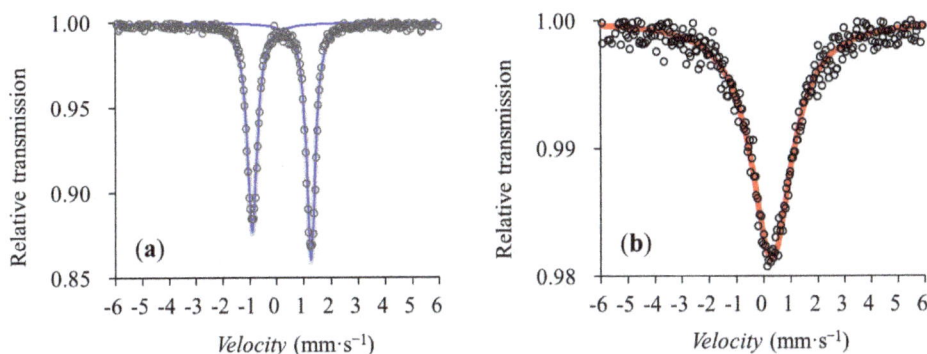

Figure 3. Mössbauer spectra of **1-Cl** at (**a**) 123 K and (**b**) 300 K. Open circles and lines are measurement data and fitting curves, respectively.

2.4. Magnetic Property

The magnetic susceptibility measurements of **1-Cl** and **1-Br** were performed over 100–400 K, whereas those of **2-Cl** and **2-Br** were performed over 150–400 K (Figure 4). At 123 K, the $\chi_m T$ value of **1-Cl** was 0.32 cm^3·K·mol^{-1}, indicating that the Fe(III) complex in **1-Cl** was in the LS state at 123 K. This result is in agreement with the spin-state predictions from the crystal structure and Mössbauer spectra at the same temperature. Additionally, the $\chi_m T$ values of **1-Br**, **2-Cl**, and **2-Br** at 150 K were 0.40, 0.32, and 0.36 cm^3·K·mol^{-1}, respectively; these values also show that **1-Br**, **2-Cl**, and **2-Br** were in the LS state at 150 K. The $\chi_m T$ value of **1-Cl** gradually increased during the heating process as a result of the SCO and ultimately reached 3.72 cm^3·K·mol^{-1} at 400 K. Meanwhile, **1-Br** showed a gradual two-step SCO between 150 and 240 K and finally reached 4.07 cm^3·K·mol^{-1} at 400 K. The spin-transition temperature $T_{1/2}$ in the first and second steps were 188 and 213 K, respectively. However, there was no clear plateau in the two-step SCO in **1-Br**, so the crystal structure of the intermediate phase could not be determined.

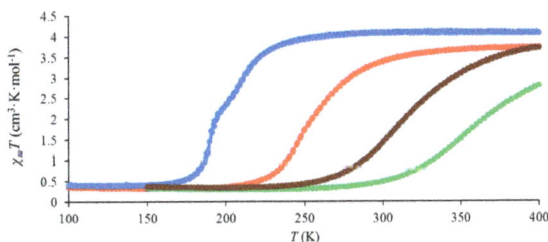

Figure 4. $\chi_m T$–T plots of **1-Br** (blue dots), **1-Cl** (red dots), **2-Br** (brown dots), and **2-Cl** (green dots).

2-Cl and **2-Br** showed gradual SCO near room temperature and finally reached 2.80 and 3.72 cm^3·K·mol^{-1}, respectively. The $\chi_m T$ value for **2-Cl** at 400 K was clearly lower than the expected value for the HS Fe(III) state (S = 5/2). Additionally, the χT–T plot of **2-Br** still did not reach the plateau region, even at 400 K. For this reason, the SCO of **2-Cl** and **2-Br** were considered incomplete at 400 K. In addition, **2-Cl** and **2-Br** showed a $T_{1/2}$ dependence on the halogen substituent in the dicarboxylic acid, similar to **1-Cl** and **1-Br**. The $\chi_m T$ values of **2-Cl** and **2-Br** in the HT phase were assumed to reach the $\chi_m T$ values of **1-Cl** (3.72 cm^3·K·mol^{-1}) and **1-Br** (4.07 cm^3·K·mol^{-1}) in the HS state, respectively. It was estimated that the $T_{1/2}$ of **2-Cl** and **2-Br** were 373 and 320 K, respectively. The difference of $T_{1/2}$ between **1-Cl** and **1-Br**, or **2-Cl** and **2-Br** clearly showed that the $T_{1/2}$ in SCO complexes coexisting with halogen-substituted dicarboxylic acids can be controlled by changing the halogen substituent in the dicarboxylic acid while keeping the molecular crystal packing. The effect of the halogen substituent change in salicylaldehyde 2-pyridyl hydrazone-type ligands to $T_{1/2}$ was also observed. The $T_{1/2}$ values of **2-Cl** and **2-Br** were clearly shifted to higher temperatures compared with those of **1-Cl** and **1-Br** as the halogen substituent changed from Cl to Br. The shift of $T_{1/2}$ to higher temperatures with the change of halogen substituent in the ligand from Cl to Br was considered to have originated from the electron withdrawing effect of halogen substituents. It is reported that the $T_{1/2}$ of [Fe(X-sal$_2$trien^R)]$^+$ that possesses the phonoxide part, which is similar to our ligands, tends to increase with the change of the substituent on the phenoxide part from the OCH$_3$ substituent possessing electron donor character to the NO$_2$ possessing electron withdrawing character [31]. It is known that the σ_p Hammett parameter of Br is larger than that of Cl. Thus, the larger electron withdrawing effect of Br compared with that of Cl is considered to induce the shift of $T_{1/2}$ to a higher temperature, which is the same as the case with [Fe(X-sal$_2$trien^R)]$^+$.

3. Materials and Methods

3.1. Compound Synthesis

3.1.1. Synthesis of Ligand HL1

The 5′-Chloro-2′-hydroxy-3′-nitroacetophenone (2.2 g; 10 mmol) and 2-hydrazinopyridine (1.1 g; 10 mmol) were dissolved in ethanol (100 mL). The mixture was refluxed for 5 h and left to stand at room temperature. Consequently, yellow needle-shaped crystals were obtained. Anal. C$_{13}$H$_{11}$ClN$_4$O$_3$ (306.05); calcd. C 50.91, H 3.62, N 18.27; found C 50.98, H 3.57, N 18.28.

3.1.2. Synthesis of Ligand HL2

The synthesis procedure for HL2 was identical to that for HL1, except that 5′-bromo-2′-hydroxy-3′-nitroacetophenone was used instead of 5′-chloro-2′-hydroxy-3′-nitroacetophenone. Anal. C$_{13}$H$_{11}$BrN$_4$O$_3$ (350.00); calcd. C 44.46, H 3.16, N 15.96; found C 44.53, H 3.09, N 15.99.

3.1.3. Synthesis of **1-Cl**

H$_2$L1 (72 mg; 0.20 mmol) was dissolved in methanol (100 mL), and FeCl$_3$·6H$_2$O (27 mg; 0.10 mmol) was added to the solution. The mixture was stirred for 30 min. Subsequently, H$_2$Cl$_4$TPA (30 mg; 0.10 mmol) was added, and the mixture was stirred for 30 s. Then, trimethylamine (14 μL; 0.10 mmol) was added to the solution. The solution was left to stand for several days. Consequently, plate-shaped crystals were obtained. Anal. C$_{34}$H$_{21}$Cl$_6$FeN$_8$O$_{10}$ (970.13); calcd. C 42.09, H 2.18, N 11.55; found C 42.18, H 2.22, N 11.52.

3.1.4. Synthesis of **1-Br**

The synthesis procedure for **1-Br** was identical to that for **1-Cl**, except that H$_2$Br$_4$TPA was used instead of H$_2$Cl$_4$TPA. Plate-shaped crystals were obtained. Anal. C$_{34}$H$_{21}$Cl$_2$Br$_4$FeN$_8$O$_{10}$ (1147.95); calcd. C 35.57, H 1.84, N 9.76; found C 35.69, H 1.88, N 9.79.

3.1.5. Synthesis of **2-Cl**

The synthesis procedure for **2-Cl** was identical to that for **1-Cl**, except that H_2L2 was used instead of H_2L1. Plate-shaped crystals were obtained. Anal. $C_{34}H_{21}Cl_4Br_2FeN_8O_{10}$ (1059.04); calcd. C 38.56, H 2.00, N 10.58; found C 38.53, H 2.04, N 10.51.

3.1.6. Synthesis of **2-Br**

The synthesis procedure for **2-Br** was identical to that for **1-Br**, except that HL2 was used instead of HL1. Plate-shaped crystals were obtained. Anal. $C_{34}H_{21}Br_6FeN_8O_{10}$ (1236.86); calcd. C 33.02, H 1.71, N 9.06; found C 33.02, H 1.73, N 9.01.

3.2. X-ray Structure Determination

Diffraction data from **1-Cl** at 123 and 330 K and **1-Br**, **2-Cl**, and **2-Br** at 123 K were collected on a Rigaku charge-coupled device diffractometer (Rigaku, Tokyo, Japan). A crystal was glued onto a nylon loop and enveloped in a temperature-controlled stream of dry N_2 gas during data collection. The structures were solved by direct methods and refined with full-matrix least-squares procedures using the *SHELXS-97* program. Hydrogen atoms involved in the hydrogen bonds were determined from the Fourier difference map, and other hydrogen atoms were generated by calculations and refined using a riding model. All non-hydrogen atoms were refined anisotropically. CCDC 1543784 (**1-Cl 123 K**), 1543783 (**1-Cl 330 K**), 1543785 (**1-Br 123 K**), 1543786 (**2-Cl 123 K**), and 1543787 (**2-Br 320 K**) contain the supplementary crystallographic data for this paper. These data can be obtained free of charge via http://www.ccdc.cam.ac.uk/conts/retrieving.html or from the CCDC (12 Union Road, Cambridge CB2 1EZ, UK; Fax: +44-1223-336033; E-mail: deposit@ccdc.cam.ac.uk).

3.3. ^{57}Fe Mössbauer Spectroscopy

^{57}Fe Mössbauer spectra were recorded using a CryoMini/CryoStat cryogenic refrigerator set (Iwatani Industrial Gases, Osaka, Japan) and a conventional Mössbauer spectrometer (Topologic Systems, Kanagawa, Japan) with a $^{57}Co/Rh$ source. The spectra were calibrated by using the six lines of a body-centered cubic iron foil (α-Fe), the center of which was taken as the zero isomer shift. The Mössbauer spectra were fitted with the least-squares fitting program MossWinn 4.0 [32].

3.4. Magnetic Property Measurements

Magnetic measurements were performed on a Quantum Design MPMS-5S superconducting quantum-interference device magnetometer (Quantum Design, Inc., San Diego, CA, USA), and data were corrected for the diamagnetic contribution, as calculated from Pascal's constants. Temperature sweeping mode was used to collect the data at a sweep rate of 1 K·min^{-1} with a magnetic field of 5 kOe.

4. Conclusions

In brief, we synthesized four Fe(III) SCO compounds that contained a halogen-substituted salicylaldehyde 2-pyridyl hydrazone-type ligand complex with a dicarboxylic acid as the anion. Crystal structures confirmed that all the compounds were isostructural. Furthermore, the dicarboxylic acid anion formed a 1D hydrogen-bonded column, and the Fe(III) complex formed weak hydrogen bonds with the 1D hydrogen-bonded column. Mössbauer spectroscopy for **1-Cl** demonstrated that the valence of the iron center was Fe(III), and these compounds showed SCO. The magnetic property measurements of these compounds showed that they exhibited gradual SCO, and the spin-transition temperature shifted to higher values without changing the molecular crystal packing when the halogen substituent in the dicarboxylic acid changed from Br to Cl. Additionally, it was confirmed that the change of halogen substituent in the salicylaldehyde 2-pyridyl hydrazone-type ligands from Cl to Br shifts $T_{1/2}$ to a higher temperature while still maintaining the molecular crystal packing. The shift of

$T_{1/2}$ when a halogen substituent is changed within the in the ligand is considered to have originated from the difference in the electron-withdrawing effect of Cl and Br. In conclusion, the spin-transition temperature of Fe(III) complexes bearing salicylaldehyde 2-pyridyl hydrazone-type ligands can be adjusted by introducing a halogen-substituted dicarboxylic acid and changing the halogen substituent in the dicarboxylic acid or the ligand.

Supplementary Materials: The following are available online at www.mdpi.com/2304-6740/5/3/53/s1. Cif and cif-checked files.

Author Contributions: Takumi Nakanishi and Osamu Sato designed the compounds and measured the magnetic properties and crystal structures. Atsushi Okazawa operated the [57]Fe Mössbauer spectrometer and analyzed the Mössbauer spectra.

Conflicts of Interest: The authors declare no conflicts of interest.

References

1. Gülich, P.; Hauser, A. Thermal and light-induced spin crossover in iron(II) complexes. *Coord. Chem. Rev.* **1990**, *97*, 1–22. [CrossRef]
2. Sato, O. Optically switchable molecular solids: Photoinduced spin-crossover, photochromism, and photoinduced magnetization. *Acc. Chem. Res.* **2003**, *36*, 692–700. [CrossRef] [PubMed]
3. Real, J.A.; Gaspar, A.B.; Munoz, M.C. Thermal, pressure and light switchable spin-crossover materials. *Dalton Trans.* **2005**, 2062–2079. [CrossRef] [PubMed]
4. Kershaw Cook, L.J.; Kulmaczewski, R.; Mohammed, R.; Dudley, S.; Barrett, S.A.; Little, M.A.; Deeth, R.J.; Halcrow, M.A. A unified treatment of the relationship between ligand substituents and spin state in a family of iron(II) complexes. *Angew. Chem. Int. Ed.* **2016**, *55*, 4327–4331. [CrossRef] [PubMed]
5. Halcrow, M.A. The effect of ligand design on metal ion spin state-lessons from spin crossover complexes. *Crystals* **2016**, *6*, 20. [CrossRef]
6. Sato, O. Dynamic molecular crystals with switchable physical properties. *Nat. Chem.* **2016**, *8*, 644–656. [CrossRef] [PubMed]
7. Lemercier, G.; Brefuel, N.; Shova, S.; Wolny, J.A.; Dahan, F.; Verelst, M.; Paulsen, H.; Trautwein, A.X.; Tuchagues, J.P. A range of spin-crossover temperature $T_{1/2}$ > 300 k results from out-of-sphere anion exchange in a series of ferrous materials based on the 4-(4-imidazolylmethyl)-2-(2-imidazolylmethyl)imidazole (trim) ligand, [Fe(trim)$_2$]X$_2$ (X=F, Cl, Br, I): Comparison of experimental results with those derived from density functional theory calculations. *Chem. Eur. J.* **2006**, *12*, 7421–7432. [PubMed]
8. Paulsen, H.; Duelund, L.; Zimmermann, A.; Averseng, F.; Gerdan, M.; Winkler, H.; Toftlund, H.; Trautwein, A.X. Substituent effects on the spin-transition temperature in complexes with tris(pyrazolyl) ligands. *Mon. Chem.* **2003**, *134*, 295–306. [CrossRef]
9. Wei, R.-J.; Tao, J.; Huang, R.-B.; Zheng, L.-S. Anion-dependent spin crossover and coordination assembly based on [Fe(tpa)]$^{2+}$[tpa = tris(2-pyridylmethyl)amine] and [N(CN)$_2$]$^-$: Square, zigzag, dimeric, and [4 + 1]-cocrystallized complexes. *Eur. J. Inorg. Chem.* **2013**, *2013*, 916–926. [CrossRef]
10. Zhang, W.; Zhao, F.; Liu, T.; Yuan, M.; Wang, Z.M.; Gao, S. Spin crossover in a series of iron(II) complexes of 2-(2-Alkyl-2H-tetrazol-5-yl)-1,10-phenanthroline: Effects of alkyl side chain, solvent, and anion. *Inorg. Chem.* **2007**, *46*, 2541–2555. [CrossRef] [PubMed]
11. Zhu, Y.Y.; Li, H.Q.; Ding, Z.Y.; Lu, X.J.; Zhao, L.; Meng, Y.S.; Liu, T.; Gao, S. Spin transitions in a series of [Fe(pybox)$_2$]$^{2+}$ complexes modulated by ligand structures, counter anions, and solvents. *Inorg. Chem. Front.* **2016**, *3*, 1624–1636. [CrossRef]
12. Ivanova, T.A.; Ovchinnikov, I.V.; Turanov, A.N. Influence of the outersphere anion on the properties of the spin transition in Fe(4-OCH3-SalEen)$_2$ Y (Y = PF$_6$, NO$_3$). *Phys. Solid State* **2007**, *49*, 2132–2137. [CrossRef]
13. Nemec, I.; Herchel, R.; Travnicek, Z. The relationship between the strength of hydrogen bonding and spin crossover behaviour in a series of iron(III) Schiff base complexes. *Dalton Trans.* **2015**, *44*, 4474–4484. [CrossRef] [PubMed]
14. Phonsri, W.; Harding, D.J.; Harding, P.; Murray, K.S.; Moubaraki, B.; Gass, I.A.; Cashion, J.D.; Jameson, G.N.; Adams, H. Stepped spin crossover in Fe(III) halogen substituted quinolylsalicylaldimine complexes. *Dalton Trans.* **2014**, *43*, 17509–17518. [CrossRef] [PubMed]

15. Phonsri, W.; Macedo, D.S.; Vignesh, K.R.; Rajaraman, G.; Davies, C.G.; Jameson, G.N.L.; Moubaraki, B.; Ward, J.S.; Kruger, P.E.; Chastanet, G.; et al. Halogen substitution effects on N2O Schiff base ligands in unprecedented abrupt Fe^{II} spin crossover complexes. *Chem. Eur. J.* **2017**, *23*, 7052–7065. [CrossRef] [PubMed]
16. Ueno, T.; Miyano, K.; Hamada, D.; Ono, H.; Fujinami, T.; Matsumoto, N.; Sunatsuki, Y. Abrupt spin transition and chiral hydrogen-bonded one-dimensional structure of iron(III) complex [Fe^{III}(Him)$_2$(hapen)]SbF$_6$ (Him = imidazole, H$_2$hapen = *N*,*N'*-bis(2-hydroxyacetophenylidene)ethylenediamine). *Magnetochemistry* **2015**, *1*, 72. [CrossRef]
17. Nakanishi, T.; Sato, O. Synthesis, structure, and magnetic properties of new spin crossover Fe(II) complexes forming short hydrogen bonds with substituted dicarboxylic acids. *Crystals* **2016**, *6*, 8. [CrossRef]
18. Koningsbruggen, P.J.; Maeda, Y.; Oshio, H. Iron(III) spin crossover compounds. *Spin Crossover Transit. Metal Compd. I* **2004**, *233*, 259–324.
19. Nihei, M.; Shiga, T.; Maeda, Y.; Oshio, H. Spin crossover iron(III) complexes. *Coord. Chem. Rev.* **2007**, *251*, 2606–2621. [CrossRef]
20. Hayami, S.; Gu, Z.; Yoshiki, H.; Fujishima, A.; Sato, O. Iron(III) spin-crossover compounds with a wide apparent thermal hysteresis around room temperature. *J. Am. Chem. Soc.* **2001**, *123*, 11644–11650. [CrossRef] [PubMed]
21. Li, Z.Y.; Dai, J.W.; Shiota, Y.; Yoshizawa, K.; Kanegawa, S.; Sato, O. Multi-step spin crossover accompanied by symmetry breaking in an Fe(III) complex: Crystallographic evidence and DFT studies. *Chem. Eur. J.* **2013**, *19*, 12948–12952. [CrossRef] [PubMed]
22. Phonsri, W.; Davies, C.G.; Jameson, G.N.; Moubaraki, B.; Murray, K.S. Spin crossover, polymorphism and porosity to liquid solvent in heteroleptic iron(III) {quinolylsalicylaldimine/thiosemicarbazone-salicylaldimine} complexes. *Chem. Eur. J.* **2016**, *22*, 1322–1333. [CrossRef] [PubMed]
23. Fujinami, T.; Ikeda, M.; Koike, M.; Matsumoto, N.; Oishi, T.; Sunatsuki, Y. Syntheses, hydrogen-bonded assembly structures, and spin crossover properties of [Fe^{III}(Him)$_2$(n-MeOhapen)]PF$_6$ (Him = imidazole and n-MeOhapen = *N*,*N'*-bis(n-methoxy-2-oxyacetophenylidene)ethylenediamine); n = 4, 5, 6). *Inorg. Chim. Acta* **2015**, *432*, 89–95. [CrossRef]
24. Tang, J.; Sanchez Costa, J.; Smulders, S.; Molnar, G.; Bousseksou, A.; Teat, S.J.; Li, Y.; van Albada, G.A.; Gamez, P.; Reedijk, J. Two-step spin-transition iron(III) compound with a wide [high spin-low spin] plateau. *Inorg. Chem.* **2009**, *48*, 2128–2135. [CrossRef] [PubMed]
25. Anderson, R.G.; Nickless, G. Co-ordinating properties of some ligand systems related to 4-(2-pyridylazo)resorcinol. *Talanta* **1967**, *14*, 1221–1228. [CrossRef]
26. Basu, C.; Chowdhury, S.; Banerjee, R.; Evans, H.S.; Mukherjee, S. A novel blue luminescent high-spin iron(III) complex with interlayer O–HCl bridging: Synthesis, structure and spectroscopic studies. *Polyhedron* **2007**, *26*, 3617–3624. [CrossRef]
27. Liu, H.Y.; Gao, F.; Niu, D.Z. Synthesis and structure of iron(III) complex with N,N,O-donor aroylhydrazones: The chloride anion as hydrogen bond acceptor forming infinite chains. *Asian J. Chem.* **2011**, *23*, 2014–2016.
28. Zhang, J.; Campolo, D.; Dumur, F.; Xiao, P.; Fouassier, J.P.; Gigmes, D.; Lalevee, J. Iron complexes as photoinitiators for radical and cationic polymerization through photoredox catalysis processes. *J. Polym. Sci. Pol Chem.* **2015**, *53*, 42–49. [CrossRef]
29. Mohan, M.; Gupta, N.S.; Chandra, L.; Jha, N.K. Synthesis, characterization and antitumor properties of some metal–complexes of 3-substituted and 5-substituted salicylaldehyde 2-pyridinylhydrazones. *J. Inorg. Biochem.* **1987**, *31*, 7–27. [CrossRef]
30. Dega-Szafran, Z.; Katrusiak, A.; Szafran, M. Short and symmetrical OHO hydrogen bond in bis(quinuclidine betaine) hydrochloride. *J. Mol. Struct.* **2010**, *971*, 1–7. [CrossRef]
31. Tweedle, M.F.; Wilson, L.J. Variable spin iron(III) chelates with hexadentate ligands derived from triethylenetetramine and various salicylaldehydes—Synthesis, characterization, and solution state studies of a new ^{2}T reversible ^{6}A spin equilibrium system. *J. Am. Chem. Soc.* **1976**, *98*, 4824–4834. [CrossRef]
32. Klencsár, Z. MossWinn Manual. 2016. Available online: http://www.mosswinn.hu/downloads/mosswinn.pdf (accessed on 26 April 2017).

inorganics

MDPI

Article

Spin-Singlet Transition in the Magnetic Hybrid Compound from a Spin-Crossover Fe(III) Cation and π-Radical Anion

Kazuyuki Takahashi [1,*] , **Takahiro Sakurai** [2], **Wei-Min Zhang** [3], **Susumu Okubo** [3], **Hitoshi Ohta** [3], **Takashi Yamamoto** [4], **Yasuaki Einaga** [4] and **Hatsumi Mori** [5]

[1]   Department of Chemistry, Graduate School of Science, Kobe University, 1-1 Rokkodai-cho, Nada-ku, Kobe, Hyogo 657-8501, Japan
[2]   Research Facility Center for Science and Technology, Kobe University, 1-1 Rokkodai-cho, Nada-ku, Kobe, Hyogo 657-8501, Japan; tsakurai@kobe-u.ac.jp
[3]   Molecular Photoscience Research Center, Kobe University, 1-1 Rokkodai-cho, Nada-ku, Kobe, Hyogo 657-8501, Japan; tmrhu@hotmail.com (W.-M.Z.); sokubo@kobe-u.ac.jp (S.O.); hohta@kobe-u.ac.jp (H.O.)
[4]   Department of Chemistry, Graduate School of Science and Technology, Keio University, 3-14-1 Hiyoshi, Kohoku-ku, Yokohama, Kanagawa 223-8522, Japan; takyama@chem.keio.ac.jp (T.Y.); einaga@chem.keio.ac.jp (Y.E.)
[5]   Institute for Solid State Physics, The University of Tokyo, 5-1-5 Kashiwanoha, Kashiwa, Chiba 277-8581, Japan; hmori@issp.u-tokyo.ac.jp
*   Correspondence: ktaka@crystal.kobe-u.ac.jp; Tel.: +81-78-803-5691

Received: 5 July 2017; Accepted: 11 August 2017; Published: 16 August 2017

Abstract: To develop a new spin-crossover functional material, a magnetic hybrid compound $[Fe(qsal)_2][Ni(mnt)_2]$ was designed and synthesized (Hqsal = N-(8-quinolyl)salicylaldimine, mnt = maleonitriledithiolate). The temperature dependence of magnetic susceptibility suggested the coexistence of the high-spin (HS) Fe(III) cation and π-radical anion at room temperature and a magnetic transition below 100 K. The thermal variation of crystal structures revealed that strong π-stacking interaction between the π-ligand in the $[Fe(qsal)_2]$ cation and $[Ni(mnt)_2]$ anion induced the distortion of an Fe(III) coordination structure and the suppression of a dimerization of the $[Ni(mnt)_2]$ anion. Transfer integral calculations indicated that the magnetic transition below 100 K originated from a spin-singlet formation transformation in the $[Ni(mnt)_2]$ dimer. The magnetic relaxation of Mössbauer spectra and large thermal variation of a g-value in electron paramagnetic resonance spectra below the magnetic transition temperature implied the existence of a magnetic correlation between d-spin and π-spin.

Keywords: spin-crossover; π-radical; spin transition; spin-singlet formation; π-stacking interaction; magnetic correlation; Fe(III) complex; Ni dithiolene complex

---

1. Introduction

Spin-crossover (SCO) between a high-spin (HS) and low-spin (LS) state in a transition metal coordination compound is one of the molecular bistable phenomena responsive to various external stimuli such as temperature, pressure, light, magnetic field, and chemicals. Significant attention has been paid to SCO in a wide field of chemical sciences [1,2]. The SCO switches not only a spin-state, but also electronic state and coordination structure in a metal complex. Thus the utilization of electronic and structural transformation accompanying SCO can lead to potential applications of display, memory, sensing and electronic devices [3,4].

One of the emergent fields of SCO research in the last decade is the development of multifunctional compounds between SCO and other solid-state electronic properties such as conductivity [5–12], magnetism [13–23], and optical properties [24,25]. The goal of this research is that one can design and synthesize a molecular solid whose electronic property can be controlled by external stimuli. However, most reports concerning functional SCO compounds described the coexistence of both SCO and other electronic property, while the achievement of a synergy between these two properties is still very limited. Thus further investigation is required to design and synthesis of a new functional SCO hybrid compound.

Various factors are involved in the realization of the synergy between SCO and electronic property. Among them we emphasized the role of intermolecular interactions either between SCO molecules or between SCO and counterfunctional molecules in a functional SCO hybrid compound. In particular, intermolecular interactions between SCO molecules can contribute to the cooperativity in SCO behavior as well as the crystal engineering of an SCO hybrid compound. According to our strategy, we successfully developed the multifunctional SCO compounds, namely SCO conductors [5,7] and SCO magnets [22,23]. Among them, [Fe(Iqsal)₂][Ni(dmit)₂]·CH₃CN·H₂O [22] is the intriguing compound which exhibits the synergistic spin transition between SCO and spin-singlet formation of the paramagnetic π-radical anions (HIqsal = *N*-(8-quinolyl)-5-iodosalicylaldimine, dmit = 4,5-dithiolato-1,3-dithiole-2-thione). The crucial point of this synergy is that the halogen-bonding interactions between iodine-substituted SCO Fe(III) cation and paramagnetic π-radical is competed with the energy gain from the spin-singlet formation in the paramagnetic π-radical dimer. In contrast, for the mother compound [Fe(qsal)₂][Ni(dmit)₂]·2CH₃CN [26] and its analogue [Fe(qsal)₂][Ni(dmise)₂]·2CH₃CN [27], the nickel dithiolene π-radical anions formed a spin-singlet dimer below room temperature and might contribute to the cooperative SCO transitions of the SCO Fe(III) cations through π-stacking interactions between π-ligand and Ni dithiolene anion (Hqsal = *N*-(8-quinolyl)salicylaldimine, dmise = 4,5-dithiolato-1,3-dithiole-2-selone).

Further efforts to develop new paramagnetic SCO compounds, we focused on a similar paramagnetic π-radical anion, [Ni(mnt)₂]⁻, which afforded magnetic compounds showing various magnetic behaviors [28] (mnt = maleonitriledithiolate). Recently, Nihei et al. reported that the Fe(II) compound from the [Ni(mnt)₂] anion exhibited a synergistic multistep spin transition [15]. However, the Fe(III) compound with the [Ni(mnt)₂] anion has never been reported. To compare the crystal structure and magnetic properties with the above [N(dmit)₂] compounds, we selected the [Fe(qsal)₂] cation as a potential SCO cation. We report herein the preparation, crystal structures, and magnetic properties of [Fe(qsal)₂][Ni(mnt)₂] **1** (Figure 1). Although the present compound did not show an SCO phenomenon, we found a spin-singlet formation transition of the [Ni(mnt)₂] dimer and a magnetic correlation between d-spin and π-spin in Mössbauer and electron paramagnetic resonance (EPR) spectra. All these features were attributed to a strong π-stacking interaction between π-ligand of the [Fe(qsal)₂] cation and [Ni(mnt)₂] anion.

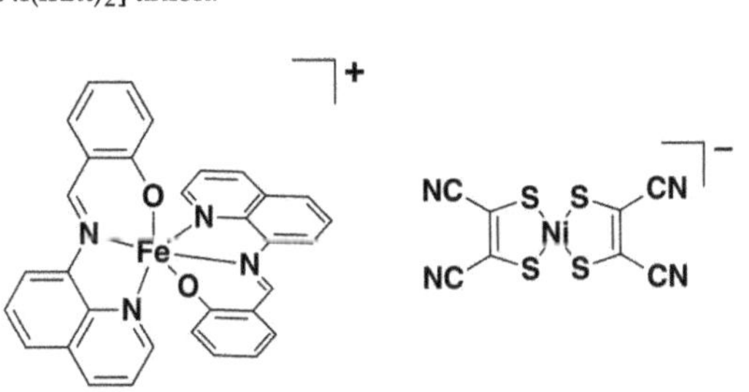

Figure 1. Chemical structural formula of compound **1**.

2. Results and Discussion

2.1. Synthesis

[Fe(qsal)$_2$][Ni(mnt)$_2$] was prepared by the metathesis reaction between [Fe(qsal)]$_2$Cl·1.5H$_2$O [29] and (TBA)[Ni(mnt)$_2$] in acetonitrile (TBA = tetrabutylammonium cation). Recrystallization from acetonitrile gave stable, tiny black crystals. The composition of the complex was confirmed by microanalysis and X-ray single crystal structural analysis described below.

2.2. Magnetic Susceptibility

The temperature variation of magnetic susceptibility for compound **1** is shown in Figure 2a. The $\chi_M T$ value for compound **1** at 300 K was 4.51 cm^3·K·mol^{-1}. Since the spin-only $\chi_M T$ values for the HS Fe(III) ion ($S = 5/2$) and paramagnetic [Ni(mnt)$_2$] anion ($S = 1/2$) are 4.375 and 0.375 cm^3·K·mol^{-1}, respectively, the $\chi_M T$ value at 300 K suggests that the compound contains the HS Fe(III) ion and paramagnetic π-radical anion. On cooling the sample, the $\chi_M T$ products were gradually decreased and reached to 4.43 cm^3·K·mol^{-1} at 100 K. Further lowering temperature, a relatively steep decrease in the $\chi_M T$ product was observed, and then a narrow plateau of the $\chi_M T$ of 4.20 cm^3·K·mol^{-1} appeared at around 35 K, suggesting compound **1** exhibited a magnetic transition. Below 15 K the $\chi_M T$ values were abruptly decreased probably due to zero-field splitting effect.

To gain an insight of the magnetic transition, the magnetic field dependence of magnetization was measured at 2.0 K. The magnetization curve is shown in Figure 2b. The magnetization at 5 T was 4.70 μ$_B$ and the curve seems to be similar to the calculated curve of $S = 5/2$ by using the Brillouin function. This implies that the [Ni(mnt)$_2$] anion might change from a paramagnetic state to a non-magnetic one in the magnetic transition.

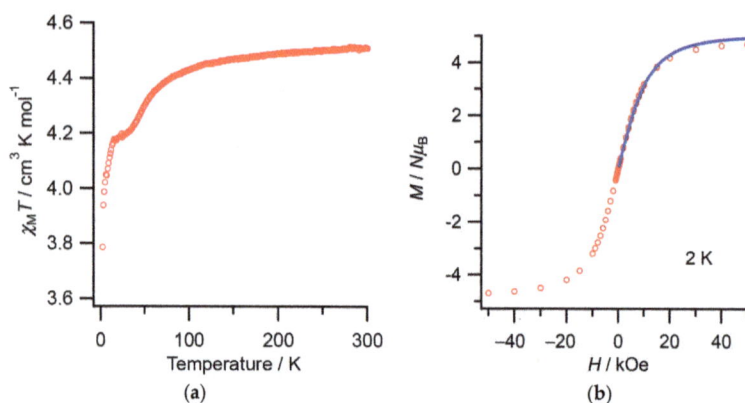

Figure 2. (a) $\chi_M T$ vs. T products for **1** under a magnetic field of 0.5 T at a scan speed of 1 K·min^{-1}; (b) Magnetic field dependence of magnetizations for **1** at 2 K. Blue curve indicates the calculated magnetization curve with $S = 5/2$ and $g = 2$.

2.3. Crystal Structural Analysis

To investigate the structural changes below and above the magnetic transition, variable temperature single-crystal X-ray structural analyses for **1** were performed using a Rigaku CMF007 Mercury CCD diffractometer attached with a Japan Thermal Engineering helium gas-flow temperature controller. Crystallographic data are listed in Table 1. The crystal structures for **1** at all temperatures were isostructural and belonged to triclinic system with $P\bar{1}$. All asymmetric units contained one [Fe(qsal)$_2$] molecule and one [Ni(mnt)$_2$] molecule.

Table 1. Crystallographic data for **1**.

	1		
Formula	$C_{40}H_{22}FeN_8NiO_2S_4$		
Formula Weight	889.45		
Color	black		
Dimension/mm	$0.20 \times 0.20 \times 0.05$		
T/K	293	100	25
Crystal system	triclinic	triclinic	triclinic
Space Group	$P\bar{1}$	$P\bar{1}$	$P\bar{1}$
$a/\text{Å}$	11.7638(6)	11.5777(7)	11.4218(11)
$b/\text{Å}$	13.5893(5)	13.6154(4)	13.7207(6)
$c/\text{Å}$	14.1000(10)	13.9687(11)	13.8041(17)
$\alpha/°$	65.962(11)	66.134(8)	67.123(12)
$\beta/°$	85.184(14)	84.758(12)	84.938(18)
$\gamma/°$	65.815(9)	65.272(9)	65.324(14)
$V/\text{Å}^3$	1868.5(3)	1821.0(3)	1804.0(4)
Z	2	2	2
$\rho_{calcd.}/\text{g·cm}^{-3}$	1.581	1.622	1.637
μ (Mo-Kα)	1.165	1.195	1.206
$2\theta_{max}/°$	55.13	54.99	55.02
No. Reflections (R_{int})	20124 (0.0214)	17468 (0.0365)	17099 (0.0347)
No. Observations ($I > 2.00\,\sigma(I)$)	8382 (6421)	7972 (6636)	7888 (6604)
No. Variables	505	505	505
R1 ($I > 2.00\,\sigma(I)$)	0.0468	0.0481	0.0553
R (all data)	0.0600	0.0567	0.0646
wR2 (all data)	0.1427	0.1309	0.1231
Residual electron density/eÅ^{-3}	0.90 −0.66	1.14 −0.93	1.05 −1.22
Goodness of fit	1.062	1.115	1.076

2.3.1. Molecular Structures in **1** at 293 K

The π-ligand molecule in the [Fe(qsal)$_2$] molecule was coordinated to a central Fe atom as a tridentate chelate ligand and thus two coordinated ligand molecules were arranged in an almost perpendicular manner (Figure 3a). The selected coordination bond lengths, distortion parameters, and angles for **1** along with those of the HS and LS [Fe(qsal)$_2$] cations in the [Ni(dmise)$_2$] compound [27] are listed in Table 2. As compared with the coordination bond lengths between **1** and [Fe(qsal)$_2$] cation, the Fe–O and Fe–N bond lengths for **1** were similar to those of the HS [Fe(qsal)$_2$] cation, indicating that the [Fe(qsal)$_2$] molecule in **1** was in the HS Fe(III) state at 293 K. On the other hand, the distortion parameters Σ and Θ for **1** were much larger than those of the HS [Fe(qsal)$_2$] cation. The dihedral angles between the quinolyl and phenyl rings in two qsal ligands of **1** were 1.37 (η_1) and 24.76° (η_2). Thus one qsal ligand was planar, whereas the other one was nonplanar in **1**. The dihedral angle η_2 was much larger than those of the HS and LS [Fe(qsal)$_2$] cations [27]. This suggests that the large distortion of a coordination structure may arise from the distortion of the nonplanar qsal ligand in the [Fe(qsal)$_2$] cation.

The molecular structure of the [Ni(mnt)$_2$] molecule is shown in Figure 3b. The selected bond lengths of the [Ni(mnt)$_2$] molecule for **1** along with the [Ni(mnt)$_2$] monoanion and dianion [30] are listed in Table 3. The mnt ligands were coordinated to a central Ni atom to form a square planar coordination structure. It is know that the charge of the [Ni(mnt)$_2$] molecule can be estimated by the Ni–S bond lengths. As compared with the Ni–S bond lengths of the [Ni(mnt)$_2$] monoanion and dianion, the [Ni(mnt)$_2$] molecule in **1** was ascribed to the monoanion. Therefore, compound **1** at 293 K consists of the HS Fe(III) cation ($S = 5/2$) and π-radical anion ($S = 1/2$). This is in good agreement with the suggestion of the existence of a π-anion radical from the magnetic susceptibility measurement.

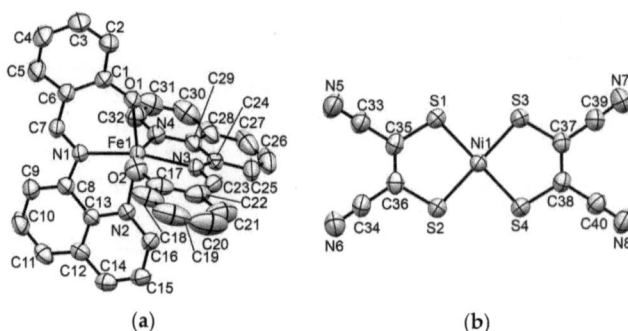

Figure 3. ORTEP drawings of 50% probability with atomic numberings for compound **1** at 293 K. (a) [Fe(qsal)$_2$] molecule; (b) [Ni(mnt)$_2$] molecule. Hydrogen atoms are omitted for clarity.

Table 2. Selected coordination bond lengths and distortion parameters of the [Fe(qsal)$_2$] molecule.

		1		HS [Fe(qsal)$_2$]$^+$	LS [Fe(qsal)$_2$]$^+$
T/K	293	100	25	273	200
Fe1–O1/Å	1.9278(19)	1.9371(18)	1.942(2)	1.918(2)	1.8806(19)
Fe1–O2/Å	1.908(2)	1.9131(18)	1.914(2)	1.909(3)	1.882(2)
Fe1–N1/Å	2.097(2)	2.101(2)	2.100(3)	2.114(3)	1.950(3)
Fe1–N2/Å	2.195(2)	2.196(2)	2.195(3)	2.151(2)	1.979(2)
Fe1–N3/Å	2.131(2)	2.132(2)	2.132(3)	2.126(3)	1.948(3)
Fe1–N4/Å	2.136(2)	2.129(2)	2.128(3)	2.133(3)	1.972(3)
Σ [1]/°	83.1(3)	85.7(3)	86.1 (4)	71.5(4)	44.6(4)
Θ [2]/°	149.1(4)	151.0(4)	151.6(5)	121.1(5)	54.9(5)
ϕ [3]/°	162.17(8)	161.97(8)	162.16(11)	166.35(11)	176.59(11)
η_1 [4]/°	1.37	1.79	2.26	7.49	6.76
η_2 [5]/°	24.76	26.66	27.46	7.94	5.88
	this work	this work	this work	Ref. [27]	Ref. [27]

[1] The sum of the absolute differences of 12-bite angles of a first coordination sphere from 90°. [2] The sum of the absolute differences of 24 angles of 8 triangle surfaces of a coordination octahedron from 60°. [3] The angles of N1–Fe1–N3. [4] Dihedral angles between the quinolyl (C24–C32, N4) and phenyl (C17–C22) rings. [5] Dihedral angles between the quinolyl (C8–C16, N2) and phenyl (C1–C6) rings.

Table 3. Selected bond lengths of the [Ni(mnt)$_2$] molecule.

		1		[Ni(mnt)$_2$]$^-$	[Ni(mnt)$_2$]$^{2-}$
T/K	293	100	25		
Ni1–S1/Å	2.1466(8)	2.1509(7)	2.1466(10)	2.147(3) [1]	2.176(1) [1]
Ni1–S2/Å	2.1383(8)	2.1432(7)	2.1459(10)	2.151(3) [1]	2.173(1) [1]
Ni1–S3/Å	2.1340(8)	2.1391(7)	2.1376(10)	2.148(3) [1]	–
Ni1–S4/Å	2.1421(8)	2.1469(7)	2.1458(10)	2.149(3) [1]	–
S1–C35/Å	1.718(3)	1.718(3)	1.708(4)	1.727(10) [1]	1.732(4) [1]
S2–C36/Å	1.715(3)	1.721(3)	1.712(4)	1.722(9) [1]	1.725(5) [1]
S3–C37/Å	1.715(3)	1.723(2)	1.722(4)	1.705(10) [1]	–
S4–C38/Å	1.720(3)	1.726(3)	1.729(4)	1.724(9) [1]	–
C35–C36/Å	1.349(4)	1.360(4)	1.374(5)	1.367(12) [1]	1.360(7) [1]
C37–C38/Å	1.358(4)	1.361(4)	1.358(5)	1.370(12) [1]	–
	this work	this work	this work	Ref. [30]	Ref. [30]

[1] The bond lengths corresponding to the atomic numberings of **1**.

2.3.2. Molecular Arrangement of **1** at 293 K

The molecular arrangement for **1** at 293 K is shown in Figure 4a–e and the selected intermolecular distances are listed in Tables 4 and 5. The planar qsal ligand (C17–C32, N3, N4) of a [Fe(qsal)$_2$] cation was stacked with that of another neighboring [Fe(qsal)$_2$] cation related with the inversion symmetry (p in Figure 4a), to form a π-stacking [Fe(qsal)$_2$] dimer with a mean π-plane distance of 3.41 Å (Table 4). On the other hand, there were π-stacking interactions both between the quinolyl rings (C8–C16, N2) (q in Figure 4b) and between the phenyl rings (C1–C6) (r in Figure 4b), where both rings belong to the nonplanar qsal ligand. Their mean π-plane distances were 3.35 and 3.50 Å, respectively (Table 4). Therefore, the [Fe(qsal)$_2$] dimers formed a two-dimensional (2D) π-stacking array parallel to the *bc* plane (Figure 4a).

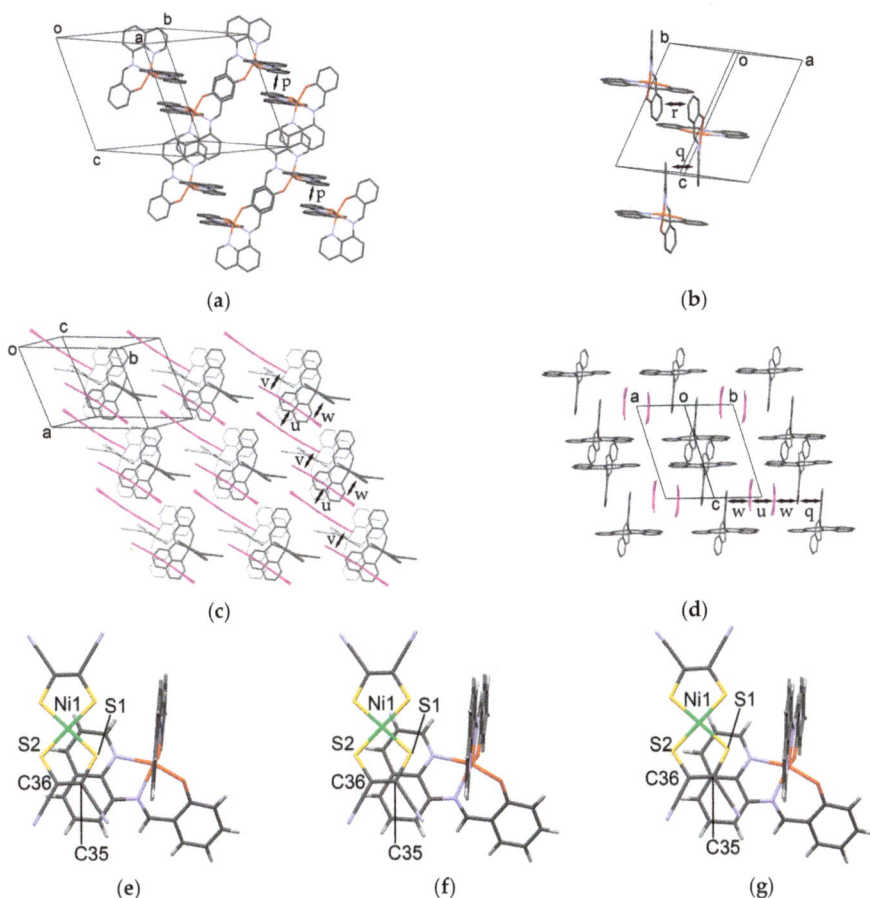

Figure 4. Crystal structure of **1**. (**a**) 2D molecular arrangement of the [Fe(qsal)$_2$] moleucle viewed along the direction perpendicular to the overlapped quinolyl plane at 293 K; (**b**) π-stacking mode between the [Fe(qsal)$_2$] dimers viewed along the direction parallel to the overlapped quinolyl plane at 293 K; (**c**) Molecular arrangement of the [Ni(mnt)$_2$] molecule (magenta sticks) along with the [Fe(qsal)$_2$] molecule layer just below the [Ni(mnt)$_2$] layer at 293 K; (**d**) Side view of π-stacking structure from the [Ni(mnt)$_2$] dimer and quinolyl rings perpendicular to the *a–b* direction at 293 K; Top views of the π-stacking between the [Ni(mnt)$_2$] and nonplanar qsal molecules at 293 K (**e**); at 100 K (**f**); at 25 K (**g**);. Letters with double-headed arrows indicate the positions listed in Tables 4 and 5.

The [Ni(mnt)$_2$] anions also formed a face-to-face π-stacking dimer with a mean π-plane distance of 3.51 Å (u in Figure 4c). The [Ni(mnt)$_2$] dimers were arranged along the *a* axis by a small π-overlap with a π-plane separation of 3.20 Å (v in Figure 4c). With respect to the intermolecular interaction between the [Fe(qsal)$_2$] cation and [Ni(mnt)$_2$] anion, the five-membered dithiolene ring (Ni1, S1, S2, C35, C36) in the [Ni(mnt)$_2$] anion overlapped the quinolyl ring (C8–C16, N2) in the nonplanar qsal ligand with a distance of 3.29–3.44 Å (w in Figure 4c and d). Although very similar π-overlaps were found in the [Fe(qsal)$_2$][Ni(dmise)$_2$] compound [27], the distances from the quinolyl plane to the S1 and C35 atoms in **1** (Figure 4e) were much shorter than the corresponding distances in the [Ni(dmise)$_2$] compounds (Table 5). This suggests a much stronger π-stacking interaction between the quinolyl ring and [Ni(mnt)$_2$] anion in **1**. Since the molecular distortion in the nonplanar qsal ligand would originate from the present strong π-stacking interaction, the π-stacking interaction may induce the HS state in the [Fe(qsal)$_2$] cation. This π-stacking interaction gave a 1D alternate π-stacking column along the *a*–*b* direction (Figure 4d). Accordingly, the crystal packing of the present compound **1** comprised 3D π-stacking interactions.

Table 4. Selected intermolecular distances (Å) in **1**.

	Position [1]	1		
T/K		293	100	25
Fe(qsal)$_2$ intradimer	p			
π-plane (qsal⋯qsal)		3.405	3.346	3.331
Fe1⋯Fe1		6.672	6.640	6.671
Fe(qsal)$_2$ interdimer	q, r			
π-plane (quinolyl⋯quinolyl)	q	3.347	3.297	3.302
Fe1⋯Fe1		8.904	8.837	8.791
π-plane (phenyl⋯phenyl)	r	3.501	3.388	3.372
Fe1⋯Fe1		9.972	9.962	10.038
Ni(mnt)$_2$ intradimer	u			
π-plane		3.505	3.450	3.400
Ni1⋯Ni1		4.028	4.002	4.142
Ni(mnt)$_2$ interdimer	v			
π-plane		3.196	3.106	3.072
Ni1⋯Ni1		8.578	8.394	7.945

[1] The positions corresponding to letters are shown in Figure 4.

Table 5. Intermolecular distances (Å) between the nickel dithiolene ring and qsal ligand (w [1]).

	1			[Ni(dmise)$_2$]	
T/K	293	100	25	273	200
Quinolyl⋯Ni1	3.348	3.309	3.297	3.419	3.325
Quinolyl⋯S1	3.288	3.244	3.301	3.579 [2]	3.519 [2]
Quinolyl⋯C35	3.402	3.348	3.367	3.494 [2]	3.498 [2]
Quinolyl⋯C36	3.448	3.388	3.372	3.365 [2]	3.404 [2]
Quinolyl⋯S2	3.448	3.393	3.362	3.297 [2]	3.278 [2]
Fe1⋯Ni1	6.214	6.223	6.433	6.591	6.282
	this work	this work	this work	Ref. [27]	Ref. [27]

[1] The position is shown in Figure 4. [2] The bond lengths corresponding to the atomic numberings of **1**.

2.3.3. Thermal Variations of the Crystal Structure of **1**

The thermal variations of the bond lengths, angles, and intermolecular distances are listed in Tables 2–5. There was no remarkable structural change in both the [Fe(qsal)$_2$] cation and the [Ni(mnt)$_2$] anion, suggesting that the HS state of the [Fe(qsal)$_2$] cation and valence state of the [Ni(mnt)$_2$] anion were retained in the whole temperature range measured. On the other hand, all intermolecular π-plane distances were shortened on lowering temperature from 293 to 100 K, whereas some of the intermolecular π-plane distances were slightly lengthened on cooling from 100 to 25 K (Tables 4 and 5). Note that the Ni⋯Ni (u) and Fe⋯Ni (w) distances at 25 K were 0.14 and 0.21 Å longer than those at

100 K despite the shrinking of its π-plane distance. These structural variations may be related to the magnetic transition.

2.4. Transfer Integrals Between the [Ni(mnt)$_2$] Molecules

To provide an insight of the thermal variations in the magnetic exchange interaction between paramagnetic [Ni(mnt)$_2$] anions, transfer integrals were calculated based on the extended Hückel method [31]. The transfer integrals of the intradimer and interdimer for **1** at 293 K were 51.5 and 22.9 meV, respectively (Table 6). The overlap of LUMOs of the intradimer is two times larger than that of the interdimer, which rationalized the definition of the [Ni(mnt)$_2$] dimer. The transfer integrals of the interdimer were almost unchanged in spite of temperature variations. It should be noted that the transfer integral of the intradimer at 25 K is almost three times larger than that at 293 and 100 K. Since the exchange coupling energy J is known to be proportional to the square of a transfer integral [32,33], this exchange coupling energy at 25 K is about 10 times larger than that at 100 K. The temperature range of the enhancement of exchange coupling energy was in good agreement with that of the magnetic transition. This clearly indicates that the magnetic transition below 100 K would originate from a spin-singlet formation of the paramagnetic [Ni(mnt)$_2$] dimer. Furthermore, this transition suggests the existence of an intermolecular interaction corresponding to the energy gain of the spin-singlet formation. As described in the previous section, on the spin-singlet formation transition, two remarkable intermolecular structural changes were observed in the [Ni(mnt)$_2$] dimer (u) and the π-stacking structure between the [Ni(mnt)$_2$] anion and quinolyl ring in the nonplanar qsal ligand (w). The former arises from the spin-singlet formation, the latter implies the variation of the intermolecular π-stacking interaction. The comparison between Figure 4f and g revealed that the [Fe(qsal)$_2$] cation and [Ni(mnt)$_2$] anion separate each other, indicating that the π-stacking interaction is weakened at 25 K. Hence, the energy gain of the spin-singlet formation competed with the π-stacking interaction between the [Ni(mnt)$_2$] anion and quinolyl ring in the nonplanar qsal ligand.

Table 6. Transfer integrals (m·eV) between the [Ni(mnt)$_2$] molecules in **1**.

Position [1]		1		
T/K		293	100	25
Intradimer	u	51.53	66.62	172.0
Interdimer	v	22.91	28.49	25.17

[1] The positions corresponding to letters are shown in Figure 4.

2.5. Electron Paramagnetic Resonance (EPR) Spectra

To confirm the electronic states of the [Fe(qsal)$_2$] and [Ni(mnt)$_2$] for compound **1**, the temperature dependence of electron paramagnetic resonance spectra for a polycrystalline sample of **1** was recorded in the temperature range of at 4.2–280 K (Figure 5a). The spectrum mainly consisted of a broad absorption at around 2 kOe, which can be ascribed to the HS [Fe(qsal)$_2$] cation with a strong rhombic distortion [34]. On the other hand, there were no absorptions assigned to the LS [Fe(qsal)$_2$] cation and paramagnetic [Ni(mnt)$_2$] anion in the magnetic field range reported in the literature [34,35]. Note that no absorption ascribed to the LS [Fe(qsal)$_2$] cation appeared and moreover the g-values of the HS [Fe(qsal)$_2$] cation were largely shifted from $g = 3.3$ to $g = 4.0$ on cooling the sample from 100 to 40 K (Figure 5b). Thus, the magnetic transition did not arise from the SCO transition and the change in a g-value will be related to the spin-singlet formation in the [Ni(mnt)$_2$] dimer. Although the change in a coordination structure may lead to the shift of a g-value in a metal coordination complex, the structural variation of the [Fe(qsal)$_2$] cation from 100 to 25 K was smaller than that from 293 to 100 K and moreover the g-values were almost constant from 293 K to 100 K. This means that the coordination structural change cannot account for the present large transformation in a g-value. Previously the g-value shifts were reported in the spin-Peierls compounds [36,37]. One possible explanation to shift a g-value in the present compound is that a magnetic exchange correlation between d-spin and π-spin might disappear as π-spins forms the spin-singlet. Additionally, it should be noted that no absorption assigned to the LS [Fe(qsal)$_2$] cation was observed even at 4.2 K, indicating that the [Fe(qsal)$_2$] cation was in the HS state above 4.2 K.

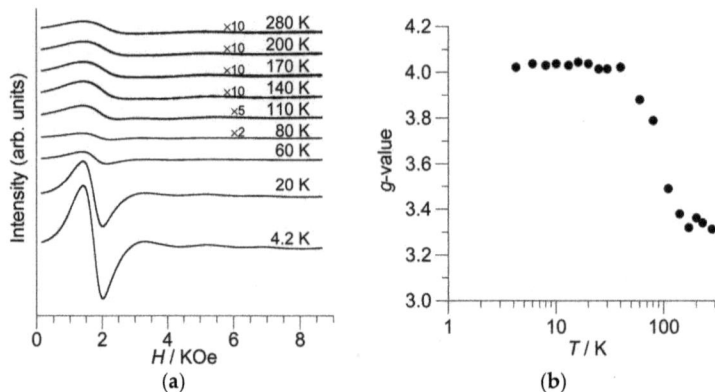

Figure 5. (a) Temperature dependence of electron paramagnetic resonance spectra for **1**; (b) Temperature dependence of g-values for the absorption from the high-spin (HS) [Fe(qsal)$_2$] cation.

2.6. Mössbauer Spectra

To confirm the valence and spin states of the Fe ion for compound **1**, the temperature valuable Mössbauer spectra for **1** were recorded at 15, 40, 100, and 293 K (Figure 6 and Table 7). The spectrum at 293 K consisted of mainly a broad asymmetric quadrupole doublet with isomer shift (IS) of 0.307(16) mm·s^{-1} and quadrupole splitting (QS) of 0.91(2) mm·s^{-1}. As compared with the [Fe(qsal)$_2$] complexes in the literature [29,38–40], the IS value is very similar to those of the HS [Fe(qsal)$_2$] species, whereas the QS value is larger than those of the HS species. This large QS value may arise from the distortion of a coordination octahedron. On cooling to 100 K, the doublet spectrum was broadening. The IS shifted to a larger value of 0.48(4) mm·s^{-1}, whereas the QS was increased slightly. Further lowering temperatures very similar broad wing-like spectra without the splitting were recorded at 40 and 15 K.

Note that the broad wing-like spectra below 100 K were indicative of a magnetic relaxation. Although a rapid exchange between the HS and LS states is known to one of the relaxation mechanisms in SCO complexes [41,42], the distorted and elongated HS coordination structure at 25 K and the absence of the LS signal in the EPR spectrum at 4.2 K would exclude this relaxation mechanism and thus all the spectra would be ascribed to the HS Fe(III) centers. As for another possible mechanism, the magnetic relaxation in HS Fe(III) compounds is known to arise from spin-spin relaxation related to the distance between the HS Fe(III) centers [43]. The adjacent distances between the Fe(III) centers in **1** were similar to those exhibiting spin-spin relaxation in the literature [43], but were almost temperature-independent (Table 4). This means that the spin-spin relaxation originating from the lengthening of the Fe(III) center distances would be excluded. Interestingly, since the spectrum broadening became remarkable in the temperature range of the formation of a spin-singlet [Ni(mnt)$_2$] dimer, the spin-spin relaxation on the Fe center might correlate with π-spin of the [Ni(mnt)$_2$] anion in the present compound. Further investigation needs to clarify the relaxation mechanism.

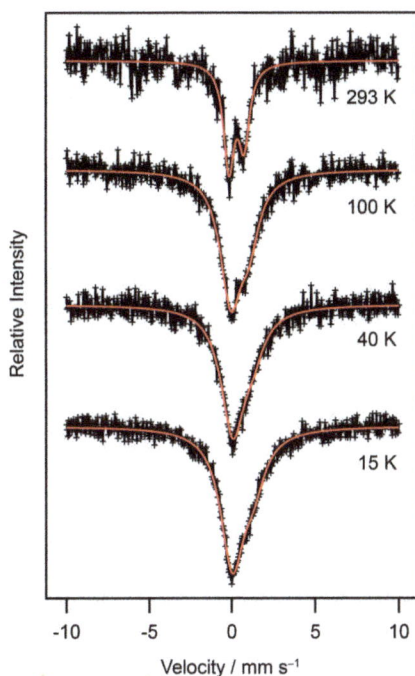

Figure 6. Mössbauer spectra for **1**. Red curves indicate spectra simulated by using the parameters listed in Table 7.

Table 7. Temperature variations of Mössbauer parameters for **1**.

T/K	Spin-State	Ratio	IS [1]/mm·s^{-1}	QS [2]/mm·s^{-1}	LW [3]/mm·s^{-1}
293	HS	100%	0.307(16)	0.91(2)	0.79(4)
100	HS	100%	0.48(3)	1.07(3)	1.54(5)
40	HS	100%	0.65(4)	1.18(6)	1.73(5)
15	HS	100%	0.65(3)	1.24(4)	1.73(4)

[1] Isomer shift. [2] Quadrupole splitting. [3] Linewidth.

3. Materials and Methods

All chemicals were purchased and used without further purification. [Fe(qsal)$_2$]Cl·1.5H$_2$O was prepared according to the literature [29].

3.1. Synthesis of [Fe(qsal)$_2$][Ni(mnt)$_2$] (1)

To a solution of [Fe(qsal)$_2$]Cl·1.5H$_2$O (122 mg, 0.2 mmol) in 100 mL of acetonitrile was added a solution of (TBA)[Ni(mnt)$_2$] (119 mg, 0.2 mmol) in 20 mL of acetonitrile in one portion. After standing for several hours, black crystals were precipitated. Filtration and dry in air gave 84 mg of compound **1** as black platelets (47% yield). The crystalline samples for the physical measurements below were obtained by recrystallization from acetonitrile.

Anal. Calcd. For C$_{40}$H$_{22}$FeN$_8$NiO$_2$S$_4$: C, 54.01%; H, 2.49%; N, 12.60%. Found: C, 53.91%; H, 2.69%; N, 12.40%.

3.2. Physical Measurements

Variable-temperature direct-current magnetic susceptibilities of polycrystalline samples (ca. 20 mg) fold in an aluminum foil were measured on a Quantum Design MPMS-XL magnetometer (San Diego, CA, USA) under a field of 0.5 T at a sweep speed of 1 K·min^{-1} in the temperature range of 2–300 K. The sample magnetization data were obtained by the subtraction of background magnetization data for the aluminum foil from the measured data including the aluminum foil, and then the magnetic susceptibilities were corrected for diamagnetic contributions estimated by Pascal constants.

The Mössbauer spectra were recorded on a constant acceleration spectrometer with a source of ^{57}Co/Rh in the transmission mode. The measurements at low temperature were performed with a closed-cycle helium refrigerator (Iwatani Co., Ltd. (Osaka, Japan)). Velocity was calibrated by using an α-Fe standard. The obtained Mössbauer spectra were fitted with asymmetric Lorentzian doublets by the least squares fitting program (MossWinn (Version 4.0, Zoltán Klencsár, Budapest, Hungary)).

X-band (~9.5 GHz) electron paramagnetic resonance measurements have been performed on polycrystalline sample using a Bruker EMX081 ESR spectrometer (Yokohama, Japan) in the temperature region from 4.2 to 280 K.

3.3. Crystal Structure Determinations

A platelet crystal was mounted in a polyimide loop. A Japan Thermal Engineering helium gas-flow temperature controller (Sagamihara, Japan) was used for the temperature variable measurements. All data were collected on a Rigaku CMF007 Mercury CCD diffractometer (Tokyo, Japan) using graphite monochromated Mo-*Kα* radiation. All data collections and calculations were performed using CrystalClear (Rigaku (Tokyo, Japan)). The data were collected to a maximum 2θ value of about 55°. A total of 720 oscillation images were collected at 293 K, whereas a total of 640 oscillation images were collected at 100 and 25 K due to the effect on the cold helium gas flow. Data were integrated and numerical absorption correction was applied. The CrystalStructure crystallographic software program package (Rigaku) was used to determine the unit cell parameters and crystal structures. The structure at 293 K was solved by direct methods and refined by full-matrix least-squares methods based on F^2 by using *SHELXL* Version 2016/6 (George Sheldrick, Göttingen, Germany). The atomic positions of the refined structure at 293 K were used as the initial structures at 100 and 25 K and refined by the same methods. All non-hydrogen atoms were refined anisotropically. Hydrogen atoms were generated by calculation and refined using the riding model. CCDC 1559560–1559562 contains the supplementary crystallographic data for this paper. These data can be obtained free of charge via http://www.ccdc.cam.ac.uk/conts/retrieving.html (or from the CCDC, 12 Union Road, Cambridge CB2 1EZ, UK; Fax: +44 1223 336033; E-mail: deposit@ccdc.cam.ac.uk).

3.4. Transfer Integral Calculations

The transfer integrals (*t*) were calculated by the extended Hückel molecular orbital calculation method by using the lowest unoccupied molecular orbital (LUMO) of Ni(mnt)$_2$ as the basis function within the tight-binding approximation [31]. The semiempirical parameters of Ni and C [44], S [45], N [46] for Slater-type atomic orbitals were taken from the literature. The *t* between each pair of molecules was assumed to be proportional to the overlap integral (*S*), *t* = *ES*, where *E* is a constant value of −10.0 eV.

4. Conclusions

We reported the synthesis and characterization of π-d multi-spin magnetic compound [Fe(qsal)$_2$][Ni(mnt)$_2$]. Unfortunately the Fe(III) center in the present compound was in the HS state and did not exhibit an SCO phenomenon, whereas a spin-singlet transition of the [Ni(mnt)$_2$] dimer was clearly evidenced. These two observations may arise from the stronger π-stacking interaction between the [Ni(mnt)$_2$] anion and [Fe(qsal)$_2$] cation than those in the [Ni(dmit)$_2$] and [Ni(dmise)$_2$] compounds [26,27]. Recently we found strong intermolecular interactions such as hydrogen-bonding and Coulomb interactions prevented Fe(III) complexes from exhibiting SCO [47,48]. As these observations are compared with functional SCO hybrid compounds [5,7,22,23], the competition of intermolecular interactions between SCO and functional units play a key role in the development of a synergistic functional SCO compound. Moreover, magnetic relaxation in Mössbauer spectra and large g-value shift in EPR spectra were unexpectedly observed in the same temperature range of the spin-singlet formation. Although elucidation of their origins needs further investigation, they show the possibility that one spin-state can be probed by using the other spin in a multi-spin molecular system.

Supplementary Materials: The following are available online at www.mdpi.com/2304-6740/5/3/54/s1, Cif and cif-checked files.

Acknowledgments: This work was partially supported by a Grant-in-Aid for Scientific Research on Innovative Areas of Molecular Degrees of Freedom (No. 20110007) and JSPS KAKENHI Grant Number 25410068. Kazuyuki Takahashi is grateful to the Instrument Center of Institute for Molecular Science for the use of Rigaku CMF007 Mercury CCD system for X-ray crystal analyses.

Author Contributions: Kazuyuki Takahashi conceived, designed, performed the experiments, and analyzed the data; Takahiro Sakurai and Hitoshi Ohta contributed magnetic susceptibility measurements; Wei-Min Zhang and Susumu Okubo contributed EPR spectroscopy; Takashi Yamamoto and Yasuaki Einaga contributed Mössbauer spectroscopy; Hatsumi Mori contributed characterization of the compound; Kazuyuki Takahashi wrote the paper.

Conflicts of Interest: The authors declare no conflicts of interest.

References

1. Gütlich, P. *Spin Crossover in Transition Metal Compounds*; Goodwin, H.A., Ed.; Springer: Berlin/Heidelberg, Germany, 2004.
2. Halcrow, M.A. *Spin-Crossover Materials*; John Wiley & Sons, Ltd.: Oxford, UK, 2013.
3. Bousseksou, A.; Molnár, G.; Salmon, L.; Nicolazzi, W. Molecular spin crossover phenomenon: Recent achievements and prospects. *Chem. Soc. Rev.* **2011**, *40*, 3313–3335. [CrossRef] [PubMed]
4. Gütlich, P.; Gaspar, A.B.; Garcia, Y. Spin state switching in iron coordination compounds. *Beilstein J. Org. Chem.* **2013**, *9*, 342–391. [CrossRef] [PubMed]
5. Takahashi, K.; Cui, H.-B.; Okano, Y.; Kobayashi, H.; Einaga, Y.; Sato, O. Electrical conductivity modulation coupled to a high-spin–low-spin conversion in the molecular system [Fe^{III}(qsal)$_2$][Ni(dmit)$_2$]$_3$·CH$_3$CN·H$_2$O. *Inorg. Chem.* **2006**, *45*, 5739–5741. [CrossRef] [PubMed]
6. Faulmann, C.; Jacob, K.; Dorbes, S.; Lampert, S.; Malfant, I.; Doublet, M.-L.; Valade, L.; Real, J.A. Electrical conductivity and spin crossover: A new achievement with a metal bis dithiolene complex. *Inorg. Chem.* **2007**, *46*, 8548–8559. [CrossRef] [PubMed]
7. Takahashi, K.; Cui, H.-B.; Okano, Y.; Kobayashi, H.; Mori, H.; Tajima, H.; Einaga, Y.; Sato, O. Evidence of the chemical uniaxial strain effect on electrical conductivity in the spin-crossover conducting molecular system: [Fe^{III}(qnal)$_2$][Pd(dmit)$_2$]$_5$·acetone. *J. Am. Chem. Soc.* **2008**, *130*, 6688–6689. [CrossRef] [PubMed]
8. Djukic, B.; Lemaire, M.T. Hybrid Spin crossover conductor exhibiting unusual variable temperature electrical conductivity. *Inorg. Chem.* **2009**, *48*, 10489–10491. [CrossRef] [PubMed]
9. Nihei, M.; Takahashi, N.; Nishikawa, H.; Oshio, H. Spin-crossover behavior and electrical conduction property in iron(II) complexes with tetrathiafulvalene moieties. *Dalton Trans.* **2011**, *40*, 2154–2156. [CrossRef] [PubMed]
10. Phan, H.; Benjamin, S.M.; Steven, E.; Brooks, J.S.; Shatruk, M. Photomagnetic response in highly conductive iron(II) spin-crossover complexes with TCNQ radicals. *Angew. Chem. Int. Ed.* **2015**, *54*, 823–827. [CrossRef] [PubMed]
11. Zhang, X.; Wang, Z.-X.; Xie, H.; Li, M.-X.; Woods, T.J.; Dunbar, K.R. A cobalt(II) spin-crossover compound with partially charged TCNQ radicals and an anomalous conducting behavior. *Chem. Sci.* **2016**, *7*, 1569–1574. [CrossRef]

12. Shvachko, Y.N.; Starichenko, D.V.; Korolyov, A.V.; Yagubskii, E.B.; Kotov, A.I.; Buravov, L.I.; Lyssenko, K.A.; Zverev, V.N.; Simonov, S.V.; Zorina, L.V.; et al. The conducting spin-crossover compound combining Fe(II) cation complex with TCNQ in a fractional reduction state. *Inorg. Chem.* **2016**, *55*, 9121–9130. [CrossRef] [PubMed]

13. Clemente-León, M.; Coronado, E.; López-Jordà, M.; Soriano-Portillo, A.; Waerenborgh, J.C.; Delgado, F.S.; Ruiz-Pérez, C. Insertion of a spin crossover FeIII complex into an oxalate-based layered material: Coexistence of spin canting and spin crossover in a hybrid magnet. *Inorg. Chem.* **2008**, *47*, 9111–9120. [CrossRef] [PubMed]

14. Neves, A.I.S.; Dias, J.C.; Vieira, B.J.C.; Santos, I.C.; Branco, M.B.C.; Pereira, L.C.J.; Waerenborgh, J.C.; Almeida, M.; Belo, D.; Gama, V. A new hybrid material exhibiting room temperature spin-crossover and ferromagnetic cluster-glass behavior. *CrystEngComm* **2009**, *11*, 2160–2168. [CrossRef]

15. Nihei, M.; Tahira, H.; Takahashi, N.; Otake, Y.; Yamamura, Y.; Saito, K.; Oshio, H. Multiple bistability and tristability with dual spin-state conversions in [Fe(dpp)$_2$][Ni(mnt)$_2$]$_2$·MeNO$_2$. *J. Am. Chem. Soc.* **2010**, *132*, 3553–3560. [CrossRef] [PubMed]

16. Clemente-León, M.; Coronado, E.; López-Jordà, M.; Espallargas, G.M.; Soriano-Portillo, A.; Waerenborgh, J.C. Multifunctional magnetic materials obtained by insertion of a spin-crossover Fe^{III} complex into bimetallic oxalate-based ferromagnets. *Chem. Eur. J.* **2010**, *16*, 2207–2219. [CrossRef] [PubMed]

17. Ohkoshi, S.; Imoto, K.; Tsunobuchi, Y.; Takano, S.; Tokoro, H. Light-induced spin-crossover magnet. *Nat. Chem.* **2011**, *3*, 564–569. [CrossRef] [PubMed]

18. Clemente-León, M.; Coronado, E.; López-Jordà, M.; Waerenborgh, J.C. Multifunctional magnetic materials obtained by insertion of spin-crossover Fe^{III} complexes into chiral 3D bimetallic oxalate-based ferromagnets. *Inorg. Chem.* **2011**, *50*, 9122–9130. [CrossRef] [PubMed]

19. Roubeau, O.; Evangelisti, M.; Natividad, E. A spin crossover ferrous complex with ordered magnetic ferric anions. *Chem. Commun.* **2012**, *48*, 7604–7606. [CrossRef] [PubMed]

20. Clemente-León, M.; Coronado, E.; López-Jordà, M.; Waerenborgh, J.C.; Desplanches, C.; Wang, H.; Létard, J.-F.; Hauser, A.; Tissot, A. Stimuli responsive hybrid magnets: Tuning the photoinduced spin-crossover in Fe(III) complexes inserted into layered magnets. *J. Am. Chem. Soc.* **2013**, *135*, 8655–8667. [CrossRef] [PubMed]

21. Ababei, R.; Pichon, C.; Roubeau, O.; Li, Y.-G.; Bréfuel, N.; Buisson, L.; Guionneau, P.; Mathonière, C.; Clérac, R. Rational design of a photomagnetic chain: Bridging single-molecule magnets with a spin-crossover complex. *J. Am. Chem. Soc.* **2013**, *135*, 14840–14853. [CrossRef] [PubMed]

22. Fukuroi, K.; Takahashi, K.; Mochida, T.; Sakurai, T.; Ohta, H.; Yamamoto, T.; Einaga, Y.; Mori, H. Synergistic spin transition between spin crossover and spin-Peierls-like singlet formation in the halogen-bonded molecular hybrid system: [Fe(Iqsal)$_2$][Ni(dmit)$_2$]·CH$_3$CN·H$_2$O. *Angew. Chem. Int. Ed.* **2014**, *53*, 1983–1986. [CrossRef] [PubMed]

23. Okai, M.; Takahashi, K.; Sakurai, T.; Ohta, H.; Yamamoto, T.; Einaga, Y. Novel Fe(II) spin crossover complexes involving a chalcogen-bond and π-stacking interactions with a paramagnetic and nonmagnetic M(dmit)$_2$ anion (M=Ni, Au; dmit = 4,5-dithiolato-1,3-dithiole-2-thione). *J. Mater. Chem. C* **2015**, *3*, 7858–7864. [CrossRef]

24. Ohkoshi, S.; Takano, S.; Imoto, K.; Yoshikiyo, M.; Namai, A.; Tokoro, H. 90-degree optical switching of output second-harmonic light in chiral photomagnet. *Nat. Photonics* **2014**, *8*, 65–71. [CrossRef]

25. Wang, C.F.; Li, R.-F.; Chen, X.-Y.; Wei, R.-J.; Zheng, L.-S.; Tao, J. Synergetic spin crossover and fluorescence in one-dimensional hybrid complexes. *Angew. Chem. Int. Ed.* **2015**, *54*, 1574–1577. [CrossRef] [PubMed]

26. Takahashi, K.; Cui, H.-B.; Kobayashi, H.; Einaga, Y.; Sato, O. The light-induced excited spin state trapping effect on Ni(dmit)$_2$ salt with an Fe(III) spin-crossover cation: [Fe(qsal)$_2$][Ni(dmit)$_2$]·2CH$_3$CN. *Chem. Lett.* **2005**, *34*, 1240–1241. [CrossRef]

27. Takahashi, K.; Mori, H.; Kobayashi, H.; Sato, O. Mechanism of reversible spin transition with a thermal hysteresis loop in [Fe^{III}(qsal)$_2$][Ni(dmise)$_2$]·2CH$_3$CN: Selenium analogue of the precursor of an Fe(III) spin-crossover molecular conducting system. *Polyhedron* **2009**, *28*, 1776–1781. [CrossRef]

28. Duana, H.-B.; Ren, X.-M.; Meng, Q.-J. One-dimensional (1D) [Ni(mnt)$_2$]$^-$-based spin-Peierls-like complexes: Structural, magnetic and transition properties. *Coord. Chem. Rev.* **2010**, *254*, 1509–1522. [CrossRef]

29. Dickinson, R.C.; Baker, W.A.; Collins, R.L. The magnetic properties of bis[*N*-(8-quinolyl)-salicylaldimine] halogenoiron(III)·X hydrate, Fe(8-QS)$_2$X·xH$_2$O: A reexamination. *J. Inorg. Nucl. Chem.* **1977**, *39*, 1531–1533. [CrossRef]

30. Kobayashi, A.; Sasaki, Y. One-dimensional system of square-planar bis(1,2-dicyanovinylene-1,2-dithiolato) metal complexes. I. The crystal structure of [(C₄H₉)₄N]₂[Ni(mnt)₂] and [(C₂H₅)₄N][Ni(mnt)₂]. *Bull. Chem. Soc. Jpn.* **1977**, *50*, 2650–2656. [CrossRef]

31. Mori, T.; Kobayashi, A.; Sasaki, Y.; Kobayashi, H.; Saito, G.; Inokuchi, H. The intermolecular interaction of tetrathiafulvarene in organic metals. Calculation of orbital overlaps and models of energy-band structures. *Bull. Chem. Soc. Jpn.* **1984**, *57*, 627–633. [CrossRef]

32. Akutagawa, T.; Nakamura, T.; Inabe, T.; Underhill, A.E. Structures of Ni(dmit)₂ salts of lithium or ammonium included in crown ether assemblies. *Thin Solid Films* **1998**, *331*, 264–271. [CrossRef]

33. Akutagawa, T.; Nakamura, T. Control of assembly and magnetism of metal-dmit complexes by supramolecular cations. *Coord. Chem. Rev.* **2002**, *226*, 3–9. [CrossRef]

34. Ivanova, T.A.; Ovchinnikov, I.V.; Garipov, R.R.; Ivanova, G.I. Spin crossover [Fe(qsal)₂]X (X = Cl, SCN, CF₃SO₃) complexes: EPR and DFT study. *Appl. Magn. Reson.* **2011**, *40*, 1–10. [CrossRef]

35. Zhou, H.; Wen, L.L.; Ren, X.M.; Meng, Q.J. Novel molecular staircases constructing from H–bonding interactions based on the building blocks of [Ni(mnt)₂]⁻ ions: Syntheses, crystal structures, EPR spectra and magnetic properties. *J. Mol. Struct.* **2006**, *787*, 31–37. [CrossRef]

36. Hijmans, T.W.; Beyermann, W.P. Electron-spin-resonance study of the high-field phase of the spin-Peierls system tetrathiafulvalene-Au-bis-dithiolene. *Phys. Rev. Lett.* **1987**, *58*, 2351–2354. [CrossRef] [PubMed]

37. Yamamoto, Y.; Ohta, H.; Motokawa, M.; Fujita, O.; Akimitsu, J. The observation of *g*-shifts in spin-Peierls material CuGeO₃ by submillimeter wave ESR. *J. Phys. Soc. Jpn.* **1997**, *66*, 1115–1123. [CrossRef]

38. Oshio, H.; Kitazaki, K.; Mishiro, J.; Kato, N.; Maeda, Y.; Takashima, Y. New spin-crossover iron(III) complexes with large hysteresis effects and time dependence of their magnetism. *J. Chem. Soc. Dalton Trans.* **1987**, 1341–1347. [CrossRef]

39. Hayami, S.; Gu, Z.-Z.; Yoshiki, H.; Fujishima, A.; Sato, O. Iron(III) spin-crossover compounds with a wide apparent thermal hysteresis around room temperature. *J. Am. Chem. Soc.* **2001**, *123*, 11644–11650. [CrossRef] [PubMed]

40. Takahashi, K.; Mori, H.; Tajima, H.; Einaga, Y.; Sato, O. Cooperative spin transition and thermally quenched high-spin state in new polymorph of [Fe(qsal)₂]I₃. *Hyperfine Interact.* **2012**, *206*, 1–5. [CrossRef]

41. Bousseksou, A.; Place, C.; Linares, J.; Varret, F. Dynamic spin crossover in [Fe(2-BIK)₃](ClO₄)₂ and [Fe(Me₂-BIK)₃](BF₄)₂ investigated by Mossbauer spectroscopy. *J. Mag. Mag. Mater.* **1992**, *104–107*, 225–226. [CrossRef]

42. Maeda, Y.; Tsutsumi, N.; Takashima, Y. Examples of fast and slow electronic relaxation between ⁶A and ²T. *Inorg. Chem.* **1984**, *23*, 2440–2447. [CrossRef]

43. Wignall, J.W.G. Mössbauer line broadening in trivalent iron compounds. *J. Chem. Phys.* **1966**, *44*, 2462–2467. [CrossRef]

44. Summerville, R.H.; Hoffmann, R. Tetrahedral and other M₂L₆ transition metal dimers. *J. Am. Chem. Soc.* **1976**, *98*, 7240–7254. [CrossRef]

45. Chen, M.M.L.; Hoffmann, R. Sulfuranes. Theoretical aspects of bonding, substituent site preferences, and geometrical distortions. *J. Am. Chem. Soc.* **1976**, *98*, 1647–1653. [CrossRef]

46. Jørgensen, K.A.; Hoffmann, R. Binding of alkenes to the ligands in OsO₂X₂ (X = O and NR) and CpCo(NO)₂. A frontier orbital study of the formation of intermediates in the transition-metal-catalyzed synthesis of diols, amino alcohols, and diamines. *J. Am. Chem. Soc.* **1986**, *108*, 1867–1876. [CrossRef]

47. Murata, S.; Takahashi, K.; Sakurai, T.; Ohta, H.; Yamamoto, T.; Einaga, Y.; Shiota, Y.; Yoshizawa, K. The role of Coulomb interactions for spin crossover behaviors and crystal structural transformation in novel anionic Fe(III) complexes from a π-extended ONO ligand. *Crystals* **2016**, *6*, 49. [CrossRef]

48. Murata, S.; Takahashi, K.; Sakurai, T.; Ohta, H. Single-crystal-to-single-crystal transformation in hydrogen bond induced high spin pseudopolymorphs from protonated cation salts with a π extended spin crossover Fe(III) complex anion. *Polyhedron* **2017**. [CrossRef]

![inorganics logo] *inorganics*

MDPI

Article

Modification of Cooperativity and Critical Temperatures on a Hofmann-Like Template Structure by Modular Substituent

Takashi Kosone [1,*], Takeshi Kawasaki [2], Itaru Tomori [2], Jun Okabayashi [3] and Takafumi Kitazawa [2,4,*]

[1] Department of Creative Technology Engineering Course of Chemical Engineering, Anan College, 265 Aoki, Minobayashi, Anan, Tokushima 774-0017, Japan
[2] Department of Chemistry, Faculty of Science, Toho University, 2-2-1 Miyama, Funabashi, Chiba 274-8510, Japan; takeshi.kawasaki@sci.toho-u.ac.jp (T.K.); synapse_yf@yahoo.co.jp (I.T.)
[3] Research Center for Spectrochemistry, University of Tokyo, Bunkyo-ku, Tokyo 113-0033, Japan; jun@chem.s.u-tokyo.ac.jp
[4] Research Centre for Materials with Integrated Properties, Toho University, 2-2-1 Miyama, Funabashi, Chiba 274-8510, Japan
* Correspondence: kosone@anan-nct.ac.jp (T.K.); kitazawa@chem.sci.toho-u.ac.jp (T.K.); Tel.: +81-474-72-5077 (T.K.); +81-884-23-7195 (T.K.)

Received: 12 July 2017; Accepted: 7 August 2017; Published: 16 August 2017

Abstract: In a series of Hofmann-like spin crossover complexes, two new compounds, {Fe(3-F-4-Methyl-py)$_2$[Au(CN)$_2$]$_2$} (1) and {Fe(3-Methyl-py)$_2$[Au(CN)$_2$]$_2$} (2) (py = pyridine) are described. The series maintains a uniform 2-dimentional (2-D) layer structure of {Fe[Au(CN)$_2$]$_2$}. The layers are combined with another layer by strong aurophilic interactions, which results in a bilayer structure. Both coordination compounds 1 and 2 at 293 K crystallize in the centrosymmetric space groups $P2_1/c$. The asymmetric unit contains two pyridine derivative ligands, one type of Fe^{2+}, and two types of crystallographically distinct [Au(CN)$_2$]$^-$ units. Compound 1 undergoes a complete two-step spin transition. On the other hand, 2 maintains the characteristic of the high-spin state. The present compounds and other closely related bilayer compounds are compared and discussed in terms of the cooperativity and critical temperature. The bilayer structure is able to be further linked by substituent-substituent contact resulting in 3-dimentional (3-D) network cooperativity.

Keywords: coordination polymer; cooperative interaction; crystal engineering; spin crossover

1. Introduction

An essential part for designing spin crossover (SCO) materials is to control and optimize the crystal structure [1]. Especially, construction of the strong cooperative intermolecular interactions in the whole of a structure leads to steep transition behavior with a wide hysteresis loop, which is important for the practical materials [2]. The cooperativity is now being investigated by a variety of coordination polymers. Coordination polymers are one of the interesting materials for constructing supramolecular networks. However, systematic designing of the networks is hard because of the structural diversity of coordination polymers.

Since we reported the first Hofmann like SCO coordination polymer {Fe(py)$_2$[Ni(CN)$_4$]}$_n$ (py = pyridine) [3], many derived types of {Fe^{II}(L)$_{1\sim2}$[M^{II}(CN)$_4$]}$_n$ [4–9] and {Fe^{II}(L)$_{1\sim2}$[M^I(CN)$_2$]$_2$}$_n$ [10–21] (M^I = Cu, Ag, or Au, M^{II} = Ni, Pd, or Pt, L = pyridine derivatives) have been developed. These compounds show a template 2-dimentional (2-D) sheet structure because of their strongly determinate self-assembly process in which they link octahedral metal centers through the N atoms of the bidentate [Au(CN)$_2$]$^-$ unit. Therefore, this structural system can be modified only at the axial ligands, L, for designing the cooperative networks. Here we report and discuss new Hofmann-like 2-D compounds of the general formula {Fe(X-py)$_2$[Au(CN)$_2$]$_2$} (X = 3-F-4-Methyl (**1**) or 3-Methyl (**2**) as shown in Scheme 1).

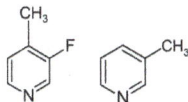

Scheme 1. Molecular structure of the ligands of 3-Furuoro-4-methyl-py and 3-Methyl-py.

2. Results

2.1. X-ray Structural Analysis

2.1.1. Structure of Compound **1** (T = 293 K)

Compound **1** at 293 K crystallizes in the monoclinic centrosymmetric space group $P2_1/c$. The asymmetric unit of the complex consists of the hetero-metal Fe^{II}Au^I unit (Figure 1a). The Fe^{II} ion is octahedrally coordinated by six N atoms. The Fe–N$_{py}$ bond lengths (Fe(1)–N(1) = 2.219(6) Å, Fe(1)–N(2) = 2.220(6) Å) are longer than the Fe–N$_{CN}$ bond lengths (Fe(1)–N(3) = 2.142(7) Å, Fe(1)–N(4) = 2.153(6) Å, Fe(1)–N(5) = 2.160(7) Å, Fe(1)–N(6) = 2.148(6) Å). The average lengths of Fe–N$_{py}$ = 2.220 Å and Fe–N$_{CN}$ = 2.151 Å (total average length of Fe–N = 2.185 Å are estimated). All Au^I atoms have linear coordination geometries with the CN substituents binding to the Fe^{II} ions. While the F(1) in the 3-F-4-Me-py ligand is disordered, the F(2) F(3) in the other 3-F-4-Me-py ligand are not disordered. Thus, the two 3-F-4-Me-py ligands in [Fe^{II}(3-F-4-Me-py)$_2$][Au^I(CN)$_2$] are not equivalent and coexist in *transoid* and *cisoid* conformations for Fe(1). The bidentate [Au^I(CN)$_2$] linear units give rise to an infinite corrugated 2-D mesh-layer formed by the assembly of –Au–N–C–Fe–C–N–Au– infinite chains (Figure 1b). In addition, the layers interact by pairs to define bilayers which stem from strong aurophilic interactions (Figure 1c). The average Au···Au distance in the bilayers is 3.142 Å, less than the sum of the van der Waals radii of Au (3.60 Å). The nearest aromatic rings form almost face-to-face superposition (dihedral angles = 8.55°). It constructs weak π-stacking interactions. The closet C$_{py}$···C$_{py}$ distances between py rings [C(2)···C(8) = 3.548(13) Å] are smaller than the sum of the van der Waals radius (ca. 3.70 Å).

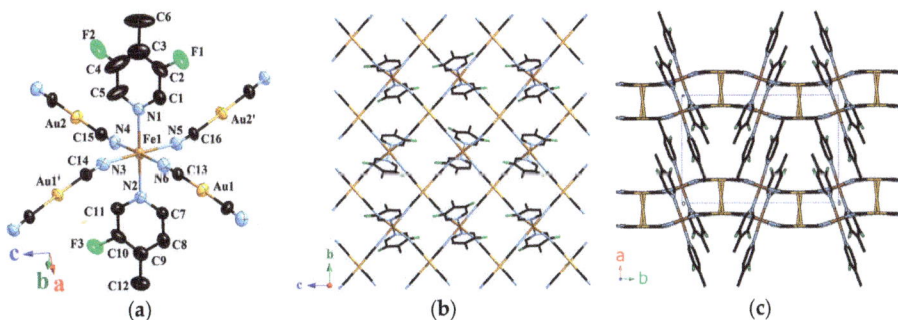

Figure 1. (a) Coordination structure of compound **1** containing its asymmetric unit at 293 K; (b) View of the 2-D layer structure of **1**; (c) Stacking of four consecutive layers of **1** at 293 K. In these pictures, hydrogen atoms are omitted for clarity.

2.1.2. Structure of Compound **1** (T = 180 K)

The crystal structure of this state is similar to that for 293 and 90 K. However, as compared to the fully high spin (HS) state, the single crystals lose their qualities due to the narrow temperature region of the half transition phase. As a result, the crystal data at 180 K is not good enough ($R = 0.1058$ [$I > 2\text{sigma}\,(I)$]) for determining the correct structure (see Table S1). So, here we discuss in detail the structure at 293 K (fully high spin (HS) state) and 90 K (fully low spin (LS) state).

2.1.3. Structure of Compound 1 (T = 90 K)

The crystal structure of **1** at 90 K is almost identical to that observed at 293 K. The Fe–N_{py} bond lengths (Fe(1)–N(1) = 2.013(4) Å, Fe(1)–N(2) = 2.002(4) Å) are longer than the Fe–N_{CN} bond lengths [Fe(1)–N(3) = 1.942(4) Å, Fe(1)–N(4) = 1.944(4) Å, Fe(1)–N(5) = 1.934(4) Å, Fe(1)–N(6) = 1.936(4) Å]. Total average length is estimated as Fe–N = 1.962 Å. The change of the average length upon spin transition is 0.223 Å, which is almost identical with the expected values for the Fe^{II} 100% LS state. The rings form more parallel superposition than that of the HS state (dihedral angles = 2.57°).

2.1.4. Structure of Compound 2 (T = 293 K)

The crystal structure of **2** at 293 K is almost similar to that of **1** at 293 K which also crystallizes in the monoclinic centrosymmetric space group $P2_1/c$. The 2-D bilayer structure is also almost same shape (Au⋯Au distance is 3.174 Å) (Figure 2). However, local differences are observed. The rings array is slightly more unparallel than that of **1** (dihedral angles = 11.97°). However, weak π-stacking interactions are also observed (C(1)⋯C(10) = 3.542(10) Å). There is a *trans* orientation of 3-Me substituents with respect to the N(1)–Fe–N(2) axis. The Fe–N_{py} bond lengths (Fe(1)–N(1) = 2.247(4) Å, Fe(1)–N(2) = 2.244(4) Å) are longer than the Fe–N_{CN} bond lengths [Fe(1)–N(3) = 2.164(4) Å, Fe(1)–N(4) = 2.162(4) Å, Fe(1)–N(5) = 2.159(4) Å, Fe(1)–N(6) = 2.158(4) Å]. The average lengths of Fe–N_{py} = 2.246 Å and Fe–N_{CN} = 2.162 Å (total average length of Fe–N = 2.204 Å are also estimated).

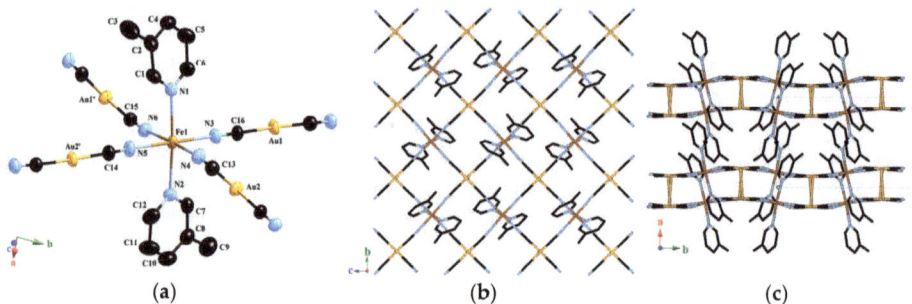

Figure 2. (**a**) Coordination structure of compound **2** containing its asymmetric unit at 293 K; (**b**) View of the 2-D layer structure of **2**; (**c**) Stacking of four consecutive layers of **2** at 293 K. In these pictures, hydrogen atoms are omitted for clarity.

2.2. Magnetic Properties

2.2.1. Thermal Dependence Magnetic Behavior of Compound 1

Figure 3a shows the thermal dependence of $\chi_M T$ for **1** with χ_M being the molar magnetic susceptibility and T the temperature. At room temperature, $\chi_M T$ is 3.65 $cm^3 \cdot K \cdot mol^{-1}$. This value is slightly higher than usual for paramagnetic Fe^{II} compounds, possibly due to oxidation of the complex. Upon cooling, $\chi_M T$ remains almost constant down to 195 K; below this temperature, $\chi_M T$ undergoes a sharp decrease. The complex displays two-step spin transition with a characteristic plateau centered at around 50% conversion, and the warming mode reveals the occurrence of a hysteresis loop (second step). The SCO for this complex causes a reversible change of color from blue (HS) to purple (LS). The critical temperature (T_c) in the first step (T_c^1) is 188.5 K and the cooling ($T_c^{2\,\text{down}}$) and warming

($T_c{}^{2\,up}$) modes in the second step are 164.5 and 171.5 K, respectively, giving an approximately 7.0 K width hysteresis loop.

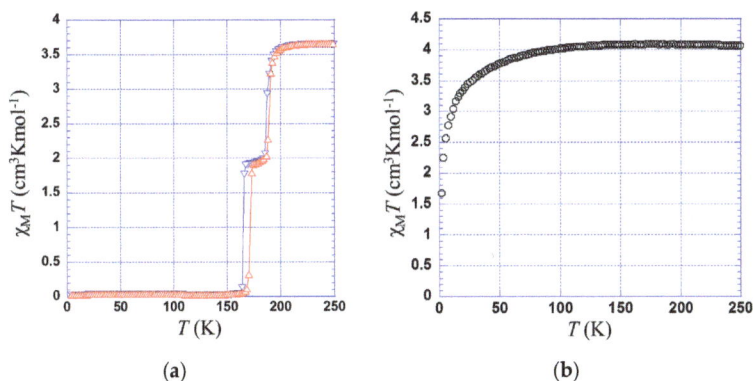

Figure 3. (a) Thermal dependence of $\chi_M T$ for compound **1**. The sample was cooled from 300 to 2 K (blue) and then warmed from 2 to 300 K (red). (b) Thermal dependence of $\chi_M T$ for compound **2**.

2.2.2. Thermal Dependence Magnetic Behavior of Compound 2

$\chi_M T$ versus T plots for **2** are shown in Figure 3b. At room temperature, the magnetic behavior of the complex **2** is characteristic of Fe(II) compounds in the HS state, $\chi_M T = 4.05$ cm^3·K·mol^{-1}. The value is slightly higher than that of a pure spin only system, whereas the value is similar to the values of other Hofmann-like SCO Fe^{II} compounds. The $\chi_M T$ value is nearly constant in the range of 80–300 K. However, it decreases steeply as the temperature is lowered to less than 30 K. The decrease in the value of $\chi_M T$ at lower temperatures is due to a typical behavior of zero-field splitting (ZFS) effects of the metallic Fe^{II} centers in the residual HS ($S = 2$) species [22].

3. Discussion

The bilayer structures of **1** and **2** are almost identical with the former reported compounds of {Fe^{II}(L)$_2$[Au^I(CN)$_2$]$_2$} (L = pyridine derivatives). In the previous papers, the synthesis and characterization of the closely related bilayer structures of {Fe(X-py)$_2$[Au(CN)$_2$]$_2$} (X = 3-F (**3**) [12,13,15], 4-Methyl (**4**) [17], 3-Br (**5**) [12], and 3-Br-4-Methyl (**6**) [16]) have been reported. These analogous compounds are also discussed in this paper.

Cell volumes, SCO behavior type, and T_c for **1–6** are summarized in Table 1. It is noted that the cell volume clearly shows an expansion with the increase of the substituent bulk. The compounds from smallest to largest are as follows: 3-F-py (1915.7 Å^3) < 3-Methyl-py (1976.3 Å^3) < 3-Br-py (1990.9 Å^3) < 4-Methyl-py (2112.5 Å^3) < 3-F-4-Methyl-py (2155.2 Å^3) < 3-Br-4-Methyl-py (2209.7 Å^3). In spite of the lattice expansion, these compounds completely maintain a bilayer structure.

In terms of the magnetic properties of **1** and **2**, **1** displays a steep two-step spin conversion with a hysteresis loop (second step), while **2** is fully HS in the whole range of temperatures. This result suggests that the T_c of **2** might be so low that no spin transition is observed in an ambient pressure. There is also no spin conversion observed for X = 3-Br. In this series, T_c increases in the following way: 3-Me-py, 3-Br-py (no transition) << 3-Br-4-Me-py (109.1 K) < 3-F-py (first step 147.9 K) < 3-F-4-Me-py (first step 188.5 K) < 4-Me-py (first step 210 K). The order of T_c must result from the difference in the ligand field strength Δ_o. It seems that the 4-Me substituent is the most effective for decreasing Δ_o (see Scheme S1). This trend of T_c is in the same order for [Fe(L)$_2$Ag(CN)$_2$] (L = 3-F-py, 3-Br-py, 3-Me-py, and 4-Me-py) and is as follows: 3-Me-py [18], 3-Br-py (no transition) [11] << 3-F-py (first step

146.3 K) [11] < 4-Mepy (first step 189 K) [18]. These orders of T_c are opposite to the expected electronic effect according to the Hammet constants [23].

Table 1. Spin crossover (SCO) behavior types, cell volumes, and critical temperatures of $\{Fe^{II}(X\text{-py})_2[Au^I(CN)_2]_2\}$ at 293 K or room temperature.

Substituents	Cell Volume (Å^3)	SCO Behavior Type	Critical Temperature T_c (K)	Ref.
3-Methyl (2)	1976.30 Å^3	None	None	–
4-Methyl (HS) (4)	2112.5 Å^3	Steep (3-step)	210 K (first step)	[17]
3-Fuluoro (HS) (3)	1915.7 Å^3	Steep (2-step)	147.9 K (first step)	[12,13,15]
3-Fuluoro-4-Methyl (HS) (1)	2155.2 Å^3	Steep (2-step)	188.5 K (first step)	–
3-Bromo (5)	1990.9 Å^3	None	None	[12]
3-Bromo-4-Methyl (HS) (6)	2204.5 Å^3	Steep (1-step, 50% transition)	109.1 K	[16]

In a previous related study, Real and co-workers discussed the effect of substituted pyridines for the analogous series of [Fe(X-py)$_2$Ag(CN)$_2$] (X = halogen atoms) [11]. The research says that T_c may be dominantly influenced by crystal packing factor and polymeric structure. We interpret the former research as meaning that the Fe–N bond length is strongly influenced by the interlayer spaces. The steric effect from substituent bulk makes the space tight. In fact, 4-position is vertical to the layer. Thus, a 4-Methyl substituent gives rise to higher chemical pressure in the Fe–N bond. Consequently, the trend of T_c can be explained by the substituent bulk. Electronic effects righteously coexist and compete with steric effects.

Both **1** and **3** have significantly similar transition behavior. However, each behavior has a different hysteresis width. In compound **3**, intermolecular distance between F⋯F is significantly shorter as compared to that of **1** (Scheme 2). The crystal structure of **3** shows the lowest cell volume which makes the interlayer spaces narrow. It results in shorter F⋯F contact (closest distance of F⋯F (HS) = 3.104 Å) between bilayers. Specifically, the half spin transition state (HS–LS) has extremely short F⋯F contact (2.955 Å). Therefore, this complex must have higher cooperative networks which stabilize the HS–LS state. On the other hand, the F⋯F contact of **1** is much longer than that of **3** (closest distance of F⋯F (HS) = 3.717 Å). This longer distance of **1** is due to the 4-Me substituent which serves as bulk for the expansion along the *a* axis between the layers. Thus it generates weakened cooperativity. In fact, the thermal trapping of the HS–LS state for **3** is observed (see Figure S1), while this process cannot be seen for **1** at the same cooling/warming rate. This result is also strongly supported by the difference in the cooperativity of **1** and **3**. SCO behavior types, hysteresis width, closest approach F⋯F distance, and presence or absence of thermal trapping for **1** and **3** are summarized in Table 2.

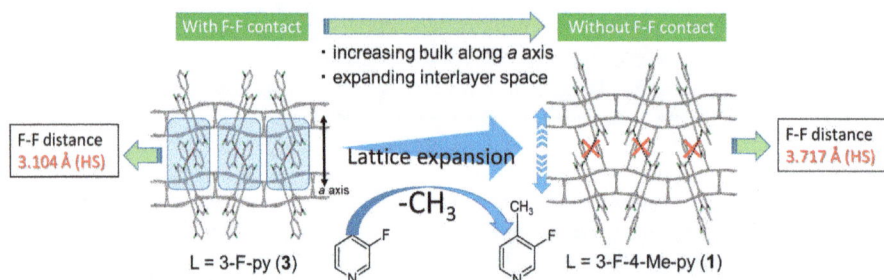

Scheme 2. Representation of the lattice expansion with the addition of the 4-position substituent. Blue highlighted box shows the F⋯F interactions (black and red dotted line).

Table 2. Comparative hysteresis width, closest approach F···F distances, and presence or absence of thermal trapping for compounds **1** and **3**. SI = supplemental information.

$Fe(X-py)_2[Au(CN)_2]_2$	Hysteresis Width (K)	Closest Approach F···F Distances [Å]	Thermal Trapping After Cooling at 2 K/min
X = 3-F (**3**)	ca. 40 K	F(1)···F(1): 3.104 Å (HS) F(1)···F(1): 2.955 Å (HS–LS) F(3)···F(3): 3.264 Å (LS)	Observed (see SI)
X = 3-F-4-Me (**1**)	ca. 7 K	F(3)···F(3): 3.717 Å (HS) F(2)···F(3): 3.305 Å (LS)	None

4. Materials and Methods

4.1. Materials

All the chemicals were purchased from commercial sources and used without any further purification.

4.2. Preparation of Compounds **1** and **2**

To 10 mL of water, $FeSO_4·(NH_4)_2SO_4·6H_2O$ (0.10 mmol, 39.2 mg) and $K[Au(CN)_2]$ (0.1 mmol, 28.8 mg) were dissolved. The vapor diffusion method using 3-F-4-Me-py or 3-Me-py as a source of ligand molecules provided colorless single crystals suitable for X-ray diffraction of **1–2** over a period of two days. Elem. Anal. Calcd. for $C_{16}H_{12}N_6F_2Au_2Fe$ (**1**): C, 25.76; H, 1.56; N, 10.83. Found: C, 25.98; H, 1.58; N, 10.90. Calcd. for $C_{16}H_{14}N_6Au_2Fe$ (**2**): C, 25.97; H, 1.91; N, 11.36. Found: C, 25.78; H, 1.92; N, 11.35.

4.3. Magnetic Measurements

Measurements of the temperature dependence of the magnetic susceptibility of the complexes **1** and **2** of the powdered samples in the temperature range of 2–300 K with a cooling and heating rate of 2 K·min^{-1} in a 1 kOe field were measured on a MPMS-XL Quantum Design SQUID magnetometer in the Cryogenic Research Center, the University of Tokyo. The diamagnetism of the samples and sample holders were taken into account.

4.4. X-ray Crystallography

Data collection was performed on a BRUKER APEX SMART CCD area-detector diffractometer at 293, 180 K, and 90 K for **1** and **2** with monochromated Mo–Kα radiation (λ = 0.71073 Å) (Bruker, Billerica, MA, USA). A selected single crystal was carefully mounted on a thin glass capillaly and immediately placed under a liquid N_2 cooled N_2 stream in each case. Crystal structures of the complexes **1** and **2** were determined using a BRUKER APEX SMART CCD area-detector diffractometer with monochrometed Mo Kα radiation (λ = 0.71073 Å). The diffraction data were treated using SMART and SAINT, and absorption correction was performed using SADABS [24]. The structures were solved by using direct methods with *SHELXTL* [25]. All non-hydrogen atoms were refined anisotropically, and the hydrogen atoms were generated geometrically. Pertinent crystallographic parameters and selected metric parameters for **1** and **2** are displayed in Tables S1 and S2. Crystallographic data have been deposited with Cambridge Crystallographic Data Centre: Deposition numbers CCDC-1559723 for compound **1** (293 K), CCDC-1559724 for **1** (90 K), and CCDC-1559832 for **2** (293 K). These data can be obtained free of charge via http://www.ccdc.cam.ac.uk/conts/retrieving.html.

5. Conclusions

In this paper, the 2-D bilayer structures of Hofmann-like coordination polymers {$Fe^{II}(X-py)_2[Au^I(CN)_2]_2$} were synthesized. It is noted that the substituents of pyridine derivatives have two steric effects in the whole of the crystal structure. One is the blocking interlayer spaces resulting in modification of the cooperativity. The other is the chemical pressure to the Fe^{II} centers resulting in modification of the critical temperature. The two effects are opposite forces to each other, like action-reaction forces with a wall.

In general, the 3-D coordination bonding types, such as $[Fe^{II}(pyrazine)[M^{II}(CN)_4]]$ (M = Ni, Pd, and Pt) display stable discontinuous spin transitions with strong cooperativity. On the other hand, most of the 2-D coordination types had weaker cooperativity than that of the 3-D type.

However, the present 2-D system shows a strong cooperativity that is as strong as the 3-D type due to the substituent-substituent interaction.

The 2-D coordination structure can be used in designs due to the unique properties, such as gate-opening behavior [26]. Therefore, the 2-D structure would create novel and unique SCO material with strong cooperativity for practical applications.

Supplementary Materials: The following are available online at www.mdpi.com/2304-6740/5/3/55/s1, Scheme S1: Critical temperature (T_c) changes depending on the substituent of {Fe(X-py)$_2$[Au(CN)$_2$]$_2$} (X = 3-F-4-Methyl (**1**), 3-Methyl (**2**), 3-F (**3**), 4-Methyl (**4**), 3-Br (**5**) and 3-Br-4-Methyl (**6**)), Figure S1: Thermal dependence of spin transition curve for compound **3**. The sample was cooled from 300 to 2 K at a rate of 2 K·min^{-1} and then warmed again to 300 K at a rate of 2 K·min^{-1}. Table S1: Crystal data and structure refinement for compounds **1** and **2**, Table S2: Selected bond lengths for compounds **1** and **2**. For CIF, check CIF files.

Acknowledgments: A part of this work was supported by a Ministry of Education, Culture, Sports, Science, and Technology, Japan (MEXT)-Supported program for the Strategic Research Foundation at Private Universities 2012–2016. This work was also supported by JSPS KAKENHI Grant Number 15K05485.

Author Contributions: Takashi Kosone, Itaru Tomori, and Takeshi Kawasaki carried out the synthesis, XRD measurements with the structural analysis, and Jun Okabayashi carried out the magnetic measurements. Takashi Kosone and Takafumi Kitazawa interpreted and discussed the result.

Conflicts of Interest: The authors declare no conflict of interest.

References

1. Real, J.A.; Andrés, E.; Munoz, M.C.; Julve, M.; Granier, T.; Bousseksou, A.; Varret, F. Spin crossover in a catenane supramolecular system. *Science* **1995**, *268*, 265–268. [CrossRef] [PubMed]

2. Gütlich, P.; Garcia, Y.; Goodwin, H.A. Spin crossover phenomena in Fe(II) complexes. *Chem. Soc. Rev.* **2000**, *29*, 419–427. [CrossRef]

3. Kitazawa, T.; Gomi, Y.; Takahashi, M.; Takeda, M.; Enomoto, M.; Miyazaki, A.; Enoki, T. Spin-crossover behaviour of the coordination polymer $Fe^{II}(C_5H_5N)_2Ni^{II}(CN)_4$. *J. Mater. Chem.* **1996**, *6*, 119–121. [CrossRef]

4. Niel, V.; Martinez-Agudo, J.M.; Muñoz, M.C.; Gaspar, A.B.; Real, J.A. Cooperative Spin Crossover Behavior in Cyanide-Bridged Fe(II)–M(II) Bimetallic 3D Hofmann-like Networks (M = Ni, Pd, and Pt). *Inorg. Chem.* **2001**, *40*, 3838–3839. [CrossRef] [PubMed]

5. Agusti, G.; Cobo, S.; Gaspar, A.B.; Molnár, G.; Moussa, N.O.; Szilágyi, P.Á.; Pálfi, V.; Vieu, C.; Carmen Muñoz, M.; Real, J.A.; et al. Thermal and light-induced spin crossover phenomena in new 3D Hofmann-like microporous metalorganic frameworks produced as bulk materials and nanopatterned thin films. *Chem. Mater.* **2008**, *20*, 6721–6732. [CrossRef]

6. Martínez, V.; Gaspar, A.B.; Muñoz, M.C.; Bukin, G.V.; Levchenko, G.; Real, J.A. Synthesis and Characterisation of a New Series of Bistable Iron(II) Spin-Crossover 2D Metal–Organic Frameworks. *Chem. Eur. J.* **2009**, *15*, 10960–10971. [CrossRef] [PubMed]

7. Bartual-Murgui, C.; Ortega-Villar, N.A.; Shepherd, H.J.; Muñoz, M.C.; Salmon, L.; Molnár, G.; Bousseksou, A.; Real, J.A. Enhanced porosity in a new 3D Hofmann-like network exhibiting humidity sensitive cooperative spin transitions at room temperature. *J. Mater. Chem.* **2011**, *21*, 7217. [CrossRef]

8. Ohtani, R.; Arai, M.; Hori, A.; Takata, M.; Kitao, S.; Seto, M.; Kitagawa, S.; Ohba, M. Modulation of Spin-Crossover Behavior in an Elongated and Flexible Hofmann-Type Porous Coordination Polymer. *J. Inorg. Organomet. Polym. Mater.* **2013**, *23*, 104–110. [CrossRef]

9. Sciortino, N.F.; Zenere, K.A.; Corrigan, M.E.; Halder, G.J.; Chastanet, G.; Létard, J.-F.; Kepert, C.J.; Neville, S.M. Four-step iron(II) spin state cascade driven by antagonistic solid state interactions. *Chem. Sci.* **2017**, *8*, 701–707. [CrossRef] [PubMed]

10. Galet, A.; Muñoz, M.C.; Martinez, V.; Real, J.A. Supramolecular isomerism in spin crossover networks with aurophilic interactions. *Chem. Commun.* **2004**, 2268–2269. [CrossRef] [PubMed]

11. Muñoz, M.C.; Gaspar, A.B.; Galet, A.; Real, J.A. Spin-Crossover Behavior in Cyanide-Bridged Iron(II)–Silver(I) Bimetallic 2D Hofmann-like Metal–Organic Frameworks. *Inorg. Chem.* **2007**, *46*, 8182–8192. [CrossRef] [PubMed]

12. Agustí, G.; Muñoz, M.C.; Gaspar, A.B.; Real, J.A. Spin-Crossover Behavior in Cyanide-bridged Iron(II)–Gold(I) Bimetallic 2D Hofmann-like Metal–Organic Frameworks. *Inorg. Chem.* **2008**, *47*, 2552–2561. [CrossRef] [PubMed]

13. Kosone, T.; Kachi-Terajima, C.; Kanadani, C.; Saito, T.; Kitazawa, T. A two-step and hysteretic spin-crossover transition in new cyano-bridged hetero-metal $Fe^{II}Au^{I}$ 2-dimensional assemblage. *Chem. Lett.* **2008**, *37*, 422–423. [CrossRef]

14. Kosone, T.; Kachi-Terajima, C.; Kanadani, C.; Saito, T.; Kitazawa, T. Isotope Effect on Spin-crossover Transition in a New Two-dimensional Coordination Polymer $[Fe^{II}(C_5H_5N)_2][Au^{I}(CN)_2]_2$, $[Fe^{II}(C_5D_5N)_2][Au^{I}(CN)_2]_2$, and $[Fe^{II}(C_5H_5{}^{15}N)_2][Au^{I}(CN)_2]_2$. *Chem. Lett.* **2008**, *37*, 754–755. [CrossRef]

15. Kosone, T.; Kanadani, C.; Saito, T.; Kitazawa, T. Synthesis, crystal structures, magnetic properties and fluorescent emissions of two-dimensional bimetallic coordination frameworks Fe^{II}(3-fluoropyridine)$_2$ $[Au^{I}(CN)_2]_2$ and Mn^{II}(3-fluoropyridine)$_2[Au^{I}(CN)_2]_2$. *Polyhedron* **2009**, *28*, 1930–1934. [CrossRef]

16. Kosone, T.; Kanadani, C.; Saito, T.; Kitazawa, T. Spin crossover behavior in two-dimensional bimetallic coordination polymer Fe^{II}(3-bromo-4-picoline)$_2[Au^{I}(CN)_2]_2$: Synthesis, crystal structures, and magnetic properties. *Polyhedron* **2009**, *28*, 1991–1995. [CrossRef]

17. Kosone, T.; Tomori, I.; Kanadani, C.; Saito, T.; Mochida, T.; Kitazawa, T. Unprecedented three-step spin-crossover transition in new 2-dimensional coordination polymer {Fe^{II}(4-methylpyridine)$_2[Au^{I}(CN)_2]_2$}. *Dalton Trans.* **2010**, *39*, 1719–1721. [CrossRef] [PubMed]

18. Rodríguez-Velamazán, J.A.; Carbonera, C.; Castro, M.; Palacios, E.; Kitazawa, T.; Létard, J.-F.; Burriel, R. Two-Step Thermal Spin Transition and LIESST Relaxation of the Polymeric Spin-Crossover Compounds Fe(X-py)$_2$[Ag(CN)$_2$]$_2$ (X=H, 3-methyl, 4-methyl, 3,4-dimethyl, 3-Cl). *Chem. Eur. J.* **2010**, *16*, 8785–8796. [CrossRef] [PubMed]

19. Kosone, T.; Kitazawa, T. Guest-dependent spin transition with long range intermediate state for 2-dimensional Hofmann-like coordination polymer. *Inorg. Chim. Acta* **2016**, *439*, 159–163. [CrossRef]

20. Chiruta, D.; Linares, J.; Garcia, Y.; Dimian, M.; Dahoo, P.R. Analysis of multi-step transitions in spin crossover nanochains. *Phys. B Condens. Matter* **2014**, *434*, 134–138. [CrossRef]

21. Okabayashi, J.; Ueno, S.; Kawasaki, T.; Kitazawa, T. Ligand 4-X pyridine (X = Cl, Br, I) dependence in Hofmann-type spin crossover complexes: Fe(4-Xpyridine)$_2$[Au(CN)$_2$]$_2$. *Inorg. Chim. Acta* **2016**, *445*, 17–21. [CrossRef]

22. Krzystek, J.; Ozarowski, A.; Telser, J. Multi-frequency, high-field EPR as a powerful tool to accurately determine zero-field splitting in high-spin transition metal coordination complexes. *Coord. Chem. Rev.* **2006**, *250*, 2308–2324. [CrossRef]

23. Nakano, K.; Suemura, N.; Yoneda, K.; Kawata, S.; Kaizaki, S. Substituent effect of the coordinated pyridine in a series of pyrazolato bridged dinuclear diiron(II) complexes on the spin-crossover behavior. *Dalton Trans.* **2005**, 740. [CrossRef] [PubMed]

24. Sheldrick, G.M. *SADABS, Program for Empirical Absorption Correction for Area Detector Data*; University of Göttingen: Göttingen, Germany, 1996.

25. Sheldrick, G.M. *SHELXL, Program for the Solution of Crystal Structures*; University of Göttingen: Göttingen, Germany, 1997.

26. Sakaida, S.; Otsubo, K.; Sakata, O.; Song, C.; Fujiwara, A.; Takata, M.; Kitagawa, H. Crystalline coordination framework endowed with dynamic gate-opening behaviour by being downsized to a thin film. *Nat. Chem.* **2016**, *8*, 377–383. [CrossRef] [PubMed]

inorganics

MDPI

Review

Control of Spin-Crossover Phenomena in One-Dimensional Triazole-Coordinated Iron(II) Complexes by Means of Functional Counter Ions

Akira Sugahara [1], Hajime Kamebuchi [2], Atsushi Okazawa [3], Masaya Enomoto [2] and Norimichi Kojima [3,4,*]

[1] Department of Chemical System Engineering, Faculty of Engineering, The University of Tokyo, Hongo 7-3-1, Bunkyo-ku, Tokyo 113-8656, Japan; a_sugahara@battery.t.u-tokyo.ac.jp

[2] Department of Chemistry, Faculty of Science Division I, Tokyo University of Science, Kagurazaka 1-3, Shinjuku-ku, Tokyo 162-8601, Japan; hkamebuchi@rs.tus.ac.jp (H.K.); menomoto@rs.kagu.tus.ac.jp (M.E.)

[3] Graduate School of Arts and Sciences, The University of Tokyo, Komaba 3-8-1, Meguro-ku, Tokyo 153-8902, Japan; cokazawa@mail.ecc.u-tokyo.ac.jp

[4] Toyota Physical and Chemical Research Institute, Yokomichi 41-1, Nagakute-shi, Aichi 480-1192, Japan

* Correspondence: kojima@toyotariken.jp or cnori@mail.ecc.u-tokyo.ac.jp; Tel.: +81-561-57-9593; Fax: +81-561-63-6302

Received: 23 June 2017; Accepted: 16 July 2017; Published: 19 August 2017

Abstract: The spin-crossover (SCO) phenomenon between a high-spin and a low-spin state has attracted much attention in the field of materials science. Among the various kinds of SCO complexes, the triazole-bridged iron(II) polymeric chain system, $[Fe(II)(R\text{-}trz)_3]X_2 \cdot xH_2O$ (where trz is triazole and X is the anion), exhibiting the SCO phenomenon with thermal hysteresis around room temperature, has been extensively studied from the viewpoint of molecular memory and molecular devices. In connection with this system, we have controlled the SCO phenomenon according to the characteristic properties of counter ions. In the case of X being $C_nH_{2n+1}SO_3^-$, the spin transition temperature ($T_{1/2}$) increases with increasing the length (n) of the alkyl chain of the counter ion and saturates above $n = 5$, which is attributed to the increase in the intermolecular interaction of the alkyl chains of $C_nH_{2n+1}SO_3^-$, called the fastener effect. The hysteresis width of $T_{1/2}$ decreases with increasing n, showing the even-odd, also known as parity, effect. In the cases where X is toluenesulfonate (tos: $CH_3C_6H_4SO_3^-$) and aminobenzenesulfonate (abs: $NH_2C_6H_4SO_3^-$), $T_{1/2}$ and its hysteresis width vary drastically with the structural isomerism (*ortho*-, *metha*-, and *para*-substitution) of counter ions, which implies the possibility of photoinduced spin transition by means of the photoisomerization of counter ions. From this strategy, we have synthesized $[Fe(II)(NH_2\text{-}trz)_3](SP150)_2 \cdot 2H_2O$ (SP150 = N-alkylsulfonated spiropyran) and investigated the SCO phenomenon. Moreover, we have developed $[Fe(II)(R\text{-}trz)_3]@$Nafion films exhibiting spin transition around room temperature, where the Nafion membrane behaves as a counter anion as well as a transparent substrate, and investigated the photogenerated high-spin state below 35 K. The lifetime of the photogenerated high-spin state strongly depends on the intensity of irradiated light.

Keywords: multifunctionality; spin-crossover; isomerization effect; fastener effect; Nafion; transition metal complex film; LIESST

---

1. Introduction

An octahedrally coordinated transition metal ion, having an electron configuration of $3d^n$ ($n = 4\text{–}7$), exhibits the possibility to change the ground spin state between the low-spin (LS) and high-spin (HS) states. Switching between the LS and HS states is known as the spin-crossover (SCO) phenomenon, which can be controlled by external stimuli such as heat, applied pressure, and light irradiation.

The first SCO phenomenon was observed in the tris(dithiocarbamato)iron(III) family, [Fe(III)(S$_2$CNR$_2$)$_3$] (R = C$_n$H$_{2n+1}$ etc.), reported by Cambi et al. in the 1930s [1]. Since then, a huge number of SCO complexes containing 3d^n (n = 4–7) metal ions have been developed [2].

From the viewpoint of molecular devices and optical information storage, the SCO phenomenon has gained renewed importance since the discovery in 1984 of photoinduced LS–HS transition for [Fe(II)(ptz)$_6$](BF$_4$)$_2$ (ptz is 1-propyltetrazole) [3]. This photoinduced SCO phenomenon has been named LIESST (Light-Induced Excited Spin State Trapping). The LIESST mechanism is a light-induced conversion from the ground LS state (^{1}A$_{1g}$) to the lowest HS state (^{5}T$_{2g}$), which was proposed by Decurtins et al., involving the first intersystem crossing step with ΔS = 1 from the initially excited ^{1}T$_{1g}$ state to the lower lying ^{3}T$_{1g}$ state. In the second intersystem crossing step with ΔS = 1, the system has a possibility to drop into the lowest HS state of ^{5}T$_{2g}$, and it remains trapped at sufficiently low temperatures because of the potential barrier between the ^{5}T$_{2g}$ and ^{1}A$_{1g}$ states [4]. Being stimulated by the discovery of LIESST for [Fe(II)(ptz)$_6$](BF$_4$)$_2$, various kinds of SCO complexes showing LIESST have been studied for mainly iron(II) and iron(III) complexes [5].

For the application of the SCO phenomenon to molecular devices, an SCO complex, [Fe(II)(H-trz)$_{2.85}$(NH$_2$-trz)$_{0.15}$](ClO$_4$)$_2$·H$_2$O (R-trz is 4-R-1,2,4-triazole), exhibiting a thermally induced spin transition with large thermal hysteresis around room temperature, was synthesized in 1993 [6]. Over the past two decades, one-dimensional (1D) iron(II) coordination polymers bridged by 4-substituted-1,2,4-triazole (R-trz) (Figure 1) have provided many kinds of extended systems [7]. The general formula of these complexes is described as [Fe(II)(R-trz)$_3$]Anion·xH$_2$O, where Anion is a mono- or divalent anion. The materials are applicable to molecular electronics or sensors [8,9] because their SCO phenomena are abrupt and tunable, and sometimes associated with wide thermal hysteresis around room temperature. In this system, these SCO characteristics vary depending on three factors: (i) substituents (R) on triazole ligands; (ii) non-coordinated anions located in the voids in 1D chains; and (iii) interstitial crystal water molecules which form hydrogen bonds with the ligands. An excellent example of controlling the transition temperature ($T_{1/2}$) was demonstrated by Kahn et al. based on a concept called "molecular alloy" [6]. The alloy compounds [Fe(II)(H-trz)$_{3-3x}$(NH$_2$-trz)$_{3x}$](ClO$_4$)$_2$·yH$_2$O show a linear variation of $T_{1/2}\uparrow$ and $T_{1/2}\downarrow$ as a function of x, where $T_{1/2}\uparrow$ and $T_{1/2}\downarrow$ stand for the temperatures at the HS and LS ratio of 50:50 on the heating and cooling processes, respectively. Another kind of alloy involves mixed counter anions, the formula of which is [Fe(II)(NH$_2$-trz)$_3$](NO$_3$)$_{2x}$(BF$_4$)$_{2-2x}$·yH$_2$O [10]. Thus, the spin transition behavior is finely tunable to serve molecular devices working at room temperature due to optical and magnetic bistabilities. Furthermore, the LIESST effect was also reported for [Fe(II)(R-trz)$_3$] complexes as seen in typical SCO complexes [11].

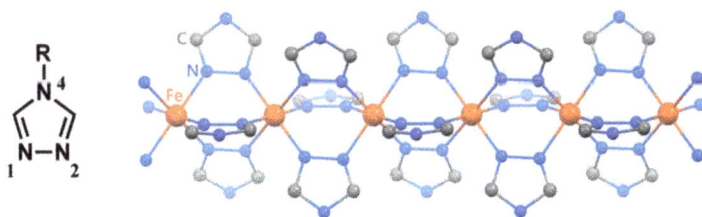

Figure 1. Molecular structure of 4-R-1,2,4-triazole (**left**) and one-dimensional (1D) coordination polymer chains of [Fe(II)(R-trz)$_3$] (**right**). Hydrogen atoms and substituents (R) are omitted for clarity.

In order to create molecular devices, a preparation of single crystals or a transparent film is more suitable for memory or display, respectively, than powder conditions. However, single crystals for the [Fe(II)(R-trz)$_3$]X$_2$·xH$_2$O family have never been created, except for the single crystal X-ray diffraction study on [Fe(II)(NH$_2$-trz)$_3$](NO$_3$)$_2$ in 2011 [12]. Furthermore, the [Fe(II)(R-trz)$_3$]X$_2$·xH$_2$O family has low solubility. By increasing the solubility of the organic solvent, the [Fe(II)(R-trz)$_3$] complex is able to be applied to solution processes, such as the spin coating method, the spray method, and so on. Therefore, various approaches to manipulating the processes of using triazole complexes were reported. One of the approaches is the modification of triazole ligands by substituting the long alkyl groups. Substituting the triazole complexes with long alkyl groups (octadecyl [13], 3-dodecyloxy propyl [14]) was reported. Another modification of the triazole ligand is the construction of the dendrimer by substituting it with the dendrons of a benzyl ether-type unit [14–17]. Another approach is embedding the complex in a polymer matrix, such as polystyrene [18], polymethyl methacrylate [19,20], or Nafion [21–23].

In connection with the [Fe(II)(R-trz)$_3$]X$_2$ system, we controlled the SCO phenomena by using the characteristic properties of counter anions. In the case when X is toluenesulfonate (tos: CH$_3$C$_6$H$_4$SO$_3{}^-$) and aminobenzenesulfonate (abs: NH$_2$C$_6$H$_4$SO$_3{}^-$), $T_{1/2}$ and its hysteresis width vary drastically with the structural isomerism (substitution at the *o*-, *m*-, and *p*-positions on the benzene ring of tos and abs) [24,25], which implies the possibility of photoinduced spin transition by means of the photoisomerization of counter ions. Based on this strategy, we synthesized [Fe(II)(NH$_2$-trz)$_3$](SP150)$_2$·2H$_2$O (SP150 is *N*-alkylsulfonated spiropyran) and investigated the SCO phenomenon [26]. In the case where X is C$_n$H$_{2n+1}$SO$_3{}^-$, the $T_{1/2}$ increases with increasing the length (*n*) of the alkyl chain of the counter ion and saturates above *n* = 5. This is attributed to the increase in the intermolecular interaction of the alkyl chains of C$_n$H$_{2n+1}$SO$_3{}^-$, called the fastener effect [27]. The hysteresis width of $T_{1/2}$ decreases with increasing *n*, showing the even-odd (parity) effect, which is known for 1D iron(II) SCO coordination polymers based on bridging di-tetrazoles [28]. Moreover, we developed [Fe(II)(R-trz)$_3$]@Nafion films exhibiting spin transition around room temperature, in which the Nafion membrane behaves as a counter anion as well as a transparent substrate [21], and investigated the condensed photogenerated HS state below 35 K [29]. The lifetime of the photogenerated HS state strongly depends on the intensity of irradiated light. In this paper, we review the effect of the characteristic properties of counter anions on the SCO behavior and the photoinduced SCO phenomenon for [Fe(II)(R-trz)$_3$]X$_2$·*x*H$_2$O (R = H, NH$_2$, etc.; X = anion), which has been mainly contributed by our group.

2. The Effect of Crystal Solvent on SCO for the [Fe(II)(R-trz)$_3$] System

In the [Fe(II)(R-trz)$_3$] chain system, not only are various kinds of counter anions acceptable, but also various crystal solvents, which remarkably influences the SCO phenomenon as well as other SCO complexes. As for the crystal solvent effect, there are some prominent reports on tris(2-picolylamine)iron(II) dichlorides, where the SCO behavior significantly depends on of the type of alcohol solvent [30] and the amount of crystal water [31]. In this section, we describe the effect of crystal water on the SCO for [Fe(II)(NH$_2$-trz)$_3$](*p*-tos)$_2$·*x*H$_2$O. In connection with [Fe(II)(NH$_2$-trz)$_3$](*p*-tos)$_2$·1.5H$_2$O, Codjvi and Kahn et al. reported the SCO transition with wide thermal hysteresis by means of optical spectroscopy [32]. According to them, in the first heating process [Fe(II)(NH$_2$-trz)$_3$](*p*-tos)$_2$·1.5H$_2$O shows an abrupt spin transition accompanied by an apparent optical absorption change at 361 K, and a wide thermal hysteresis width at 82 K was observed in the subsequent cooling process. In the second cycle, however, the spin transition in the heating process takes place at the much lower temperature of 296 K. The thermogravimetric analysis revealed that the abrupt spin transition at 361 K in the heating process is induced by the dehydration of crystal water molecules. The reproducible spin transition of $T_{1/2}$↑ = 296 K and $T_{1/2}$↓ = 279 K with the hysteresis width of 17 K is attributed to that of anhydrous [Fe(II)(NH$_2$-trz)$_3$](*p*-tos)$_2$.

In order to investigate the effect of crystal water on the SCO phenomenon for the [Fe(II)(R-trz)$_3$] system, we prepared [Fe(II)(NH$_2$-trz)$_3$](*p*-tos)$_2$·*x*H$_2$O for different values of *x*: 1.5, 3.5, 5.5, and 8 [33]. Although the number (*x*) of crystal water nearly equals 1.5 under ambient humidity, *x* increases from 1.5 under more humid conditions. To keep *x* in the desired range above 1.5, the hydrated sample was sealed into an aluminum capsule after the determination of *x* by means of thermogravimetry. For the extended X-ray absorption fine structure (EXAFS) measurements of *x* = 1.5, the sample was diluted with boron nitride (BN) to create a pellet. In order to prevent the desorption of crystal water in a vacuum, the pellet was completely sealed with an adhesive substrate (Stycast No. 1266).

To prove the formation of a 1D Fe chain structure, [Fe(II)(NH$_2$-trz)$_3$]$_\infty$, in [Fe(II)(NH$_2$-trz)$_3$] (p-tos)$_2 \cdot$1.5H$_2$O, we performed the Fe K-edge EXAFS spectra [33]. Figure 2 shows the Fourier transform of the spectra at 295 and 365 K. The most intense peak, around 1.7 Å in the Fourier transform, is derived from the Fe–N scattering. We found a remarkable peak at 7 Å, while in general, peaks are hardly observed in this region. The result indicates the existence of a linear arrangement of the Fe atoms, where enhanced multiple scattering can be observed as a significant peak in the long radius region [34]. The peak at about 7 Å corresponds to the Fe–Fe–Fe multiple scattering, which proves the existence of a straight 1D Fe chain structure such as [Fe(II)(H-trz)$_2$(trz)](BF$_4$), where trz is 1,2,4-triazolate [35]. Reflecting the SCO transition, the peak positions at 365 K are clearly shifted to the longer side than those at 295 K. The Fourier transform of EXAFS for [Fe(II)(NH$_2$-trz)$_3$](p-tos)$_2 \cdot$1.5H$_2$O at 295 K and 365 K corresponds to the LS and HS states, respectively.

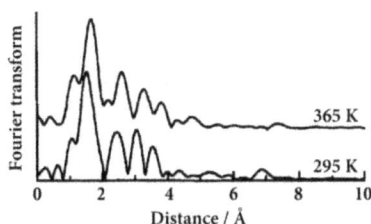

Figure 2. Fourier transforms of Fe K-edge extended X-ray absorption fine structure (EXAFS) oscillation function, $k^3\chi(k)$, for [Fe(II)(NH$_2$-trz)$_3$](p-tos)$_2 \cdot$1.5H$_2$O at 295 K and 365 K [33].

Figure 3 shows the temperature dependence of the magnetic susceptibility multiplied by temperature ($\chi_M T$) for [Fe(II)(NH$_2$-trz)$_3$](p-tos)$_2 \cdot$1.5H$_2$O, indicating the stepwise SCO behavior. The intermediate plateau between 310 and 360 K increases with repeating the heating cycles, and the saturated $\chi_M T$ corresponds to the HS state. The disappearance of the intermediate state by repeating the heating cycles is thought to arise from the rearrangement of crystal water molecules by annealing. A similar irreversible thermal history of SCO behavior was reported on a Langmuir film of an amphiphilic iron(II) complex [36]. Owing to the rearrangement of crystal water molecules by annealing, we investigated the effect of crystal water on the SCO behavior for [Fe(II)(NH$_2$-trz)$_3$](p-tos)$_2 \cdot x$H$_2$O, where x was 1.5, 3.5, 5.5, and 8, under the second heating cycle.

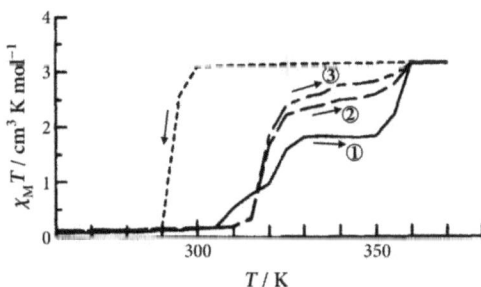

Figure 3. Temperature dependence of the magnetic susceptibility multiplied by temperature ($\chi_M T$) for [Fe(II)(NH$_2$-trz)$_3$](p-tos)$_2 \cdot$1.5H$_2$O. The solid line represents the first heating process, the broken line represents the second heating process, the dot-dashed line represents the third heating process, and the dotted line represents the cooling process. Reproduced with permission from Reference [33].

Figure 4 shows the SCO behavior for $[Fe(II)(NH_2\text{-trz})_3](p\text{-tos})_2 \cdot xH_2O$, where x equals 1.5, 3.5, 5.5, and 8, under the second heating cycle. The samples where x equals 1.5, 3.5, and 5.5 show a stepwise LS to HS transition, whereas the sample where x is 8 shows a simple LS to HS transition with negligible thermal hysteresis. In the case where x is 3.5, the temperature dependence of $\chi_M T$ is equivalent to the superposition of $\chi_M T$ between x at 1.5 and x at 5.5. As shown in Figure 4, with increasing the amount of crystal water, the transition temperature lowers and the thermal hysteresis width becomes narrow. This can be explained by the water molecules in $[Fe(II)(NH_2\text{-trz})_3](p\text{-tos})_2 \cdot xH_2O$ having two opposite functions in the SCO phenomenon. One is the formation of hydrogen bonding, which makes the lattice harder and stabilizes the LS state. The other function is behaving as a spacer between the $[Fe(II)(NH_2\text{-trz})_3]$ chains, which makes the lattice softer and stabilizes the HS state. Therefore, with increasing the amount of crystal water, the function as a spacer becomes more dominant, which decreases $T_{1/2}$ and the hysteresis width. In the case where x is 1.5, the effect of the hydrogen bonding of crystal water molecules in the SCO phenomenon is the most dominant among the tests amounts of x at 1.5, 3.5, 5.5, and 8. In connection with this, it should be mentioned that the effect of crystal water on the SCO transition is similar to that of the counter anions for $[Fe(II)(NH_2\text{-trz})_3]Anion \cdot xH_2O$ [37–39]. With increasing the radius of the counter anion, the $T_{1/2}$ of the complex decreases. The results indicate that the counter anion acts as a spacer for the Fe chains. In addition, the anion with a larger radius makes the crystal lattice softer. Thus, the crystal water molecules and the counter anions both play an important role in the SCO behavior.

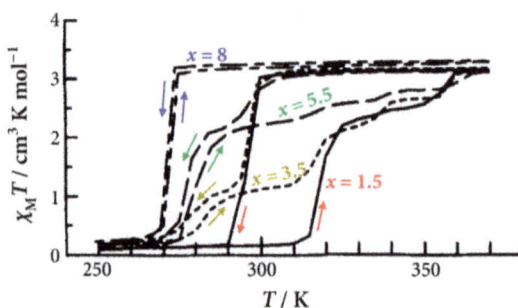

Figure 4. Temperature dependence of the magnetic susceptibility multiplied by temperature ($\chi_M T$) for $[Fe(II)(NH_2\text{-trz})_3](p\text{-tos})_2 \cdot xH_2O$. x = 1.5 is represented by the solid line, x = 3.5 is represented by the dotted line, x = 5.5 is represented by the broken line, and x = 8 is represented by the dot-dashed line. The arrows represent the sweeping directions of temperature. Reproduced with permission from Reference [33].

3. The Effect of Structural Isomerism of Counter Ions on SCO for the $[Fe(II)(R\text{-trz})_3]$ System

In the case of spherical anions, isotropic structural changes affect the interchain interaction for $[Fe(II)(R\text{-trz})_3]X_2 \cdot xH_2O$. Anisotropic structural changes, such as the structural isomerism of organic molecules, also affect the cooperative interaction among $[Fe(II)(NH_2\text{-trz})_3]$ chains. One of the structural isomers is an *ortho-*, *meta-*, and *para-*substituted benzene. In order to investigate the effect of structural isomerism of counter ions on the SCO phenomenon for the $[Fe(II)(R\text{-trz})_3]$ system, we prepared $[Fe(II)(NH_2\text{-trz})_3](tos)_2 \cdot xH_2O$ and $[Fe(II)(NH_2\text{-trz})_3](abs)_2 \cdot xH_2O$ [24,25]. The amount of crystal water in $[Fe(II)(NH_2\text{-trz})_3](tos)_2 \cdot xH_2O$ was determined as x = 2, 1.5, and 1.5 for *o-*, *m-*, and *p-*tos salts, respectively. The amount of crystal water in $[Fe(II)(NH_2\text{-trz})_3](abs)_2 \cdot xH_2O$ was x = 2 for all the *o-*, *m-*, and *p-*abs salts, obtained by means of thermogravimetric analysis. Figure 5 shows the molecular structures of *o-*, *m-*, *p-*tos and *o-*, *m-*, *p-*abs.

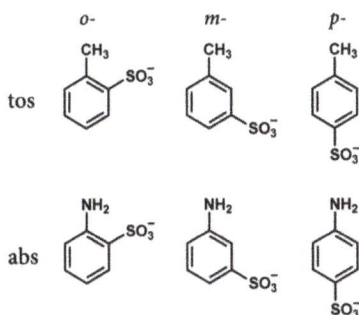

Figure 5. Molecular structures of *o*-, *m*-, and *p*-toluenesulfonate (tos) and *o*-, *m*-, and *p*- aminobenzenesulfonate (abs).

The temperature dependence on the magnetic susceptibility of $[Fe(II)(NH_2\text{-trz})_3](tos)_2 \cdot xH_2O$ is shown in Figure 6 [24]. All the complexes show the spin transition around 300 K. In the case of the *o*-tos salt, the spin transition occurs at 250 K with a thermal hysteresis width of 80 K. For the *m*-tos salt, the abrupt spin transition was observed at 319 K, where hysteresis was negligible. The spin transition of the *p*-tos salt occurs abruptly at 295 K in the cooling process, while stepwise spin transitions were observed on heating. It should be noted that the $T_{1/2}\downarrow$ strongly depends on the structural-isomerized counter anions compared to $T_{1/2}\uparrow$. Such a remarkable structural isomerism effect should be related to the arrangement of counter anions among $[Fe(II)(NH_2\text{-trz})_3]$ chains.

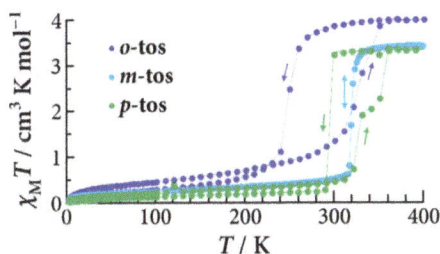

Figure 6. Temperature dependence of the magnetic susceptibility multiplied by temperature ($\chi_M T$) for $[Fe(II)(NH_2\text{-trz})_3](o\text{-}, m\text{-}, \text{and } p\text{-tos})_2 \cdot xH_2O$. The arrows represent the sweeping directions of temperature. Reproduced with permission from Reference [24].

The SCO behavior of $[Fe(II)(NH_2\text{-trz})_3](tos)_2 \cdot xH_2O$ was also confirmed by ^{57}Fe Mössbauer spectroscopy. Figure 7 shows the ^{57}Fe Mössbauer spectra of $[Fe(II)(NH_2\text{-trz})_3](tos)_2 \cdot xH_2O$ at 290 and 77 K [24]. For ^{57}Fe Mössbauer spectroscopy, the powder samples were dominant in the complexes. The LS state remained dominant upon heating to 290 K, which is apparently different from those in the heating process. This is indicative of thermal hysteresis. For the *m*-tos salt, the ^{57}Fe Mössbauer spectra was almost unchanged below 290 K, both in the cooling and heating processes, indicating a stable ground LS state. The ^{57}Fe Mössbauer study is consistent with the magnetic sealed in an acrylate resin sample holder to avoid the desorption of crystal water during the heating process. Two doublets were found in the ^{57}Fe Mössbauer spectra of the *o*- and *p*-tos salts at 290 K. The doublets with small and large quadrupole splitting are assigned to the LS and HS states, respectively. On cooling down to 77 K, the HS signals considerably decreased, and the LS state became measurable.

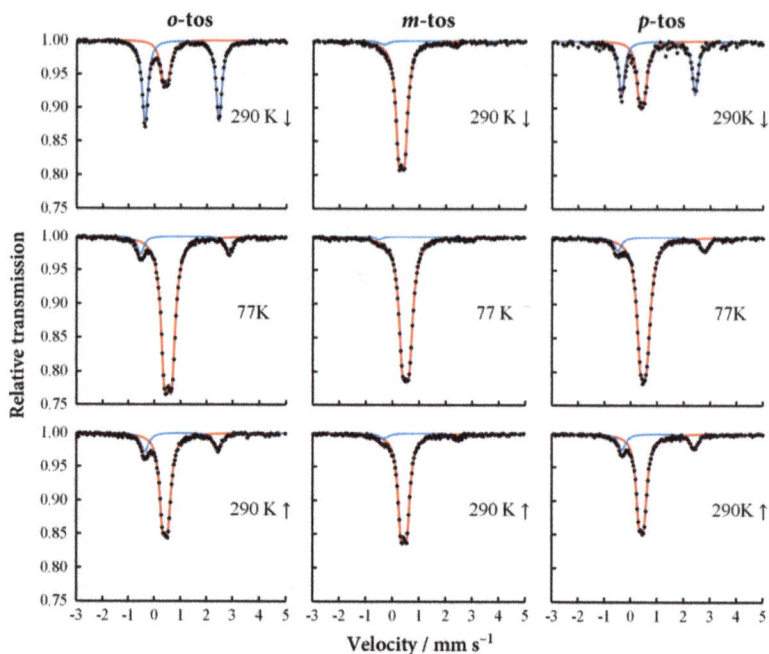

Figure 7. ^{57}Fe Mössbauer spectra for [Fe(II)(NH$_2$-trz)$_3$](o-, m-, p-tos)$_2$·xH$_2$O at 290 K and 77 K. Reproduced with permission from Reference [24].

Figure 8 shows the Fourier transform of Fe K-edge EXAFS oscillation function, $k^3\chi(k)$, for [Fe(II)(NH$_2$-trz)$_3$](abs)$_2$·2H$_2$O at 30 K [25]. The dominant peaks at about 1.7 Å are attributed to the N atoms of the FeN$_6$ core with six-fold coordination. The first-nearest neighbor Fe–N distances at 30 K were estimated at 1.974 Å, 1.974 Å, and 1.978 Å, for the o-abs, m-abs, and p-abs salts, respectively, which is typical of Fe–N distances in the LS state. On the other hand, those at 350 K were estimated at 2.165 Å, 2.158 Å, and 2.167 Å, for the o-abs, m-abs, and p-abs salts, respectively, which are typical of Fe–N distances in the HS state. All the salts show a noticeable peak at about 7 Å, corresponding to the Fe–Fe–Fe multiple scattering, which proves the existence of a straight 1D Fe chain structure.

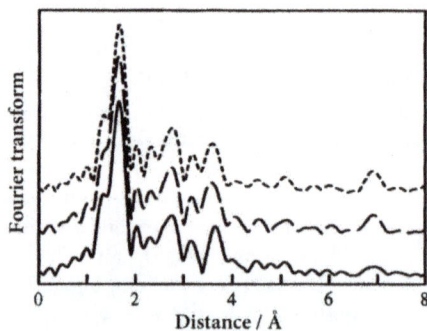

Figure 8. Fourier transform of Fe K-edge EXAFS oscillation function, $k^3\chi(k)$, for [Fe(II)(NH$_2$-trz)$_3$](o-, m-, p-abs)$_2$·2H$_2$O at 30 K. The dotted line represents o-abs salt, the broken line denotes the m-abs salt, and the solid line represents the p-abs salt [25].

The temperature dependence of the magnetic susceptibility of $[Fe(II)(NH_2\text{-}trz)_3](abs)_2 \cdot 2H_2O$ is shown in Figure 9 [25]. In the first cycle of the heating process, the spin transition of the *o*-, *m*-, and *p*-abs salts occurred at 350 K, 320 K, and 240 and 340 K, respectively. These transition temperatures are typical of $[Fe(II)(NH_2\text{-}trz)_3]$ complexes showing SCO around or above room temperature. Unlike the case of the tos salts, the spin transitions of all the abs salts in the second cycle remain practically unchanged. This may be related to the fact that the abs molecules have both the amino (hydrogen bond donor) and sulfonate (hydrogen bond acceptor) groups which can form a tight hydrogen bond network with itself, while the tos molecules lack a hydrogen bond donor site.

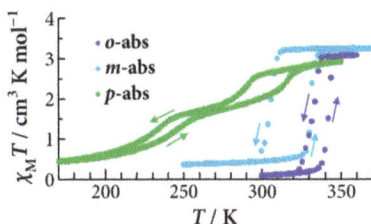

Figure 9. Temperature dependence of the magnetic susceptibility multiplied by temperature ($\chi_M T$) of $[Fe(II)(NH_2\text{-}trz)_3](abs)_2 \cdot 2H_2O$. The arrows represent the sweeping directions of temperature. Reproduced with permission from Reference [25].

In connection with the effect of structural isomerism of counter ions on the SCO behavior for the $[Fe(II)(R\text{-}trz)_3]$ system, it should be mentioned that, as an example of using organic counter ions, Koningsbruggen et al. reported $[Fe(II)(NH_2\text{-}trz)_3]$ complexes with naphthalenesulfonates and its derivatives [40], allowing for systematic variation of the anions. The sulfonate group can be placed in two different positions on the naphthalene ring where the anions have a potential to be modified with further functional groups, such as the hydroxyl and amino groups, forming hydrogen bonds. According to them, the SCO temperatures of $[Fe(II)(NH_2\text{-}trz)_3]X_2 \cdot xH_2O$ (where X was 1-naphthalenesulfonate, 4-hydroxy-1-naphthalenesulfonate, 4-amino-1-naphthalenesulfonate, 2-naphthalenesulfonate, or 6-hydroxy-2-naphthalenesulfonate) were determined from the optical measurements. All the compounds show roughly the same SCO behavior, where the spin transition occurred irreversibly in the first heating process, before reproducible transitions with thermal hysteresis widths of about 10 K, in the subsequent measuring cycles. In the first heating process, the complexes show an abrupt spin transition at 330 to 340 K accompanied by an apparent absorption change. A wide thermal hysteresis width of approximately 100 K was observed in the subsequent cooling process. However, the spin transition on the subsequent heating process occurred at much lower temperatures, ranging from 229 to 297 K, depending on the anion. The reproducible hysteresis width was about 10 K for all the complexes.

As mentioned above, in the $[Fe(II)(NH_2\text{-}trz)_3]X_2 \cdot xH_2O$ system, $T_{1/2}$ and its hysteresis width vary drastically with the structural isomerism of counter ions, which implies the possibility of photoinduced spin transition by means of the photoisomerization of the counter ions. Based on this strategy, we synthesized $[Fe(II)(NH_2\text{-}trz)_3](SP150)_2 \cdot 2H_2O$ by using SP150 as a photochromic anion, and investigated the SCO behavior [26]. Figure 10 shows the molecular structure of SP150.

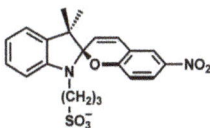

Figure 10. Molecular structure of a photochromic anion, SP150.

To prove the formation of a 1D Fe chain structure, $[Fe(II)(NH_2\text{-}trz)_3]_\infty$, in $[Fe(II)(NH_2\text{-}trz)_3]$ $(SP150)_2 \cdot 2H_2O$, we performed the Fe K-edge EXAFS spectra [26]. The Fourier transform at 30 K is shown in Figure 11. The peak at about 6.8 Å corresponds to the Fe–Fe–Fe multiple scattering, which proves the formation of straight 1D Fe chain structure in $[Fe(II)(NH_2\text{-}trz)_3](SP150)_2 \cdot 2H_2O$.

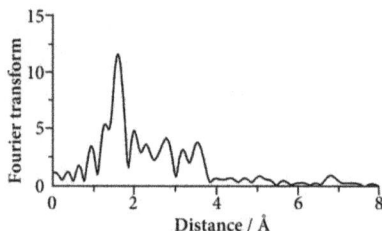

Figure 11. Fourier transform of Fe K-edge EXAFS oscillation function, $k^3\chi(k)$, for $[Fe(II)(NH_2\text{-}trz)_3]$ $(SP150)_2 \cdot 2H_2O$ at 30 K [26].

Figure 12 shows the temperature dependence of $\chi_M T$ for $[Fe(II)(NH_2\text{-}trz)_3](SP150)_2 \cdot 2H_2O$. At first, $\chi_M T$ was measured from 298 K to 10 K. After cooling down to 10 K, $\chi_M T$ was measured from 10 K to 370 K during the heating process, where a partial and gradual SCO phenomenon appears at around 340 K. After heating up to 370 K, $\chi_M T$ versus T during the cooling process, from 370 K to 10 K, shows a completely different line from that during the first heating process, and a gradual SCO appears at around 260 K. The $\chi_M T$ versus T during the cooling process from 370 K to 10 K was reproduced on the third cycle. The irreversible SCO behavior is considered to be the rearrangement of crystal water molecules and/or SP150 molecules by annealing.

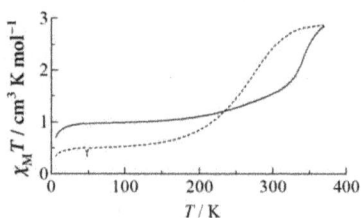

Figure 12. Temperature dependence of the magnetic susceptibility multiplied by temperature ($\chi_M T$) for $[Fe(II)(NH_2\text{-}trz)_3](SP150)_2 \cdot 2H_2O$ [26]. The solid line denotes the process from 298 K → 10 K → 370 K, and the dotted line represents the process from 370 K → 10 K → 298 K.

4. The Effect of Counter Ion Size on SCO for the [Fe(II)(R-trz)$_3$] System

The SCO characteristics are strongly influenced by counter ions in the triazole-bridged iron(II) systems as well as in a number of SCO materials. One of the importance instances of this is the counter ion dependence on the transition temperature and thermal hysteresis in a $[Fe(II)(NH_2\text{-}trz)_3]$-type system with monovalent inorganic anions, such as Cl^-, Br^-, NO_3^-, I^-, BF_4^-, and ClO_4^-. An example where the ion radius directly influences the transition temperature is the SCO phenomena of $[Fe(II)(NH_2\text{-}trz)_3]X_2$ (X = Cl^-, Br^-, NO_3^-, I^-, BF_4^-, ClO_4^-) [37–39]. The transition temperature and hysteresis width linearly decreased from 355 to 180 K depending on the ionic radii of the spherical anions, denoted by the filled triangles in Figure 13. A similar tendency was discovered in other triazole-based SCO complexes. Garcia et al. reported this in $[Fe(II)(hye\text{-}trz)_3]X_2 \cdot xH_2O$ (hye-trz = 4-(2'-hydroxyethyl)-1,2,4-triazole, X = PF_6^-, ClO_4^-, BF_4^-, I^-, Br^-, NO_3^-, and Cl^-) [41]. The transition temperatures, determined by optical measurement, linearly decrease in correlation with ionic radii from 314 to 205 K, represented by the open triangles in Figure 13.

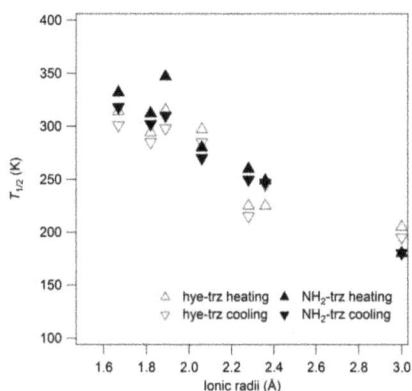

Figure 13. Variation of the transition temperatures versus the ionic radii for [Fe(II)(NH$_2$-trz)$_3$]X$_2$·xH$_2$O and [Fe(II)(hye-trz)$_3$]X$_2$·xH$_2$O [37–39,41].

Among the same series of anions, the Fe–N coordination bond length will likely decrease according to the enhancement of the anion–cation interactions, which have been investigated using EXAFS techniques in combination with [57]Fe Mössbauer [42] and X-ray fluorescence spectroscopy [38]. The anion–cation interaction can be regarded as a kind of chemical pressure, which originates in intermolecular interactions and would provide an effect similar to physical hydrostatic pressure on SCO phenomena, except for the changes in chemical properties.

Moreover, Garcia et al. reported on the correlation between $T_{1/2}$ and the volume of the counter anion in the series of fluorinated inorganic divalent anions (TiF$_6^{2-}$, ZrF$_6^{2-}$, SnF$_6^{2-}$, GeF$_6^{2-}$, and TaF$_7^{2-}$) salts of [Fe(II)(NH$_2$-trz)$_3$] complexes, in addition to the related monovalent complexes [43]. These divalent anions are expected to favor the engineering of hydrogen-bonding networks. They considered the variation of $T_{1/2}$ with the volume of the counter anion, instead of considering only the radius determined from Kapustinskii's equation [44]. This allows us to take into account the dimensional volume (V) of the anions. Two linear regimes were found in the regions of 0.04 nm$^3 \leq V \leq$ 0.09 nm^3 and $V \geq$ 0.11 nm^3. In the former, corresponding with the monovalent anions, $T_{1/2}$ decreases linearly with increasing volume, and spin transition requires less energy. Above a V of 0.11 nm^3, corresponding with the divalent anions, it seems that $T_{1/2}$ is almost constant at about 200 K. These complexes seem to have enough chain distance to freely expand the Fe–N coordination bonds, so that the counter anions hardly affect the spin transition. Thus, a larger counter anion induces the reduction of the relative energy of the potential wells between the HS and LS states, resulting in a negative pressure effect. Indeed, [Fe(II)(NH$_2$-trz)$_3$](B$_{10}$H$_{10}$)·H$_2$O shows the HS state over the temperature range of 300 to 2 K, which contains the largest counter anion in reported systems, with V being approximately 0.23 nm^3 [45].

5. Fastener Effect of Counter Ions on SCO for the [Fe(II)(R-trz)$_3$] System

As described in Section 4, $T_{1/2}$ increases with decreasing the ionic radii of counter anions in [Fe(II)(NH$_2$-trz)$_3$]X$_2$·xH$_2$O. The charge density of the anion increases with decreasing the ionic radius, leading to incremental anion-cation interaction between the [Fe(II)(NH$_2$-trz)$_3$] unit, the anion, and the compression of the FeN$_6$ coordination sphere, which induces two effects: the incremental intrachain (short-range) interaction and ligand field strength. Furthermore, $T_{1/2}$ increases due to both of these effects. We also observed an effect on uniaxial chemical pressure along the direction of the [Fe(II)(NH$_2$-trz)$_3$]$_\infty$ chain by van der Waals force between alkanesulfonate anions (C$_n$H$_{2n+1}$SO$_3^-$, n = 1–9) in the series of [Fe(II)(NH$_2$-trz)$_3$](C$_n$H$_{2n+1}$SO$_3$)$_2$·xH$_2$O. Magnetostructural correlation between $T_{1/2}$ and the Fe–Fe distance is described in this section.

To determine the local structure around Fe ions for these complexes, Fe K-edge EXAFS measurements were conducted. This measurement is a useful tool to reveal the local structure around iron(II) ions in [Fe(II)(R-trz)$_3$]Anion·xH$_2$O; however, the crystal structure has become clearer in recent studies [12,46,47].

The Fourier-transformed Fe K-edge EXAFS spectra at 35 K (LS state) and 370 K (HS state) are shown in Figure 14. The most intense peak around 1.7 Å is attributed to the N atoms in the six-fold FeN$_6$ coordination environment. The peak obviously moved to the longer side according to the change from the HS to LS state. Multiple peaks were found in the range of 2–4 Å, which were attributed to the Fe–N, Fe–C, and Fe–Fe coordination shells and multiple-scattering paths. For the methanesulfonate compound (n = 1), the closest Fe–N distance of the LS state is estimated to be 1.977 to 1.984 Å at 35 K. For the other compounds (n = 2 to 9), the Fe–N distances were estimated to be 1.965 to 1.983 Å (LS state), which are close to that of the methanesulfonate compound. Accordingly, the Fe–N distances are insensitive to the chain length of alkanesulfonates. As shown in Figure 15, we found that the estimated Fe–Fe distances decreased from 3.34 to 3.29 Å accompanied by the even-odd effect with increasing alkyl chain length. This tendency is caused by the fastener effect that was derived from van der Waals interactions between self-assembled alkyl chains. It is worth noting that we can apply uniaxial pressure to the triazole-bridged Fe(II) system along the 1D chain direction by causing the fastener effect. Figure 16 shows a schematic illustration of the compression of the [Fe(II)(NH$_2$-trz)$_3$] chain associated with the fastener effect. The Fe chain can be assumed to follow the self-assembling interaction between alkyl chains, where the Fe chain would be compressed along the direction of Fe–Fe array with increasing the fastener effect. At this time, the linear arrangement of iron atoms could likely be maintained with compensatory in N–Fe–N bridging angle. This concept is supported by the Fe–Fe–Fe multiple scattering peaks around 7 Å in EXAFS spectra (Figure 14a) [27]. [Fe(II)(NH$_2$-trz)$_3$](C$_n$H$_{2n+1}$SO$_3$)$_2$·xH$_2$O is a unique system which has a varying Fe–Fe interaction.

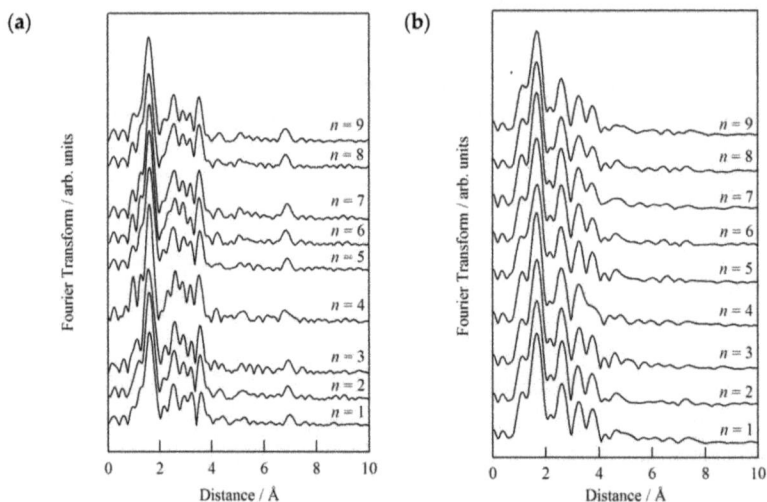

Figure 14. Fourier transform of Fe K-edge EXAFS oscillation functions, $k^3\chi(k)$, for [Fe(II)(NH$_2$-trz)$_3$] (C$_n$H$_{2n+1}$SO$_3$)$_2$·xH$_2$O (n = 1–9) at (**a**) 35 K, low-spin (LS) state; and (**b**) 370 K, high-spin (HS) state [27].

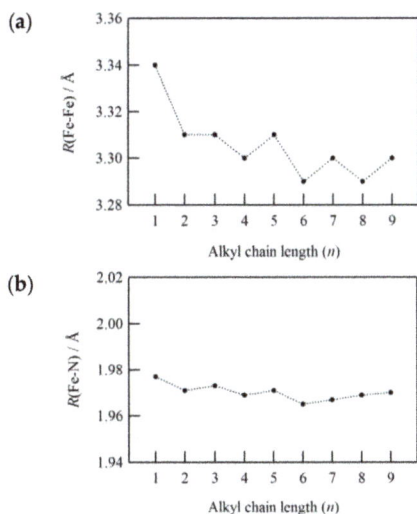

Figure 15. The nearest-neighbor (**a**) Fe–Fe and (**b**) Fe–N distances as a function of the alkyl chain length (n) at 35 K for [Fe(II)(NH$_2$-trz)$_3$](C$_n$H$_{2n+1}$SO$_3$)$_2$·xH$_2$O (n = 1–9) [27].

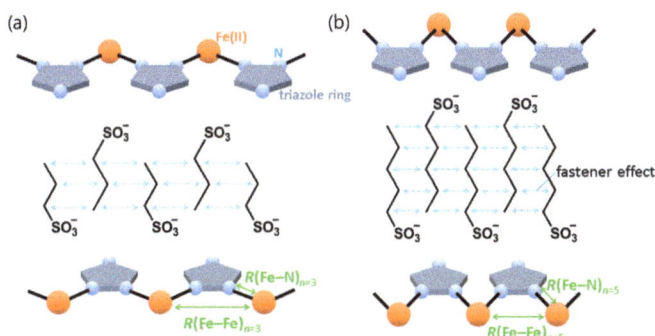

Figure 16. Schematic representation of the fastener effect on R(Fe–Fe) and R(Fe–N) for [Fe(II)(NH$_2$-trz)$_3$](C$_n$H$_{2n+1}$SO$_3$)$_2$·xH$_2$O; (**a**) n = 3 and (**b**) n = 5.

Figure 17 shows the temperature dependence of $\chi_M T$ in the temperature range of 250 to 370 K. In the case of n = 1, the $\chi_M T$ value observed at 320 K was 3.42 cm^3·K·mol^{-1} and is quite close to the theoretical value for the HS state (S = 2). On cooling, the value started to decrease abruptly around 280 K ($T_{1/2}\downarrow$ = 276 K) and then arrived a value of 0.21 cm^3·K·mol^{-1}. In the subsequent heating process, the spin transition occurred at $T_{1/2}\uparrow$ of 295 K and the $\chi_M T$ value was restored completely. The thermal hysteresis width ($\Delta T_{1/2} = T_{1/2}\uparrow - T_{1/2}\downarrow$) is 19 K.

As can be seen in Figure 18a, $T_{1/2}$ increases as the number of carbons in the alkyl chain increase up to n = 4 accompanied by the even-odd effect. However, it remains almost constant above n = 5. In connection with this, $T_{1/2}$ has relevant, significant impacts on the nearest-neighbor Fe–Fe distances (R(Fe–Fe)), estimated from EXAFS to be 35 K. Taking into account that the alkyl chain length has negligible influence on R(Fe–N), $T_{1/2}$ is considered to be mainly related to R(Fe–Fe). The displacement of iron(II) ions in a 1D arrangement move coherently with counter anions due to the fastener effect, leading to the squeeze of the neighboring Fe–Fe distances. Consequently, the elastic energy of [Fe(II)(NH$_2$-trz)$_3$]$_\infty$ chains is considered to be increased. This situation is analogous to SCO behavior under external hydrostatic pressure [48–50], except that hydrostatic pressure is isotropic. Increasing external pressure provides narrow magnetic hysteresis width and a high $T_{1/2}$ value. For the 1D systems, SCO behavior affected by external pressure has been successfully described by using a model that takes into account the competing short- and long-range interactions, where the former represents the interaction of molecules inside the chain and the latter is assumed to arise from the coupling of the iron(II) d electrons with the deformation accompanied by the HS to LS transitions. When the short-range interaction overcomes the long-range one, the LS state becomes stable and leads the hysteresis loop to disappear [51]. This concept can be also applied to the present results. Increasing the alkyl chain length corresponds to the enhancement of the short-range interaction in the system. In fact, we found a close correlation between the $T_{1/2}$ value and the R(Fe–Fe) estimated from EXAFS as shown in Figure 18a, where R(Fe–Fe) is directly connected to the short-range interaction. In the range of $n \geq 5$, the fastener effect seems to reach a critical limit in shortening the Fe–Fe distance. To our knowledge, this is the first system with systematic control of the Fe–Fe interaction in the 1D triazole-coordinated iron(II) SCO complexes. Moreover, there is a noticeable difference between this and other studies that investigated the dependence of SCO behavior by spherical counter anions, as described in numerous previous reports and our compounds [24,25,37–39,41–43,45,52–56].

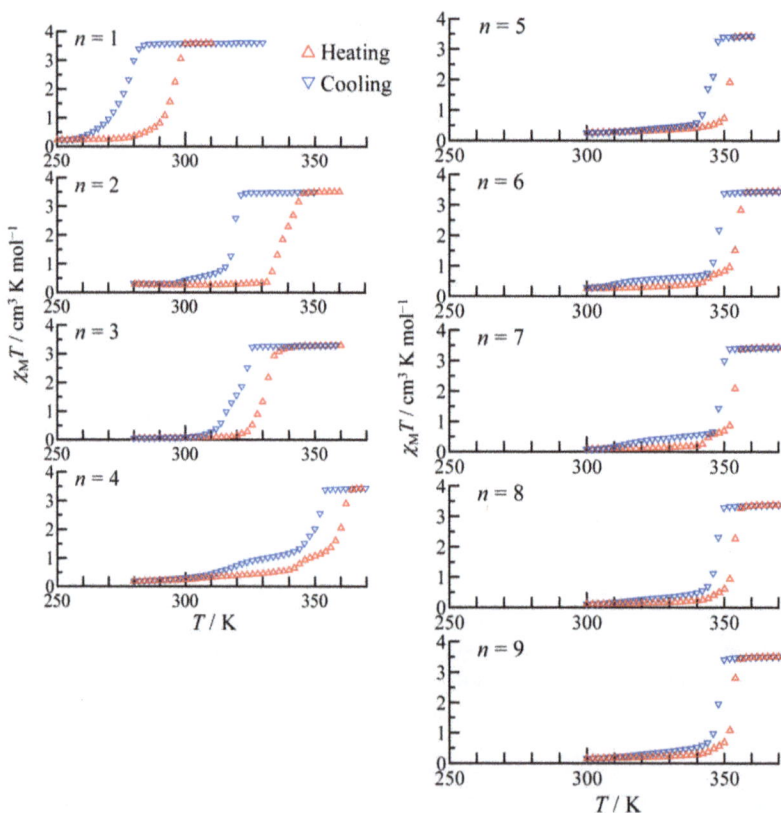

Figure 17. Temperature dependence of molar magnetic susceptibility multiplied by temperature for [Fe(II)(NH$_2$-trz)$_3$](C$_n$H$_{2n+1}$SO$_3$)$_2$·xH$_2$O (n = 1–9). Blue and red triangles represent the first cooling and the subsequent heating processes, respectively [27].

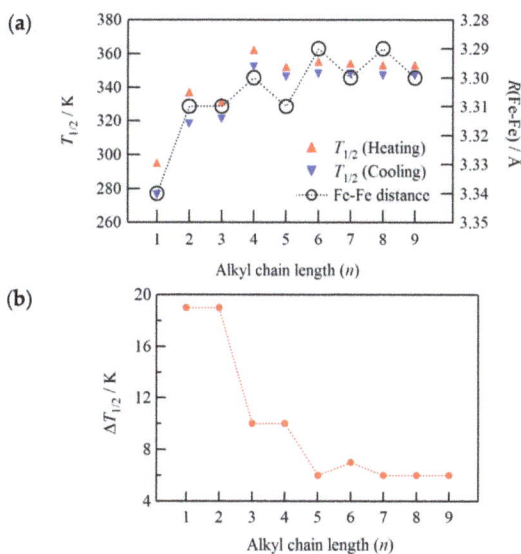

Figure 18. (a) Correlation between $T_{1/2}$ and Fe–Fe distance at 35 K for [Fe(II)(NH$_2$-trz)$_3$] (C$_n$H$_{2n+1}$SO$_3$)$_2$·xH$_2$O (n = 1–9); and (b) hysteresis width of spin-crossover (SCO) transition ($\Delta T_{1/2}$) as a function of alkyl chain length (n) for [Fe(II)(NH$_2$-trz)$_3$](C$_n$H$_{2n+1}$SO$_3$)$_2$·xH$_2$O (n = 1–9) [27].

The relationship between $\Delta T_{1/2}$ and the alkyl chain length is shown in Figure 18b. A clear even-odd effect was also observed in the hysteresis width, where $\Delta T_{1/2}$ was slightly wider for even n than for odd n. The magnitude of $\Delta T_{1/2}$ became smaller with increasing n and saturated over n = 5 at $\Delta T_{1/2}$ ~6 K. This tendency, except for the even-odd effect, is the same as with [Fe(II)(NH$_2$-trz)$_3$] with spherical counter anions. The counter anions are accommodated among the cationic iron(II) chains and behave as spacers. Accordingly, the incremental increase of anion size leads to detachment from the adjacent cationic SCO chains, and thereby interchain interaction, i.e., the cooperative effect, would decrease.

6. Photoinduced SCO for the [Fe(II)(R-trz)$_3$] System

Since the first LIESST phenomenon was discovered with the SCO complex [Fe(II)(ptz)$_6$](BF$_4$)$_2$ in 1984 [3], varied iron(II) and iron(III) compounds showing LIESST have been developed [5]. Being able to switch between non-magnetic and magnetically ordered states by light irradiation at room temperature will open a new field of photonic molecular devices. However, there still remains a limitation on the critical temperature in LIESST. The highest critical temperature is 130 K for [Fe(II)(L)(CN)$_2$]·H$_2$O (L is the Schiff base macrocyclic ligand derived from the condensation of 2,6-diacetylpyridine with 3,6-dioxaoctane-1,8-diamine) [57]. Therefore, another approach has been eagerly anticipated. Recently, the photoinduced SCO phenomenon has been demonstrated for [Fe(II)(pz){Pt(CN)$_4$}] (pz is pyrazine) [58] and [Fe(II)(NH$_2$-trz)$_3$](Br)$_2$·3H$_2$O by photoirradiation within the thermal hysteresis loop of the SCO transition [59]. Such a strategy is one of the possible approaches to control the spin states in SCO compounds at room temperature.

In the [Fe(II)(pz){Pt(CN)$_4$}] complex, where thermal SCO transitions occur at $Tc{\uparrow}$ = 284 K and $Tc{\downarrow}$ = 308 K with a hysteresis width of 24 K, light-induced spin transition at room temperature was achieved by applying one-shot laser-pulsed irradiation [58]. The photoinduced spin conversion was confirmed by investigating Raman bands at 650 and 682 cm^{-1} assigned to the in-plane bending mode of the pyrazine ring for the HS and LS states, respectively. Note that the photoinduced spin transition is reversible in the reported system. [Fe(II)(NH$_2$-trz)$_3$]Br$_2$·3H$_2$O, [Fe(II)(NH$_2$-trz)$_3$](NO$_3$)$_2$·H$_2$O, and [Fe(II)(H-trz)$_2$(trz)](BF$_4$)$_2 \cdot$ H$_2$O are the second systems for the photoinduced LS to HS conversion in the thermal hysteresis region by a single shot of laser irradiation [60].

As for the triazole-bridged iron(II) SCO system, the development of single crystal or transparent film has been desired for the detailed investigation of the optical properties and the LIESST effect at low temperatures. However, the single crystal of triazole-bridged polymeric iron(II) complex has rarely been obtained, except for in a few complexes [12]. The triazole-bridged iron(II) chain system accepts various kinds of counter anions and exhibits a wide variety of spin transitions. From this viewpoint, we have developed [Fe(II)(R-trz)$_3$] films exhibiting the SCO phenomenon around room temperature and observed the LIESST effect below 30 K, in which Nafion behaves as a counter anion as well as a transparent substrate [21,29].

To confirm the 1D chain structure in Nafion 117 film, the Fe K-edge EXAFS measurements were performed [21]. As shown in Figure 19, the dominant peak of the Fourier transform was observed around 1.7 Å at 65 K, corresponding to the Fe–N scattering. Furthermore, a noticeable peak can be found at 7 Å, indicating the formation of a 1D chain structure consisting of a linear Fe arrangement, as schematically shown in Figure 20.

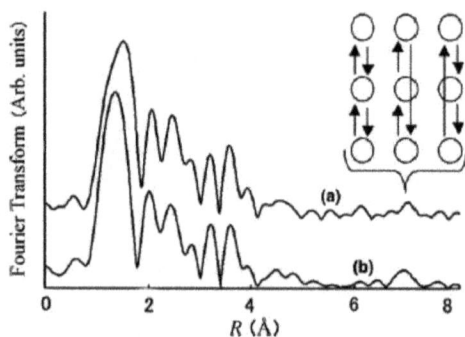

Figure 19. Fourier transforms of the Fe K-edge EXAFS oscillation function, $k^3\chi(k)$, for (a) [Fe(II)(H-trz)$_3$]@Nafion and (b) [Fe(II)(NH$_2$-trz)$_3$]@Nafion at 65 K [21]. The inset shows the schematic representation of double-scattering (1,2) and triple-scattering (3) from the next nearest neighbor Fe–Fe–Fe shell, which is responsible for the EXAFS peak at around 7 Å.

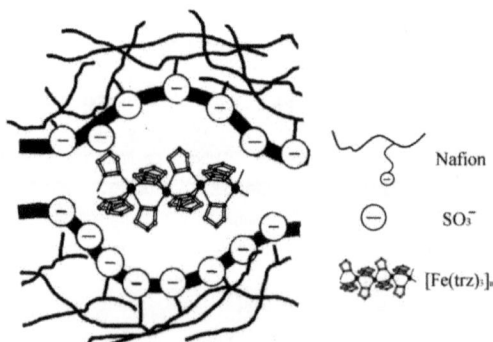

Figure 20. Schematic representation of the oligomer of [Fe(R-trz)$_3$]$_n{}^{2n+}$ embedded in Nafion 117.

The magnetic properties of [Fe(II)(H-trz)$_3$]@Nafion and [Fe(II)(NH$_2$-trz)$_3$]@Nafion were measured at 0.1 T for the samples sealed in an aluminum capsule to prevent water loss [21]. Figure 21 shows the temperature dependence of χT for both materials. For [Fe(II)(H-trz)$_3$]@Nafion, the reversible spin transition occurred around a $T_{1/2}$ of 260 K, accompanied with a small thermal hysteresis width of 3 K. The SCO transition for [Fe(II)(NH$_2$-trz)$_3$]@Nafion was observed around a $T_{1/2}$ of 198 K without thermal hysteresis. Therefore, it seems that the hysteresis width ($\Delta T_{1/2} = 3$ K) of the spin transition for [Fe(II)(H-trz)$_3$]@Nafion is attributed to the cooperative effect due to the intrachain interaction in [Fe(II)(H-trz)$_3$]$_n$.

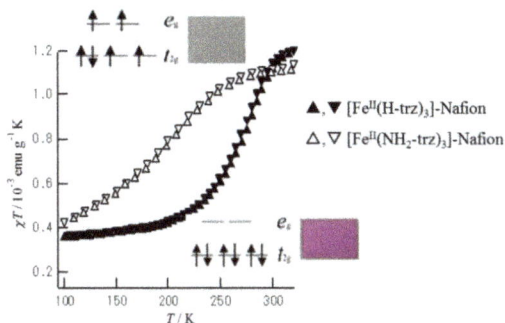

Figure 21. Temperature dependence of the magnetic susceptibility multiplied by temperature (χT) for [Fe(II)(H-trz)$_3$]@Nafion and [Fe(II)(NH$_2$-trz)$_3$]@Nafion [21]. Upright and inverted triangles represent the heating and cooling processes, respectively. The inset demonstrates the thermochromism due to the SCO phenomenon for [Fe(II)(H-trz)$_3$]@Nafion 117. Purple film in the LS state (t_{2g}^6: $S = 0$) at 77 K; Colorless film in the HS state ($t_{2g}^4 e_g^2$: $S = 2$) at 300 K.

Figure 21 also indicates residual paramagnetic fraction below 200 K, which is attributed to the HS state of iron(II). The residual HS fraction of iron(II) is attributed to the terminal iron(II) site in the oligomer of [Fe(II)(H-trz)$_3$]$_n^{2n+}$ in Nafion 117. It should be mentioned that in the iron(II) trimer complex, [Fe(II)$_3$(Et-trz)$_6$(H$_2$O)$_6$](CF$_3$SO$_3$)$_6$, the central iron(II) site undergoes the LS–HS transition at about 200 K, while the spin state of the terminal iron(II) sites is the HS state between 2 K and 300 K [61].

As shown in Figure 21, we demonstrated the thermochromism induced by the LS–HS transition for [Fe(II)(H-trz)$_3$]@Nafion. The colors of the HS state ($t_{2g}^4 e_g^2$, $S = 2$) and the LS state (t_{2g}^6, $S = 0$) of [Fe(II)(R-trz)$_3$]@Nafion (R = H, NH$_2$) are colorless and purple, respectively. In fact, the LS state of [Fe(II)(R-trz)$_3$]@Nafion shows a broad band in the visible region, at about 18,500 cm^{-1}, corresponding to the $^1A_{1g} \rightarrow {}^1T_{1g}$ transition, while the HS state shows a broad band in the near-infrared region, at about 12,500 cm^{-1}, corresponding to the $^5T_{2g} \rightarrow {}^5E_g$ transition.

In order to investigate the LIESST effect for [Fe(II)(H-trz)$_3$]@Nafion, we used the LS absorption band corresponding to the $^1A_{1g} \rightarrow {}^1T_{1g}$ transition as a sensitive monitor of the density of the photoexcited HS molecules, n_{HS} (= $1 - n_{LS}$), and investigated the variation of n_{HS} against the excitation power density (P), temperature (T), and the delay time (t) [29]. The 514.5 nm line of Ar$^+$ ion laser was used as the source of excitation and probe lights. Figure 22 shows the LIESST effect for [Fe(II)(H-trz)$_3$]@Nafion. The photoexcitation was performed at 5 K (the thick line in Figure 22a) with increasing P stepwise every 255 s. For the weak excitation ($P \leq 100$ W/cm^2), n_{HS} was essentially unchanged ($\leq 10\%$). In the range of 100 to 400 mW/cm^2, n_{HS} sharply increased from approximately 10% to about 60%. According to the variation of P, n_{HS} promptly changed and reached the steady state under each photoexcitation. In the photostationary state, the conversion to the HS sites balances with their thermal and/or quantum relaxation. Stronger excitation above $P = 400$ mW/cm^2 hardly affected n_{HS}, which is almost constant from about 60% to 70%. Such a dependency of n_{HS} as a function of P becomes insensitive with increasing the measuring temperature, as shown by the thin line in Figure 22a.

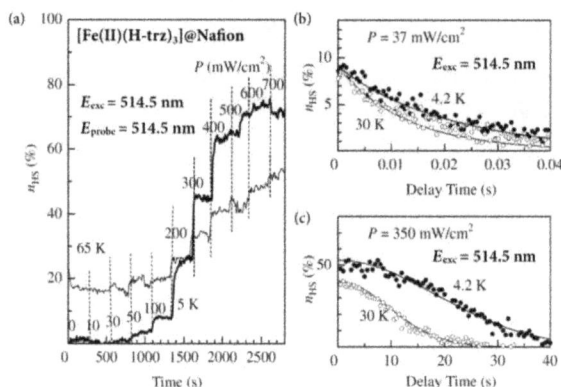

Figure 22. (a) Variation of the density, n_{HS}, of the photoexcited HS molecules in the [Fe(II)(H-trz)$_3$]@Nafion at 5 K (thick curve) and 65 K (thin curve) under photoexcitation. Relaxation dynamics of the photogenerated HS molecules for [Fe(II)(H-trz)$_3$]@Nafion at 4.2 K and 30 K after turning off the excitation light: (b) P = 37 mW/cm^2 and (c) P = 350 mW/cm^2. Solid curves are least-squares-fitting results [29].

At low temperatures, the photogenerated HS fraction as a function of P shows a discontinuous jump when P reaches a threshold value (P_c), which is found at ~60 mW/cm^2 at 4.2 K. Such threshold behavior is ascribed to the critical nucleation of the photogenerated HS clusters, and the discontinuous jump from the low-n_{HS} phase to the high-n_{HS} one corresponds to the phase transition into the condensed state [29]. The lifetime of HS molecules in the high-n_{HS} region becomes three orders of magnitude longer than that of the isolated molecules, reflecting the elastic interaction between neighboring HS molecules, which looks like a condensation process. Actually, the relaxation process from the condensed state is non-exponential, making a sharp contrast with the conventional exponential decay for non-interacting particles, which is shown in Figure 22b. At low temperatures (≤35 K) and in the high-P region, as shown in Figure 22, the nonlinear increase in n_{HS}, as well as the non-exponential decay of n_{HS}, suggest the dynamical phase transition into the photoexcited condensed phase, which is schematically shown in Figure 23. The hatched region represents the condensed phase of the photoexcited HS state, which is observed only under photoexcitation, and vanishes if the excitation light is switched off. In other words, this condensed phase is stabilized by the attractive interaction between the HS sites and not by the transformation into a metastable structure.

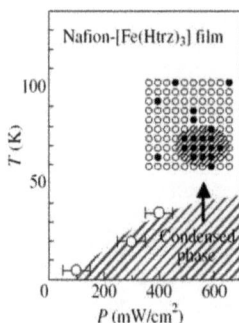

Figure 23. Phase diagram of the [Fe(II)(Htrz)$_3$]@Nafion against excitation power density (P) and temperature (T) [29]. The hatched region represents the condensed phase of the photoexcited HS molecules. The inset of closed circles shows the distribution pattern of the photoexcited HS molecules.

7. Experimental Procedure

7.1. Syntheses

7.1.1. [Fe(II)(NH$_2$-trz)$_3$](o-, m-, p-tos)$_2$·xH$_2$O Synthesis

Iron powder was dissolved in an aqueous solution of toluenesulfonic acid. Adding an aqueous solution of NH$_2$-trz to the above solution, [Fe(II)(NH$_2$-trz)$_3$](o-, m-, p-tos)$_2$·xH$_2$O was precipitated immediately. The amount of hydrated water, x, was determined by thermogravimetric analysis as x = 2, 1.5, and 1.5 for the o-, m-, p-isomers. [Fe(II)(NH$_2$-trz)$_3$](o-, m-, p-abs)$_2$·2H$_2$O was also synthesized likewise. The number of hydrated water molecules was determined by thermogravimetric analysis as 2 for all the o-, m-, p-isomers.

7.1.2. [Fe(II)(NH$_2$-trz)$_3$](SP150)$_2$·2H$_2$O Synthesis

An aqueous solution of SP150 (Japan Photopigment Institute Co., Okayama, Japan) was mixed with the cation-exchange resin (TOYOPEARL 650) adsorbing iron(II) ions, stirred overnight, then filtered off. After evaporating water, Fe(SP150)$_2$ was obtained. The methanol solution of Fe(II)(SP150)$_2$ was mixed with a methanol solution of NH$_2$-trz. Then, [Fe(II)(NH$_2$-trz)$_3$](SP150)$_2$ precipitated immediately. [Fe(II)(NH$_2$-trz)$_3$](SP150)$_2$ absorbs two crystal water molecules per the formula unit under ambient humidity, which is determined by means of thermogravimetric analysis.

7.1.3. [Fe(II)(NH$_2$-trz)$_3$](C$_n$H$_{2n+1}$SO$_3$)$_2$·xH$_2$O Synthesis

In the cases of n = 1 and n = 2, iron powder was dissolved in an aqueous solution of the alkanesulfonic acid C$_n$H$_{2n+1}$SO$_3$H. After evaporating the water, the desired compound precipitated as a bluish powder. Ascorbic acid was used to prevent the partial oxidation of iron(II). The crude product was recrystallized from methanol. In the cases of n = 3–9, an aqueous solution of FeSO$_4$·7H$_2$O was mixed with cation-exchange resin, stirred overnight, then filtered off. The aqueous solution of C$_n$H$_{2n+1}$SO$_3$Na was mixed with the cation-exchange resin absorbing iron(II) ions, stirred overnight, then filtered off. After evaporating the water, the desired compound precipitated. The crude product was recrystallized from methanol. The methanol solution of Fe(C$_n$H$_{2n+1}$SO$_3$)$_2$·xH$_2$O was mixed with the methanol solution of NH$_2$-trz, then [Fe(II)(NH$_2$-trz)$_3$](C$_n$H$_{2n+1}$SO$_3$)$_2$·xH$_2$O precipitated immediately as a white powder, and was then filtered off. Because of the hydration, the color of the compound turned to red-purple after a while. The number of crystal water was determined by the thermogravimeter. The value of x is 0.5 for n = 1, 2.5 for n = 2, and 1.5 for the other compounds.

7.1.4. SCO Complex Film [Fe(II)(R-trz)$_3$]@Nafion (R = H, NH$_2$) Synthesis

The acid form of Nafion 117 was immersed in an aqueous solution of FeSO$_4$·7H$_2$O. The mobile anions (SO$_4{}^{2-}$) were excluded from the polymer matrix because their ionic charge was identical to that of the fixed ions. After being immersed in the aqueous solution of FeSO$_4$ for 2 h, the film was rinsed in methanol, then it was immersed in a methanol solution of R-trz. One hour later, the film was picked up and washed with methanol, then dried under a nitrogen atmosphere.

7.2. Measurements

7.2.1. Fe K-Edge EXAFS Spectroscopy

Fe K-edge EXAFS spectra were measured in the conventional transmission mode at BL-10B in the Photon Factory (operation energy of 2.5 GeV and stored current of 400–200 mA) at the High Energy Accelerator Research Organization in Tsukuba, Japan. A water-cooled Si(311) channel-cut crystal was employed as a monochromator. The intensities of the incident and transmitted X-ray were recorded using ionization chambers filled with pure N$_2$. Although the absolute photon energy was not calibrated, the relative photon energies never changed during the measurements within the employed energy step (0.2 eV), judging from several glitches appearing in the I_0 function. The sample was diluted with BN to give a pellet for EXAFS measurements. In order to prevent the desorption of crystal water in the vacuum, the pellet was completely sealed with an adhesive substrate (STYCAST 1266). The Fe K-edge jump was found to be approximately 0.1, while the total absorption coefficient was estimated to be less than 4.0, most of which originated from BN and the adhesive substrate. For the low-temperature measurements, a closed cycle He refrigerator was used and the temperature was measured with a Si diode placed close to the sample. In order to evaluate the back-scattering amplitude and the phase shift, we employed the theoretical standards given by FEFF6 [62]. A model structure constituted of a $Fe_5(NH_2$-trz$)_{12}$ chain cluster was presumed in FEFF6 calculations, where the central iron atom is an X-ray-absorbing atom and all the H atoms are neglected, therefore there are 77 atoms in total.

7.2.2. Magnetic Susceptibility Measurements

The temperature dependence of $\chi_M T$ was measured with an MPMS-5S SQUID magnetometer (Quantum Design, San Diego, CA, USA). Powder samples were sealed in an aluminum capsule to prevent the loss of crystal water on the heating process. The applied field was 0.1 T or 0.5 T. The temperature was swept in heating and cooling modes in order to examine the thermal hysteresis effect. In compounds showing the thermal hysteresis, the temperature was cycled more than twice.

7.2.3. ^{57}Fe Mössbauer Spectroscopy

For ^{57}Fe Mössbauer spectroscopic measurement (Topologic Systems Co., Kanagawa, Japan), ^{57}Co in Rh matrix was used as a Mössbauer source. Powder samples were sealed in an acrylate resin sample holder with silicon grease to prevent the loss of crystal water. The spectra were calibrated by using the six lines of a body-centered cubic iron foil (α-Fe), the center of which was taken as zero isomer shift. A cryogenic refrigerator set (Iwatani Industrial Gases Co., Osaka, Japan), Cryomini and MiniStat, was used in the temperature range between 10 and 300 K. The spectra were fitted with a MossWinn 3.0 program [63].

7.2.4. Optical Spectroscopy

In order to investigate the LIESST effect for [Fe(II)(H-trz)$_3$]@Nafion, we utilized the LS absorption band corresponding to the $^1A_{1g} \rightarrow {}^1T_{1g}$ transition as a sensitive monitor of the density of photoexcited HS molecules, n_{HS} (= $1 - n_{LS}$), and investigated the variation of n_{HS} against the excitation power density (P), temperature (T), and the delay time (t). The 514.5 nm line of Ar$^+$ ion laser was used as the source of excitation and probe lights [29]. The power density of the probe light was kept below 0.5 mW/cm^2. The excitation light, which selectively excites the LS state into the HS state, was focused on the sample so that the excitation light (3 mm φ) completely overlaps the probe light (1 mm φ).

8. Conclusions

In the triazole-bridged iron(II) polymeric chain system, [Fe(II)(R-trz)$_3$]X$_2 \cdot x$H$_2$O, where X is the anion, we have controlled the SCO phenomenon by means of characteristic properties of counter anions and crystal water molecules.

With regard to the effect of crystal water on the SCO phenomenon, we prepared [Fe(II)(NH$_2$-trz)$_3$] (p-tos)$_2 \cdot x$H$_2$O (x = 1.5, 3.5, 5.5, and 8) and investigated the LS to HS transition. In this system, with increasing the amount of crystal water, x, the transition temperature lowers and the thermal hysteresis width becomes narrow.

In order to investigate the effect of counter anions on the SCO phenomenon, we have synthesized [Fe(II)(R-trz)$_3$]X$_2 \cdot x$H$_2$O (X = tos and abs). In these systems, $T_{1/2}$ and its hysteresis width vary drastically with the structural isomerism, which implies the possibility of photoinduced spin transition by means of the photoisomerization of counter ions. Based on this strategy, we synthesized

[Fe(II)(NH$_2$-trz)$_3$](SP150)$_2$·2H$_2$O and investigated the SCO phenomenon. In the case where X is C$_n$H$_{2n+1}$SO$_3^-$, the spin transition temperature ($T_{1/2}$) increases with increasing the length of the alkyl chain of the counter ion and saturates above n = 5, which is attributed to the increase in the intermolecular interaction between the alkyl chains of C$_n$H$_{2n+1}$SO$_3^-$, called the fastener effect. The hysteresis width of $\Delta T_{1/2}$ decreases with increasing n, showing the even-odd effect. Moreover, we developed [Fe(II)(R-trz)$_3$]@Nafion films exhibiting SCO phenomena around room temperature, in which the Nafion membrane behaves as a counter anion as well as a transparent substrate. In this preparation process, it is possible to produce the homogeneous SCO complex film of 300 by 300 mm^2 based on Nafion. Moreover, we investigated the LIESST effect for [Fe(II)(H-trz)$_3$]@Nafion and observed the condensed photogenerated HS state below 35 K. The lifetime of the photogenerated HS state strongly depends on the intensity of irradiated light.

Acknowledgments: We wish to thank Akio Nakamoto, Yuichi Murakami and Syuji Toyazaki for the development of triazole-bridged iron(II) polymeric chain system, [Fe(II)(R-trz)$_3$]A·xH$_2$O (A = anion, ion-exchange resin). We also thank to Toshihiko Yokoyama for the measurement and analysis of Fe K-edge EXAFS, and XiaoJun Liu, Yutaka Moritomo, and Arao Nakamura for the investigation of LIESST effect. This work was supported by a Grant-in-Aid for Scientific Research from the Ministry of Education, Culture, Sports, Science and Technology, Japan.

Author Contributions: Norimichi Kojima conceived and directed the project. Akira Sugahara, Hajime Kamebuchi, and Norimichi Kojima synthesized, characterized and measured the SCO materials. Atsushi Okazawa and Masaya Enomoto also contributed to the measurements of magnetic susceptibility and ^{57}Fe Mössbauer spectra. Akira Sugahara, Hajime Kamebuchi, and Norimichi Kojima mainly wrote the manuscript, which was improved by all the authors.

Conflicts of Interest: The authors declare no conflict of interest.

References

1. Cambi, L.; Gagnasso, A. Iron dithiocarbamates and nitrosodithiocarbamates. *Atti. Accad. Naz. Lincei.* **1931**, *13*, 809–813.

2. Gütlich, P.; Goodwin, H.A. *Spin Crossover in Transition Metal Compounds I–III*; Springer: Berlin/Heidelberg, Germany, 2004.

3. Decurtins, S.; Gütlich, P.; Köhler, C.P.; Spiering, H. Light-induced excited spin state trapping in a transition-metal complex: The hexa-1-propyltetrazole-iron(II) tetrafluoroborate spin-crossover system. *Chem. Phys. Lett.* **1984**, *105*, 1–4.

4. Decurtins, S.; Gütlich, P.; Hasselbach, K.M.; Hauser, A.; Spiering, H. Light-induced excited-spin-state trapping in iron(II) spin-Crossover systems. Optical spectroscopic and magnetic susceptibility study. *Inorg. Chem.* **1985**, *24*, 2174–2178.

5. Gütlich, P.; Gaspar, A.B.; Garcia, Y. Spin state switching in iron coordination compounds. *Beilstein J. Org. Chem.* **2013**, *9*, 342–391. [CrossRef] [PubMed]

6. Krober, J.; Codjovi, E.; Kahn, O.; Groliere, F.; Jay, C. A spin transition system with a thermal hysteresis at room temperature. *J. Am. Chem. Soc.* **1993**, *115*, 9810–9811. [CrossRef]

7. Aromí, G.; Barrios, L.A.; Roubeau, O.; Gamez, P. Triazoles and tetrazoles: Prime ligands to generate remarkable coordination materials. *Coord. Chem. Rev.* **2011**, *255*, 485–546. [CrossRef]

8. Roubeau, O. Triazole-based one-dimensional spin-crossover coordination polymers. *Chem. Eur. J.* **2012**, *18*, 15230–15244. [CrossRef] [PubMed]

9. Jureschi, C.-M.; Linares, J.; Boulmaali, A.; Dahoo, P.R.; Rotaru, A.; Garcia, Y. Pressure and temperature sensors using two spin crossover materials. *Sensors* **2016**, *16*, 187. [CrossRef] [PubMed]

10. Kahn, O.; Martinez, C.J. Spin-transition polymers: From molecular materials toward memory devices. *Science* **1998**, *279*, 44–48. [CrossRef]

11. Garcia, Y.; Renz, F.; Gütlich, P. LIESST effect in Fe(II) 1,2,4-triazole chains. *Curr. Inorg. Chem.* **2016**, *6*, 4–9.

12. Grosjean, A.; Daro, N.; Kauffmann, B.; Kaiba, A.; Létard, J.-F.; Guionneau, P. The 1-D polymeric structure of the [Fe(NH$_2$trz)$_3$](NO$_3$)$_2$·nH$_2$O (with n = 2) spin crossover compound proven by single crystal investigations. *Chem. Commun.* **2011**, *47*, 12382–12384. [CrossRef] [PubMed]

13. Armand, F.; Badoux, C.; Bonville, P.; Ruaudel-Teixier, A.; Kahn, O. Langmuir–blodgett films of spin transition iron(II) metalloorganic polymers. 1. Iron(II) complexes of octadecyl-1,2,4-triazole. *Langmuir* **1995**, *11*, 3467–3472. [CrossRef]

14. Kuroiwa, K.; Shibata, T.; Sasaki, S.; Ohba, M.; Takahara, A.; Kunitake, T.; Kimizuka, N. Supramolecular control of spin-crossover phenomena in lipophilic Fe(II)-1,2,4-triazole complexes. *J. Polym. Sci. Polym. Chem.* **2006**, *44*, 5192–5202. [CrossRef]

15. Fujigaya, T.; Jiang, D.L.; Aida, T. Spin-crossover dendrimers: Generation number-dependent cooperativity for thermal spin transition. *J. Am. Chem. Soc.* **2005**, *127*, 5484–5489. [CrossRef] [PubMed]

16. Sonar, P.; Grunert, C.M.; Wei, Y.-L.; Kusz, J.; Gütlich, P.; Schlüter, A.D. Iron(II) spin transition complexes with dendritic ligands, part I. *Eur. J. Inorg. Chem.* **2008**, *10*, 1613–1622. [CrossRef]

17. Wei, Y.-L.; Sonar, P.; Grunert, M.; Kusz, J.; Schlüter, A.D.; Gütlich, P. Iron(II) spin-transition complexes with dendritic ligands, part II. *Eur. J. Inorg. Chem.* **2010**, *25*, 3930–3941. [CrossRef]

18. Rubio, M.; Hernández, R.; Nogales, A.; Roig, A.; López, D. Structure of a Spin-Crossover Fe(II)–1,2,4-triazole polymer complex dispersed in an isotactic polystyrene matrix. *Eur. Polym. J.* **2011**, *47*, 52–60. [CrossRef]

19. Lee, S.-W.; Lee, J.-W.; Jeong, S.-H.; Park, I.-W.; Kim, Y.-M.; Jin, J.-I. Processable magnetic plastics composites—Spin crossover of PMMA/Fe(II)-complexes composites. *Synth. Met.* **2004**, *142*, 243–249. [CrossRef]

20. Gural'skiy, I.A.; Quintero, C.M.; Costa, J.S.; Demont, P.; Molnár, G.; Salmon, L.; Shepherd, H.J.; Bousseksou, A. Spin crossover composite materials for electrothermomechanical actuators. *J. Mater. Chem. C* **2014**, *2*, 2949–2955. [CrossRef]

21. Nakamoto, A.; Ono, Y.; Kojima, N.; Matsumura, D.; Yokoyama, T. Spin crossover complex film, [Fe^{II}(H-trz)$_3$]-nafion, with a spin transition around room temperature. *Chem. Lett.* **2003**, *32*, 336–337. [CrossRef]

22. Nakamoto, A.; Ono, Y.; Kojima, N.; Matsumura, D.; Yokoyama, T.; Liu, X.J.; Moritomo, Y. Spin transition and its photo-induced effect in spin crossover complex film based on [Fe(II)(trz)$_3$]. *Synth. Met.* **2003**, *137*, 1219–1220. [CrossRef]

23. Nakamoto, A.; Kamebuchi, H.; Enomoto, M.; Kojima, N. Study on the spin crossover transition and glass transition for Fe(II) complex film, [Fe(II)(H-triazole)$_3$]@Nafion, by means of Mössbauer spectroscopy. *Hyperfine Interact.* **2012**, *205*, 41–45. [CrossRef]

24. Sugahara, A.; Enomoto, M.; Kojima, N. Isomerization effect of counter anion on the spin crossover transition in [Fe(4-NH$_2$trz)$_3$](CH$_3$C$_6$H$_4$SO$_3$)$_2$·nH$_2$O. *J. Phys. Conf. Ser.* **2010**, *217*, 12128. [CrossRef]

25. Toyazaki, S.; Nakanishi, M.; Komatsu, T.; Kojima, N.; Matsumura, D.; Yokoyama, T. Control of T_C by isomerization of counter anion in Fe(II) spin crossover complexes, [Fe(4-NH$_2$trz)$_3$](R-SO$_3$)$_2$. *Synth. Met.* **2001**, *121*, 1794–1795. [CrossRef]

26. Toyazaki, S. The Control of Spin-Crossover Behavior of Fe(II) Triazole Complexes. Master's Thesis, The University of Tokyo, Tokyo, Japan, 2000.

27. Kamebuchi, H.; Nakamoto, A.; Yokoyama, T.; Kojima, N. Fastener effect on uniaxial chemical pressure for one-dimensional spin-crossover system, [Fe^{II}(NH$_2$-trz)$_3$](C$_n$H$_{2n+1}$SO$_3$)$_2$·xH$_2$O: Magnetostructural correlation and ligand field analysis. *Bull. Chem. Soc. Jpn.* **2015**, *88*, 419–430. [CrossRef]

28. Absmeier, A.; Bartel, M.; Carbonera, C.; Jameson, G.N.L.; Weinberger, P.; Caneschi, A.; Mereiter, K.; Létard, J.-F.; Linert, W. Both spacer length and parity influence the thermal and light-induced properties of iron(II) α,ω-bis(tetrazole-1-yl)alkane coordination polymers. *Chem. Eur. J.* **2006**, *12*, 2235–2243. [CrossRef] [PubMed]

29. Liu, X.J.; Moritomo, Y.; Kawamoto, T.; Nakamoto, A.; Kojima, N. Dynamical phase transition in a spin-crossover complex. *J. Phys. Soc. Jpn.* **2003**, *72*, 1615–1618. [CrossRef]

30. Hostettler, M.; Törnroos, K.W.; Chernyshov, D.; Vangdal, B.; Bürgi, H.-B. Challenges in engineering spin crossover: Structures and magnetic properties of six alcohol solvates of iron(II) tris(2-picolylamine) dichloride. *Angew. Chem. Int. Ed.* **2004**, *43*, 4589–4594. [CrossRef] [PubMed]

31. Nakamoto, T.; Bhattacharjee, A.; Sorai, M. Cause for unusually large thermal hysteresis of spin crossover in [Fe(2-pic)$_3$]Cl$_2$·H$_2$O. *Bull. Chem. Soc. Jpn.* **2004**, *77*, 921–932. [CrossRef]

32. Codjovi, E.; Sommier, L.; Kahn, O.; Jay, C. A spin transition molecular material with an exceptionally large thermal hysteresis loop at room temperature. *New J. Chem.* **1996**, *20*, 503–505.

33. Toyazaki, S.; Murakami, Y.; Komatsu, T.; Kojima, N.; Yokoyama, T. Study on the spin-crossover system for [Fe(4-NH$_2$trz)$_3$](p-CH$_3$C$_6$H$_4$SO$_3$)$_2$·nH$_2$O. *Mol. Cryst. Liq. Cryst.* **2000**, *343*, 175–180. [CrossRef]

34. Stern, E.A. Theory of EXAFS. In *X-ray Absorption: Principles, Applications, Techniques of EXAFS, SEXAFS and XANES*; Koningsberger, D.C., Prins, R., Eds.; Wiley: New York, NY, USA, 1988; pp. 3–52.

35. Michalowicz, A.; Moscovici, J.; Ducourant, B.; Cracco, D.; Kahn, O. EXAFS and X-ray powder diffraction studies of the spin transition molecular materials [Fe(Htrz)$_2$(trz)](BF$_4$) and [Fe(Htrz)$_3$](BF$_4$)$_2$·H$_2$O (Htrz = 1,2,4–4*H*-triazole; trz = 1,2,4-triazolato). *Chem. Mater.* **1995**, *7*, 1833–1842. [CrossRef]

36. Soyer, H.; Mingotaud, C.; Boillot, M.-L.; Delhaes, P. Spin-crossover complex stabilized on a formamide/water subphase. *Thin Solid Films* **1998**, *327–329*, 435–438. [CrossRef]

37. Lavrenova, L.G.; Yudina, N.G.; Ikorskii, V.N.; Varnek, V.A.; Oglezneva, I.M.; Larionov, S.V. Spin-crossover and thermochromism in complexes of iron(II) iodide and thiocyanate with 4-amino-1,2,4-triazole. *Polyhedron* **1995**, *14*, 1333–1337. [CrossRef]

38. Bausk, N.V.; Érenburg, S.B.; Mazalov, L.N.; Lavrenova, L.G.; Ikorskii, V.N. Electronic and spatial structure of spin transition iron(II) tris(4-amino-1,2,4-triazole) nitrate and perchlorate complexes. *J. Struct. Chem.* **1994**, *35*, 509–516. [CrossRef]

39. Varnek, V.A.; Lavrenova, L.G. Mössbauer study of the influence of ligands and anions of the second coordination sphere in Fe(II) complexes with 1,2,4-triazole and 4-amino-1,2,4-triazole on the temperature of the $^1A_1 \leftrightarrows {}^5T_2$ spin transitions. *J. Struct. Chem.* **1995**, *36*, 104–111. [CrossRef]

40. Van Koningsbruggen, P.J.; Garcia, Y.; Codjovi, E.; Lapouyade, R.; Kahn, O.; Fournès, L.; Rabardel, L. Non-classical Fe^{II} spin-crossover behaviour in polymeric iron(II) compounds of formula [Fe(NH$_2$trz)$_3$]X$_2$·xH$_2$O (NH$_2$trz = 4-amino-1,2,4-triazole; X = derivatives of naphthalene sulfonate). *J. Mater. Chem.* **1997**, *7*, 2069–2075. [CrossRef]

41. Garcia, Y.; van Koningsbruggen, P.J.; Lapouyade, R.; Rabardel, L.; Kahn, O.; Wieczorek, M.; Bronisz, R.; Ciunik, Z.; Rudolf, M.F. Synthesis and spin-crossover characteristics of polynuclear 4-(2′-hydroxy-ethyl)-1,2,4-triazole Fe(II) molecular materials. *C. R. Acad. Sci. Ser. IIC Chem.* **1998**, *1*, 523–532. [CrossRef]

42. Erenburg, S.B.; Bausk, N.V.; Lavrenova, L.G.; Varnek, V.A.; Mazalov, L.N. Relation between electronic and spatial structure and spin-transition parameters in chain-like Fe(II) compounds. *Solid State Ion.* **1997**, *101–103*, 571–577. [CrossRef]

43. Dîrtu, M.M.; Rotaru, A.; Gillard, D.; Linares, J.; Codjovi, E.; Tinant, B.; Garcia, Y. Prediction of the spin transition temperature in Fe^{II} one-dimensional coordination polymers: An anion based database. *Inorg. Chem.* **2009**, *48*, 7838–7852. [CrossRef] [PubMed]

44. Jenkins, H.D.B.; Roobottom, H.K.; Passmore, J.; Glasser, L. Relationships among Ionic Lattice Energies, Molecular (Formula Unit) Volumes, and Thermochemical Radii. *Inorg. Chem.* **1999**, *38*, 3609–3620. [CrossRef] [PubMed]

45. Bushuev, M.B.; Lavrenova, L.G.; Shvedenkov, Y.G.; Varnek, V.A.; Sheludyakova, L.A.; Volkov, V.V.; Larionov, S.V. Complexes Fe(HTrz)$_3$B$_{10}$H$_{10}$·H$_2$O and Fe(NH$_2$Trz)$_3$B$_{10}$H$_{10}$·H$_2$O (HTrz = 1,2,4-triazole and NH$_2$Trz = 4-amino-1,2,4-triazole). The spin transition $^1A_1 \leftrightarrows {}^5T_2$ in Fe(HTrz)$_3$B$_{10}$H$_{10}$·H$_2$O. *Russ. J. Coord. Chem.* **2008**, *34*, 190–194. [CrossRef]

46. Urakawa, A.; van Beek, W.; Monrabal-Capilla, M.; Galán-Mascarós, J.R.; Palin, L.; Milanesio, M. Combined, modulation enhanced X-ray powder diffraction and raman spectroscopic study of structural transitions in the spin crossover material [Fe(Htrz)$_2$(trz)](BF$_4$). *J. Phys. Chem. C* **2011**, *115*, 1323–1329. [CrossRef]

47. Grosjean, A.; Négrier, P.; Bordet, P.; Etrillard, C.; Mondieig, D.; Pechev, S.; Lebraud, E.; Létard, J.-F.; Guionneau, P. Crystal structures and spin crossover in the polymeric material [Fe(Htrz)$_2$(trz)](BF$_4$) including coherent-domain size reduction effects. *Eur. J. Inorg. Chem.* **2013**, *2013*, 796–802. [CrossRef]

48. Gütlich, P.; Gaspar, A.B.; Ksenofontov, V.; Garcia, Y. Pressure effect studies in molecular magnetism. *J. Phys. Condens. Matter* **2004**, *16*, S1087–S1108. [CrossRef]

49. Gütlich, P.; Ksenofontov, V.; Gaspar, A.B. Pressure effect studies on spin crossover systems. *Coord. Chem. Rev.* **2005**, *249*, 1811–1829. [CrossRef]

50. Gütlich, P.; Gaspar, A.B.; Garcia, Y.; Ksenofontov, V. Pressure effect studies in molecular magnetism. *C. R. Chim.* **2007**, *10*, 21–36. [CrossRef]

51. Klokishner, S.; Linares, J.; Varret, F. Effect of hydrostatic pressure on phase transitions in spin-crossover 1D systems. *Chem. Phys.* **2000**, *255*, 317–323. [CrossRef]

52. Bausk, N.V.; Érenburg, S.B.; Lavrenova, L.G.; Mazalov, L.N. EXAFS study of spin transition effect on the spatial and electronic structure of Fe(II) complexes with triazoles. *J. Struct. Chem.* **1995**, *36*, 925–931. [CrossRef]

53. Erenburg, S.B.; Bausk, N.V.; Varnek, V.A.; Lavrenova, L.G. Influence of the electronic and spatial structure parameters on the spin-transition temperature in unusual chain iron(II) complexes. *J. Magn. Magn. Mater.* **1996**, *157–158*, 595–596. [CrossRef]

54. Kojima, N.; Murakami, Y.; Komatsu, T.; Yokoyama, T. EXAFS study on the spin-crossover system, [Fe(4-NH$_2$trz)$_3$](R-SO$_3$)$_2$. *Synth. Met.* **1999**, *103*, 2154. [CrossRef]

55. Murakami, Y.; Komatsu, T.; Kojima, N. Control of Tc and spin bistability in the spin-crossover system, [Fe(4-NH$_2$trz)$_3$](R-SO$_3$)$_2$. *Synth. Met.* **1999**, *103*, 2157–2158. [CrossRef]

56. Kojima, N.; Toyazaki, S.; Itoi, M.; Ono, Y.; Aoki, W.; Kobayashi, Y.; Seto, M.; Yokoyama, T. Search on multi-functional properties of spin-crossover system. *Mol. Cryst. Liq. Cryst.* **2002**, *376*, 567–574. [CrossRef]

57. Hayami, S.; Gu, Z.-Z.; Einaga, Y.; Kobayashi, Y.; Ishikawa, Y.; Yamada, Y.; Fujishima, A.; Sato, O. A novel LIESST iron(II) complex exhibiting a high relaxation temperature. *Inorg. Chem.* **2001**, *40*, 3240–3242. [CrossRef] [PubMed]

58. Bonhommeau, S.; Molnár, G.; Galet, A.; Zwick, A.; Real, J.-A.; McGarvey, J.J.; Bousseksou, A. One shot laser pulse induced reversible spin transition in the spin-crossover complex [Fe(C$_4$H$_4$N$_2$){Pt(CN)$_4$}] at room temperature. *Angew. Chem. Int. Ed.* **2005**, *44*, 4069–4073. [CrossRef] [PubMed]

59. Hellel, W.; Ould Hamouda, A.; Degert, J.; Létard, J.F.; Freysz, E. Switching of spin-state complexes induced by the interaction of a laser beam with their host matrix. *Appl. Phys. Lett.* **2013**, *103*, 143304. [CrossRef]

60. Gallé, G.; Etrillard, C.; Degert, J.; Guillaume, F.; Létard, J.F.; Freysz, E. Study of the fast photoswitching of spin crossover nanoparticles outside and inside their thermal hysteresis loop. *Appl. Phys. Lett.* **2013**, *102*, 63302. [CrossRef]

61. Vos, G.; De Graaff, R.A.G.; Haasnoot, J.G.; van der Kraan, A.M.; De Vaal, P.; Reedijk, J. Crystal structure at 300 and 105 K, magnetic properties and Mössbauer spectra of bis(triaquatris(4-ethyltriazole-N^1)iron(II)-N^2,$N^{2'}$,$N^{2''}$)iron(II) hexakis(trifluoromethanesulfonate). A linear, trinuclear iron(II) compound, showing a unique high-spin–low-spin transition of the central iron atom. *Inorg. Chem.* **1984**, *23*, 2905–2910.

62. Zabinsky, S.I.; Rehr, J.J.; Ankudinov, A.; Albers, R.C.; Eller, M.J. Multiple-scattering calculations of X-ray-absorption spectra. *Phys. Rev. B* **1995**, *52*, 2995–3009. [CrossRef]

63. Klencsár, Z. MossWinn—Mössbauer Spectrum Analysis and Database Software. Available online: http://www.mosswinn.com/ (accessed on 18 July 2017).

![inorganics logo] *inorganics*

MDPI

Article

Structural Dynamics of Spin Crossover in Iron(II) Complexes with Extended-Tripod Ligands

Philipp Stock [1], Dennis Wiedemann [1], Holm Petzold [2] and Gerald Hörner [1,*

[1] Institut für Chemie, Technische Universität Berlin, Straße des 17. Juni 135, 10623 Berlin, Germany; philippstock@yahoo.de (P.S.); dennis.wiedemann@chem.tu-berlin.de (D.W.)
[2] TU Chemnitz, Institut für Chemie, Anorganische Chemie, Straße der Nationen 62, 09111 Chemnitz, Germany; holm.petzold@chemie.tu-chemnitz.de
* Correspondence: gerald.hoerner@tu-berlin.de; Tel.: +49-30-314-27936

Received: 1 August 2017; Accepted: 31 August 2017; Published: 5 September 2017

Abstract: Selective manipulation of spin states in iron(II) complexes by thermal or photonic energy is a desirable goal in the context of developing molecular functional materials. As dynamic spin-state equilibration in isolated iron(II) complexes typically limits the lifetime of a given spin state to nanoseconds, synthetic strategies need to be developed that aim at inhibited relaxation. Herein we show that modulation of the reaction coordinate through careful selection of the ligand can indeed massively slow down dynamic exchange. Detailed structural analysis of $[FeL]^{2+}$ and $[ZnL]^{2+}$ (L: tris(1-methyl-2-{[pyridin-2-yl]-methylene}hydrazinyl)phosphane sulfide) with crystallographic and computational methods clearly reveals a unique trigonal-directing effect of the extended-tripod ligand **L** during spin crossover, which superimposes the ubiquitous $[FeN_6]$ breathing with trigonal torsion, akin to the archetypal Bailar twist. As a consequence of the diverging reaction coordinates in $[FeL]^{2+}$ and in the **tren**-derived complex $[Fe(tren)py_3]^{2+}$, their thermal barriers differ massively, although the spin crossover energies are close to identical. As is shown by time-resolved transient spectroscopy and dynamic ^1H-NMR line broadening, reference systems deriving from **tren** (tris-(2-aminoethyl)amine), which greatly lack such trigonal torsion, harbor very rapid spin-state exchange.

Keywords: spin crossover; Bailar twist; prismatic coordination; line broadening; chemical exchange

---

1. Introduction

Numerous first-row transition metal complexes with a d^4 to d^7 electron configuration possess energetically close lying spin states. A jump in thermal energy or admittance of alternative energy sources (X-rays, UV/Vis photons) can be used to stimulate crossover among these states (spin crossover, SCO) [1–5]. The spin isomers are sharply discretized by their magnetic properties, but in many cases they also differ substantially in complementary observables, such as UV/Vis absorption. It is the combined switching of magnetic and optical properties that renders this class of compounds highly promising as opto-magnetic actors and sensors [6,7]. The most prominent example is SCO between the low-spin (**ls**; $^1A_{1g}/t_{2g}^6$, S = 0) and high-spin (**hs**; $^5T_{2g}/t_{2g}^4e_g^2$, S = 2) states in iron(II) complexes (d^6) (Equation (1); γ_{hs} and γ_{ls} denote the spin-state molar fractions).

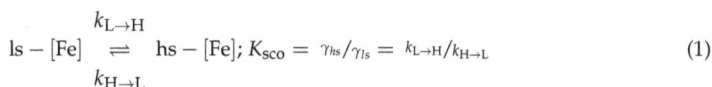

$$\text{ls} - [\text{Fe}] \underset{k_{H \to L}}{\overset{k_{L \to H}}{\rightleftharpoons}} \text{hs} - [\text{Fe}]; \, K_{sco} = \gamma_{hs}/\gamma_{ls} = k_{L \to H}/k_{H \to L} \tag{1}$$

As a prototypical example of a chemical equilibrium, the spin states underlie constant exchange with $k_{H \to L}$ and $k_{L \to H}$ as the rate constants of **hs** and **ls** decay, respectively. Thermal equilibration can be sometimes "frozen out" at cryogenic temperatures [8–12], so that minority states can be enriched as meta-stable species through irradiation with light or via rapid freezing. The former effect was coined as light-induced excited spin-state trapping, LIESST, whereas thermal induced excited spin-state trapping, TIESST, has been established as an alternative method for kinetic stabilization via rapid freezing [9]. Around ambient temperature, however, the exchange reactions are very rapid, leading to randomizing spin states within a few nanoseconds in most iron(II) complexes [13–24].

Our serendipitous observation of spin state lifetimes in the microsecond range for a number of six-coordinate iron(II) complexes therefore marks a notable exception that demands rationalization [25,26]. Ligands of the extended-tripod type with a unique thiophosphoryl capping unit (**L** in Scheme 1 and derivatives thereof) were found to harbor hindered SCO kinetics, irrespective of solvent nature and the presence of dioxygen. Through synthetic fine tuning of the ligand periphery and of the heterocycle, we could recently further extend the time domain of SCO to the millisecond range [27]. Based on extended laser-flash photolysis (LFP) studies in concert with XRD-calibrated density-functional theory modeling, we correlated the exceptionally slow SCO kinetics with the nature of SCO reaction coordinates.

In this picture, slow exchange prevails, when the reaction coordinate consists of two components; (i) trivial isotropic "breathing" of the coordination sphere due to (de)population of anti-bonding e_g-type orbitals in the (**ls**) **hs** state; (ii) anisotropic torsion of the coordination sphere along a (pseudo) C_3 axis, akin to Bailar's trigonal twist [28]. Actually, **L** and derived ligands are known to facilitate the synthesis of trigonal prismatic complexes [29–31]. This correlation acknowledges previous structure-spin state correlations [32], that are based on structural–chemical considerations going back to the early 1970s [33–42]. In particular, the idea of a taming of the SCO dynamics via ligand-imposed trigonal torsion is a recurrent motif of research. The antipode of fast exchange prevails, as long as the reaction coordinate is largely limited to isotropic "breathing". Intriguingly, the topologically related family of iron(II) complexes with **tren**-derived ligands (**tren**: tris-(2-aminoethyl)amine) falls in this second class of rapidly exchanging SCO systems. Large room temperature exchange rate constants k_{obs} of 1.1×10^7 s^{-1} and 1.4×10^7 s^{-1} have been extracted from LFP studies of [**Fe(tren)py₃**](ClO₄)₂ and [**Fe(tren)imid₃**](BF₄)₂, respectively [27,43].

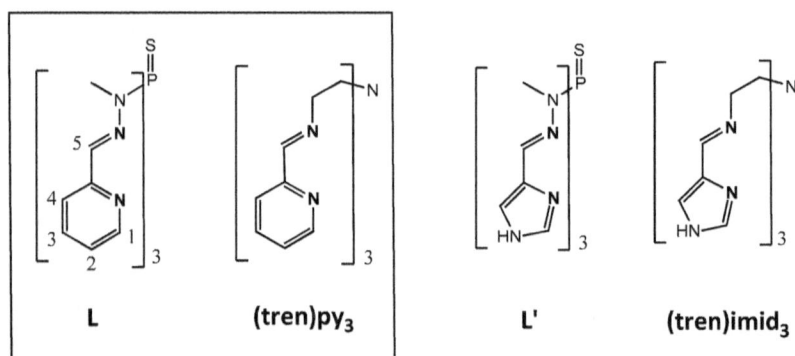

Scheme 1. Extended tripodal $\kappa^6 N$ polyimine ligands investigated in this study (proton numbering in **L** applies also for (**tren)py₃**); iron(II) complexes of the literature known imidazole ligands **L′** [31] and (**tren)imid₃** [44–48] serve as DFT reference systems.

As a matter of fact, high susceptibility towards even subtle variations in the ligand sphere is a more general feature of the SCO phenomenon. Structure–function relations have been discussed in some detail [49], mainly with respect to SCO thermodynamics and control of cooperativity among SCO manifolds [50], however. By contrast, the probably broadest experimental work on SCO dynamics dates back to 1996 [22], while the latest review in the field stems from 2004 [21]. Clearly, the knowledge on the SCO exchange kinetics in general and the ligand-borne effects on kinetics in particular require significant extension. In this work we therefore set out to provide deeper insights into the structural dynamics of spin crossover of the complexes [**FeL**]$^{2+}$ and [**Fe(tren)py$_3$**]$^{2+}$ and its interrelations with the phenomenological spin-state dynamics. X-ray crystallography of zinc(II) and iron(II) complexes of **L** and comparison with [**FeL'**]$^{2+}$ [27] served to calibrate DFT-based structure prediction of the experimentally elusive **hs** states of [**FeL**]$^{2+}$ and [**Fe(tren)py$_3$**]$^{2+}$. Based on an in-depth analysis of the calibration and prediction structure datasets, we suggest **hs**-state structures that show strong trigonal distortion in [**FeL**]$^{2+}$ but are still in line with octahedral coordination in [**Fe(tren)py$_3$**]$^{2+}$. The respective zinc(II) complexes prove to be valid real-world structure models in both cases. Accordingly, the SCO reaction coordinates of both iron(II) complexes are concluded to be distinctly different, despite the largely conserved donor set and topology. Ligand-imposed trigonal torsion akin to the Bailar twist is evident in [**FeL**]$^{2+}$, but is absent in [**Fe(tren)py$_3$**]$^{2+}$. Divergence in the reaction coordinate is manifest in the SCO dynamics, as is consistently measured by laser flash photolysis and VT-NMR spectroscopy. The latter technique allows a direct phenomenological differentiation between "slowly" and "rapidly" exchanging SCO systems through qualitative inspection of NMR linewidths.

2. Results

2.1. Structural Characteristics of [FeL]$^{2+}$ and [ZnL]$^{2+}$

2.1.1. Complex Synthesis and Solid State Structures

Ligand **L** was prepared along a published route [25,30] by reaction of the thiophosphoryl hydrazide $(S)P(N^{Me}NH_2)_3$ with three equivalents pyridine-2-carbaldehyde in methanol. In situ complexation with the hexahydrates of iron(II) tetrafluoroborate and zinc(II) perchlorate affords deep red and colorless solutions, respectively, from which the products precipitate in good yields within hours. Elemental analysis indicates 1:1 metal ligand stoichiometry and the presence of ethanol as solvate. NMR spectra reveal (averaged) C_3 symmetry in both cases. A detailed discussion of the spectra is given below. Single crystals of [**FeL**]$(BF_4)_2$ and [**ZnL**]$(ClO_4)_2$ that were suitable for XRD structure elucidation were obtained from concentrated solutions in MeCN within few days via isothermal diffusion of diethyl ether. Both compounds crystallized as MeCN solvates. Crystallographic details are summarized in Table 1, pertinent structural features are given in Table 2.

The iron(II) complex crystallizes in the orthorhombic space-group type $P2_12_12_1$ with 1.5 MeCN molecules per complex. The asymmetric unit contains two complex units of complementary helicity. While the overall packing is unexceptional, we note π-stacking between pyridine moieties of two adjacent complex units and a dispersive interaction between the capping sulfur atom and a methyl group. The zinc(II) complex also crystallizes as MeCN solvate in the monoclinic space-group type $P2_1/c$. The solid-state structure lacks intermolecular interactions between complex cations. The cation structures are found to be distinctly dependent on the nature of the central ion (Figure 1; structure details in Table 2). The iron(II) complex reveals fairly regular N_6 coordination with little variation in the Fe–N bond lengths and only minor *cis*-angle distortion. The iron ion is well centered in the N_6 coordination sphere; the displacement δ is small. Its negative value signalizes displacement towards the "P(S)-clamped" side. An average Fe–N bond length of 1.952(15) Å indicates **ls** configuration of the d^6 ion, which is in agreement with ^{1}H-NMR resonances in the range of 9.0 ppm > δ > 2 ppm, typical of diamagnetic compounds. Projection of the coordination sphere along the (pseudo) C_3 axis reveals significant distortion from a regular octahedron (right in Figure 1). The trigonal twist angle θ was introduced by Hendrickson et al. in order to quantify such Bailar-type [28,36] distortions of L_6 coordination spheres. The trigonal twist angle θ amounts to 60° for a regular octahedron and gives 0° for a trigonal prism. With a value of θ = 43.4(8)°, the coordination pattern of [**FeL**]$^{2+}$ is well within the range observed for **ls**-iron(II) complexes of N_6 ligands of the extended tripod type and closely mimics structure analogues with pyridine derivatives [26,51]. As the experimental data are closely matched by a DFT-derived vacuum structure of the isolated complex with respect to bond lengths and angles (see Table 2), only minor matrix effects may be expected. This conclusion is corroborated by crystallographic work-in-progress, which indicates conserved complex metrics of **[FeL]$^{2+}$** and **[ZnL]$^{2+}$**, even when the counter ion is varied. It is emphasized, however, that matrix effects through counter ion variation or solvates commonly and strongly affect the phenomenology of SCO. Tris(2-picolylamine)iron(II) dichloride, for instance, may be **ls**, **hs** or undergo SCO, depending on the nature of co-crystallized alcohol in an unpredictable manner [52]. Accordingly, the trigonal twist angle θ of the **hs**-complex **[FeL']$^{2+}$** was recently shown to be significantly biased by matrix effects, firstly causing some deviation of the DFT-derived and experimental values and, secondly, enhancing the susceptibility towards the counter-ion [27].

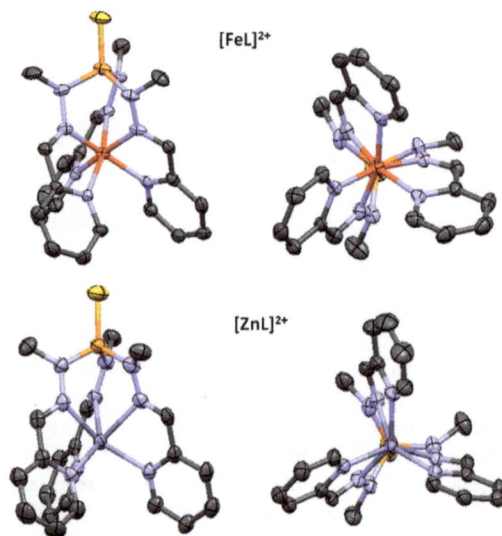

Figure 1. (**left**) Side views and (**right**) projections along M–P–S of molecular structures of (**top**) **[FeL](BF$_4$)$_2$** and (**bottom**) **[ZnL](ClO$_4$)$_2$**. Ellipsoids of 50% probability; only the dicationic complexes are shown; counterions, solvent molecules, and hydrogen atoms have been omitted for clarity. Dark grey: carbon, blue: nitrogen, yellow: sulfur, light orange: phosphorous, dark orange: iron, light grey: zinc atoms.

Table 1. Crystallographic data.

Compound	[FeL](BF$_4$)$_2$·1.5 CH$_3$CN	[ZnL](ClO$_4$)$_2$·1.5 CH$_3$CN
Sum formula	C$_{24}$H$_{28.5}$B$_2$F$_8$FeN$_{10.5}$PS	C$_{24}$H$_{28.5}$Cl$_2$N$_{10.5}$O$_8$PSZn
M (g·mol^{-1})	756.57	791.37
Shape and color	black plate	colorless column
Size (mm^3)	0.41 × 0.31 × 0.16	0.44 × 0.25 × 0.21
crystal system	orthorhombic	monoclinic
space group	$P2_12_12_1$	$P2_1/c$
a (pm)	1134.21(5)	909.34(12)
b (pm)	1532.91(7)	3165.3(3)
c (pm)	3673.55(18)	1628.1(2)
α (°)	90	90
β (°)	90	136.39(3)
γ (°)	90	90

Table 1. *Cont.*

Compound	[FeL](BF$_4$)$_2$·1.5 CH$_3$CN	[ZnL](ClO$_4$)$_2$·1.5 CH$_3$CN
V (10^6 pm^3)	6387.0(5)	3232.2(19)
μ (mm^{-1})	0.671	1.104
$\rho_{calcd.}$ (g·cm^{-3})	1.574	1.626
Z	8	4
T (K)	150(1)	150(1)
2 θ_{max} (°)	52.00	52.00
reflns. Measured	27,331	24,831
reflns. Unique	12,295	6322
parameters/restraints	932/253	475/114
R_1 ($I \geq 2\sigma(I)$)	0.0578	0.0699
R_1 (all data)	0.0810	0.0820
wR_2 ($I \geq 2\sigma(I)$)	0.1078	0.1606
wR_2 (all data)	0.1180	0.1680
u, v	0.0427, 1.3949	0.0472, 12.7098
S	1.018	1.109
Flack parameter x	0.012 (12)	n/a
ρ_{max}/ρ_{min} ($e \times 10^{-6}$ pm^{-3})	0.50/−0. 53	0.99/−0.90
CCDC number	1564278	1564279

$$w = [\sigma^2(F_o{}^2) + (uP)^2 + vP]^{-1} \text{ with } P = [\max(F_o{}^2, 0) + 2F_c{}^2]/3.$$

Table 2. Pertinent geometric parameters ((average) bond lengths in Å; angles in °; standard deviations in parentheses) of the complexes **[FeL]$^{2+}$** and **[ZnL]$^{2+}$** from XRD analysis and DFT calculations (italicized). [a]

	[FeL](BF$_4$)$_2$	*ls*-[FeL]$^{2+}$	[ZnL](ClO$_4$)$_2$	[ZnL]$^{2+}$
distances (Å)				
$d_{Fe-N(ald)}$	1.924(7)	*1.930(1)*	2.244(16)	*2.254(2)*
$d_{Fe-N(py)}$	1.979(10)	*1.974(1)*	2.133(11)	*2.134(2)*
d_{Fe-N}	1.952(15)	*1.952(11)*	2.19(2)	*2.19(3)*
bite	2.520(6)	*2.535(2)*	2.657(4)	*2.677(1)*
Fe–P	3.123(3)	*3.162*	3.482(5)	*3.509*
h [b]	2.225	*2.216*	2.587	*2.586*
δ [c]	−0.017	*−0.014*	+0.425	*+0.389*
angles (°)				
cis N$_{ald}$–Fe–N$_{ald}$	90.4(9)	*90.6(1)*	79.8(6)	*81.2(2)*
cis N$_{py}$–Fe–N$_{py}$	91.1(14)	*91.2(1)*	97(3)	*95.9(1)*
bite angle	80.4(2)	*81.0(1)*	74.7(4)	*75.1(3)*
trans N$_{py}$–Fe–N$_{ald}$	166.2(12)	*167.7(1)*	145(2)	*148.1(2)*
distortion				
$\Sigma_{cis}/°$ [d]	76.2	*57.9*	163.3	*164.4*
$\theta°$ [e]	43.4(8)	*45.0(1)*	19(2)	*22.3(1)*
S (O_h) [f]	1.464	*1.168*	8.613	*7.147*
S (TP) [f]	9.306	*10.007*	2.135	*2.700*

[a] Optimisation with BP86-D/TZVPP; [b] distance between the triangular faces of N$_{(ald)3}$ and N$_{(py)3}$; [c] metal displacement from the coordination center; [d] summed deviation from 90° of 12 N–Fe–N *cis* angles; [e] trigonal distortion ([22]); [f] continuous shape measures $S(O_h)$ and $S(TP)$ with reference to the centered octahedron and the centered trigonal prism, respectively ([40,42]).

Moderate trigonal distortion is a common structural feature when trigonally directing ligands are used [37]. This also becomes evident from an analysis in terms of the holistic continuous shape measures $S(O_h)$ and $S(TP)$. These measures, as introduced by Avnir and Alvarez et al. [40,42], approach zero when the experimental structure is in good agreement with a reference polyhedron (in our case, O_h and TP denote the centered regular octahedron and the centered regular trigonal prism, respectively). Clearly, **[FeL]**$^{2+}$ with $S(O_h) = 1.464$ is well described by an octahedron with only moderate trigonal distortion.

Distortion from O_h of the coordination sphere of **[ZnL]**$^{2+}$ is significantly stronger. Averaged trigonal twist angles amount to only 19(2)°, so that the coordination is rather describable as a moderately distorted trigonal prism. This conclusion is corroborated by the small value of the trigonal shape measure S(TP) = 2.135. In part, global distortion may be tracked to quite severe cis-angle distortion; in particular Σ_{cis} of the zinc(II) congener strongly exceeds distortion in the iron(II) complex. Concomitant with cis-angle distortion, severe bending from 180° of the nominal trans-angles (N_{py}–Fe–N_{ald}) is observed, accompanied by significant variation of the Zn–N bond-lengths. As this variation is donor-type specific (d(Zn–N_{ald}) > d(Zn–N_{py})), the result is an expulsion of the zinc ion to the "loose" end of the coordination sphere. The displacement from the center δ is large and it is positive. In this context, it is thus important to note that the displacement pattern observed for complexes of **L** (d(**[FeL]**$^{2+}$ < 0; d(**[ZnL]**$^{2+}$ > 0) is the opposite of that reported for complexes of **(tren)py$_3$** [53]. As noticed earlier [27], both ligands are members of distinctly different classes, owing to their different cap topologies. In particular, nominally hexadentate (heptadentate, if the apical nitrogen atom is considered) **(tren)py$_3$** rather consists of three largely independent diimine arms, giving 3×2 coordination [54].

hs-iron(II) [27] and zinc(II) complexes [31] with structure motifs akin to **[ZnL]**$^{2+}$ have been previously received with ligands of the extended tripod type. The quite general surrogate function of zinc(II) for **hs**-iron(II) is clearly a result of the coinciding ionic radii of Zn^{2+} and **hs**-Fe^{2+}. In consequence, the structure of the elusive **hs**-state of **[FeL]**$^{2+}$ must be expected to be strongly distorted towards a trigonal prism as well. This conclusion is fully corroborated by an XRD-calibrated DFT study (see below).

2.1.2. Complex Structures in Solution via ^{1}H-NMR Spectroscopy

The solid state structures of **[FeL]**(BF$_4$)$_2$ and **[ZnL]**(ClO$_4$)$_2$ revealed coordination units with approximate C_3 symmetry along the M–P–S vector. The occurrence of only a single set of resonances in the ^{1}H-NMR spectra indicates conserved symmetry in MeCN solution. While sharp resonances with resolved spin–spin coupling are recorded in the case of the zinc(II) complex, rather broad lines prevail at room temperature for most proton sites of the iron(II) complex (for proton assignment, see Scheme 1). Here spin–spin coupling is mostly blurred through severe line broadening, but becomes resolved upon decreasing the temperature. Low-temperature spectra recorded close to the freezing point of the solvent allowed unambiguous assignment. 2D-COSY spectra recorded in [D$_3$]MeCN at room temperature are shown in Figure 2 (region of resonances of aromatic protons, H^{1-4} and H^5). Besides the aforementioned differences in line width, the quite different spectral widths and the diverging peak positions deserve attention.

The resonances of the zinc(II) complex assemble in a very dense pattern, covering a chemical shift range Δδ of only 0.54 ppm. The spectral width is significantly smaller than the widths reported by Breher et al. [29] for both the free ligand (Δδ = 1.30 ppm) and the respective copper(I) complex (Δδ = 1.15 ppm). By contrast, the resonances of the aromatic protons (and H^5) of the iron(II) complex are spread out over a 2.11 ppm range. The increase in spectral width is mainly associated with the peak position of the ortho-proton resonance (H^1), which is located at 8.12 ppm and 6.52 ppm for the zinc and the iron complex, respectively. The substantial upfield shift, also with respect to the free ligand, points to a structure imposed origin. Actually, the crystal structure of **[FeL]**(BF$_4$)$_2$ reveals close contacts of H^1 and nitrogen atoms of a neighboring pyridine ($d(H^1 \cdots N) \approx 2.5$ Å) allowing efficient spatial overlap with its π-system. In turn, the lowfield positioned resonance of H^1 in the zinc complex indicates the lack of anisotropy effects and, therefore, the lack of close contacts of H^1 and other pyridine units. This is in agreement with the crystal structure of **[ZnL]**(ClO$_4$)$_2$, where $d(H^1 \cdots N) \approx 2.9$ Å and H^1 is fully displaced from the aromatic ring plane of neighboring pyridine. We conclude that the structures of both complexes as found in the crystal are largely conserved upon dissolving the complex salts in

MeCN. We further conclude that the solution structures of the **ls**-iron(II) complex **[FeL]²⁺** and of the **hs**-iron(II) surrogate **[ZnL]²⁺** are distinctly different.

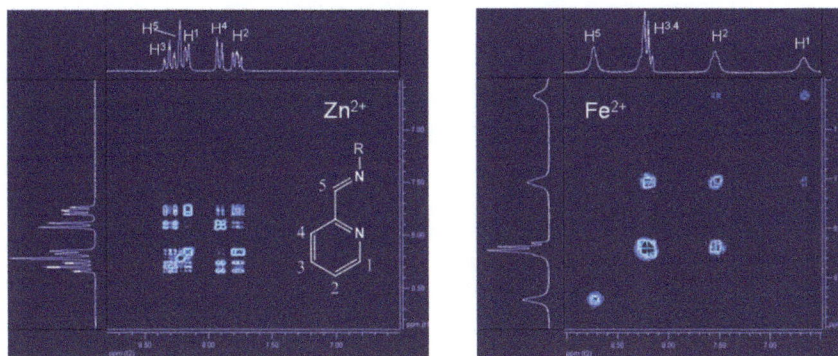

Figure 2. ¹H-NMR COSY spectra (200 MHz; D₃-MeCN) at ambient temperature; **(left)** [ZnL](ClO₄)₂; **(right)** [FeL](BF₄)₂; inset: assignment of proton sites.

2.1.3. DFT Structure Elucidation of Elusive **hs-[FeL]²⁺**

While the ls-iron(II) complex of **L** could be structurally characterized by XRD crystallography, only indirect structure information of the elusive **hs**-state is available. In particular, the trigonally distorted structure of the respective zinc(II) complex suggests similar metrics of **hs-[FeL]²⁺**. In order to validate this hypothesis, the structure variability of the iron(II) and zinc(II) complexes of **L** was investigated with DFT methods and calibrated by XRD data as far as possible. A comparison of XRD-derived data and complex metrics from DFT data (BP86-D/TZVPP) is given in Table 2. Very good agreement among both datasets is obtained with respect to all pertinent bond lengths, non-bonding distances, bond angles and several measures of complex distortion (results of a functional scan for **ls-[FeL]²⁺** are summarized in Table S1). The successful modeling of **ls-[FeL]²⁺** and **[ZnL]²** encouraged a DFT-based structure prediction of **hs-[FeL]²⁺** that has been performed by use of a variety of functionals, in order to judge the reliability of the overall prediction and the robustness of the metrical parameters. Pertinent structural data of **hs-[FeL]²⁺** are compiled in Table 3. Data for the reference systems **ls-[Fe(tren)py₃]²⁺** and **hs-[Fe(tren)py₃]²⁺** are given in Tables S2 and S3.

Table 3. Pertinent geometric parameters of DFT-optimized **hs-[FeL]²⁺**.

	B3LYP a_0 [a] =						BP86	PBE	TPSS0	TPSSh
	0.00	0.05	0.10	0.15	0.20	0.25				
Bond lengths [Å]										
$d_{Fe-N(ald)}$	2.197	2.211	2.226	2.230	2.238	2.253	2.199	2.200	2.205	2.198
$d_{Fe-N(py)}$	2.132	2.150	2.169	2.175	2.186	2.198	2.134	2.138	2.161	2.152
d_{Fe-N}	2.164	2.181	2.198	2.203	2.212	2.225	2.167	2.169	2.183	2.175
distortion										
Σ_{cis}/° [b]	150.1	154.7	160.2	161.2	163.9	171.2	150.5	151.2	159.8	155.6
θ/° [c]	24.0	22.2	20.5	20.3	20.2	17.8	24.3	24.3	20.4	21.5
$S(O_h)$ [d]	6.54	7.13	7.78	7.86	7.91	8.92	6.51	6.45	7.82	7.39
$S(TP)$ [d]	3.09	2.66	2.27	2.24	2.24	1.79	3.12	3.19	2.24	2.48

[a] amount of exact exchange; [b] summed deviation from 90° of 12 N–Fe–N *cis* angles; [c] trigonal distortion ([22]); [d] continuous shape measures $S(O_h)$ and $S(TP)$ with reference to the octahedron and the trigonal prism, respectively ([40,42]).

While a certain functional-imposed bias on the bond lengths and the distortion measures of **hs-[FeL]²⁺** cannot be denied (similar conclusions hold for the other iron(II) complexes under study), the overall appearance of the coordination sphere is independent of the functional. All optimizations of **hs-[FeL]²⁺** consistently reveal substantial trigonal tortion towards prismatic coordination (averaged across all functionals, θ amounts to 21.6(7)°). Notably, the range of the trigonal twist angles matches the respective value of the zinc(II) congener. Furthermore the averaged Fe–N bond lengths of the **hs**-iron(II) complex (d(Fe–N) = 2.188(7) Å) very well match the bond lengths of the zinc(II) complex (d(Zn–N) = 2.194 Å). Finally, the small values of the trigonal shape measures (S(TP) = 2.53(20)), again akin to **[ZnL]²⁺** with S(TP) = 2.70, clearly prove **[ZnL]²⁺** to be a convincing structure model of the **hs**-iron(II) complex and assign **hs-[FeL]²⁺** as a trigonal prismatic complex (Figure 3a).

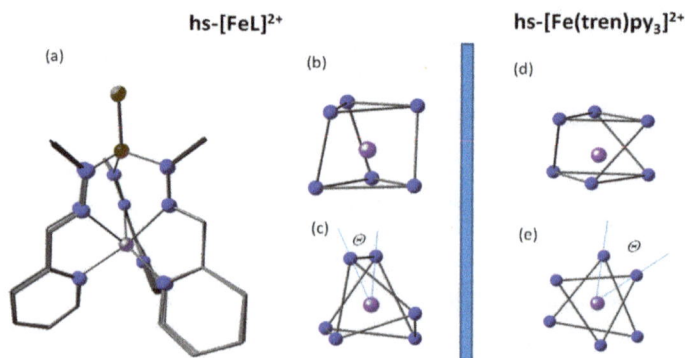

Figure 3. (**a**) Side view of the merged DFT-derived structures of **hs-[FeL]²⁺**; (**b**) side view and (**c**) top view along S–P–Fe of the [FeN₆] coordination sphere (BP86-D/TZVPP); (**d,e**) show side and top view along of the [FeN₆] coordination sphere of **hs-[Fe(tren)py₃]²⁺** (BP86-D/TZVPP).

Comparison with the optimized **hs**-structures of the well-established reference compound **[Fe(tren)py₃]²⁺** reveals a much weaker tendency towards the trigonal prism (Figure 3b; Tables S2 and S3). Averaged over all optimized structures, θ amounts to 47.1(4)°. Again there is very close agreement among the metrics of the optimized iron(II) complex and the data of the respective zinc(II) complex, both from DFT (θ = 46.3°) and XRD (θ = 45.9°; from [53]), further supporting the structure proposal for elusive **hs-[Fe(tren)py₃]²⁺**. The conclusion of a structure conserving SCO is obvious for **[Fe(tren)py₃]²⁺**: Shuttling between the **ls** and the **hs** states of this complex induces the typical expansion/contraction of the coordination sphere by ≈0.2 Å for each Fe–N bond, but leaves the overall shape and appearance of the coordination unit greatly unaffected (apart of the spin-state dependence of the apical Fe–N₇ distance). Accordingly, **[Fe(tren)py₃]²⁺** is well modeled by an ideal octahedron in both spin states, as can be read from the small values of the octahedral shape measures (**ls**: $S(O_h)$ = 0.65(5); **hs**: $S(O_h)$ = 2.12(4)).

2.2. DFT-Derived SCO Energies of [FeL]²⁺ and [Fe(tren)py₃]²⁺

Previous DFT work on the SCO energies of iron(II) complexes of some extended-tripod ligands has revealed matching **ls-hs** energy spacing within the couple **[FeL']²⁺** and **[Fe(tren)imid₃]²⁺** [27]. On the B3LYP*-D/TZVP level of theory (amount of exact exchange a_0 = 0.15), the apparent SCO energies of these complexes were identical within DFT accuracy ($\Delta_{SCO}E = E(\mathbf{hs}) - E(\mathbf{ls})$; positive values of $\Delta_{SCO}E$ indicate a **ls** ground state). Experimental results from VT-UV-Vis spectroscopy corroborated the proximity of the SCO energies. As a matter of fact, the apparent SCO energies of **[FeL]²⁺** and **[Fe(tren)py₃]²⁺** obtained in this work likewise match, when computed on the B3LYP*-D/TZVPP level of theory (Table 4). Comparison with the reference systems indicates significantly higher SCO energies, pointing to greatly favored **ls**-states and higher SCO transition temperatures, $T_{1/2}$. This qualitative conclusion is fully supported by the experimental properties of both compounds. XRD data are in line with ls-complexes in the crystal at 150 K, whereas NMR and UV-Vis spectroscopy consistently rule out more than very minor **hs**-contributions in solution around room temperature.

In order to avoid overemphasizing of the B3LYP*-derived data, we have undertaken more extensive evaluations of the SCO energies via variation of the amount of exact exchange of the parent B3LYP functional within $0.00 < a_0 < 0.20$, both for the calibration couple $\mathbf{[FeL']^{2+}}/\mathbf{[Fe(tren)imid_3]^{2+}}$ and for the prediction couple $\mathbf{[FeL]^{2+}}/\mathbf{[Fe(tren)py_3]^{2+}}$ (Table 4). Electronic energies E are reported throughout. Although harmonic frequencies have been computed for all complexes, we refrain from using DFT-derived entropies $\Delta_{SCO}S$ to adjust the scale to free energies, $\Delta_{SCO}G$. Plots of the derived apparent SCO energies of the four complexes against a_0 are convincingly linear as is generally observed [55–58], showing structure-independent slopes but structure-dependent offsets (Figure S1, in Supplementary Materials). Across the a_0 scan, we consistently find the **ls** forms of $\mathbf{[FeL]^{2+}}$ and $\mathbf{[Fe(tren)py_3]^{2+}}$ favored versus their reference compounds by 24(1) kJ·mol^{-1} and 27(1) kJ·mol^{-1}, respectively. Reasonably assuming similar SCO entropies of $\Delta_{SCO}S \approx 80$ J·(K·mol)$^{-1}$, among this set of structurally related complexes, transition temperatures $T_{1/2} > 450$ K are expected for $\mathbf{[FeL]^{2+}}$ and $\mathbf{[Fe(tren)py_3]^{2+}}$. These assumptions are corroborated by VT-UV-Vis and VT-NMR spectroscopy.

Table 4. Functional scan of apparent SCO energies of iron(II) complexes from Scheme 1. [a]

	$\Delta_{SCO}E$ (kJ·mol^{-1})				
	a_0 [b] $= 0.00$	0.05	0.10	0.15	0.20
$\mathbf{[FeL]^{2+}}$	153.4	116.7	79.4	46.6	10.3
$\mathbf{[Fe(tren)py_3]^{2+}}$	154.2	116.3	80.0	46.2	14.8
$\mathbf{[FeL']^{2+}}$	127.7	89.7	54.1	21.2	−9.0
$\mathbf{[Fe(tren)imid_3]^{2+}}$	123.3	87.3	52.8	20.5	−9.0

[a] Structures fully optimized with B3LYP-D/TZVPP; $0.00 < a_0 < 0.20$; [b] amount of exact exchange.

2.3. Dynamics of SCO in [FeL]²⁺ and [Fe(tren)py₃]²⁺

2.3.1. Optical Spectroscopy

$\mathbf{[FeL](BF_4)_2}$ and $\mathbf{[Fe(tren)py_3](ClO_4)_2}$ give deeply red and purple solutions. The colors correspond to intense absorption bands ($\varepsilon_{max} > 8000$ cm^{-1}·M^{-1}) at $\lambda = 500$ nm and $\lambda = 560$ nm, respectively. As is typical for combinations of iron(II) with π-accepting ligands, the Vis transitions can be associated with charge transfer from the iron(II) center to the ligand (MLCT). As is shown via VT-UV-Vis spectroscopy of solutions of $\mathbf{[FeL](BF_4)_2}$ in MeCN ($T_{max} = 345$ K), the MLCT transition is only marginally affected at elevated temperature (Figure S2). That is, peak position and intensity are conserved, with a very slight intensity loss of <5% at the highest available temperature. This observation points to high-lying SCO transition temperatures, in agreement with the predictions from DFT.

Previous work has shown that MLCT excitation of **ls**-iron(II) gives the meta-stable **hs**-state in a selective way. Ultra-fast depopulation of the MLCT-like Franck-Condon state concomitant with quintet-state population generally occurs on the sub-ps timescale, greatly independent of ligand nature. Accordingly, the MLCT-state in a structure analogue of $\mathbf{[FeL]^{2+}}$ is depopulated within few hundreds of femtoseconds [26]. We note in passing that there is much current interest in "taming" of MLCT states, aiming at longer MLCT-state lifetimes [24,59,60]. Laser-flash photolysis is therefore used to drive the SCO equilibrium (Equation (1)) towards the **hs**-state by means of a short-lived photochemical stimulus. Recovery of the equilibrium is recorded via transient absorption spectroscopy [13–24]. This acquisition scheme has been shown previously to be applicable for a number of iron(II) complexes [25–27]. Transient-absorption spectra recorded directly after ns-laser excitation of solutions of $\mathbf{[FeL](BF_4)_2}$ and $\mathbf{[Fe(tren)py_3](ClO_4)_2}$ in MeCN and MeOH are shown in Figure 4a. Owing to the spectral transparency of the **hs**-state in the Vis regime, the transient spectra are dominated by diagnostic bleach signals ($\Delta A < 0$), which reflect the positions and relative intensities of the peaks in the ground state absorption spectra. Additional transient absorbance ($\Delta A > 0$) appears in the spectra of **[FeL](BF$_4$)$_2$** in the near-UV region. This feature may be associated with the MLCT transition of the meta-stable **hs**-state. In the spectra of **[Fe(tren)py$_3$](ClO$_4$)$_2$** such transient absorption below 400 nm is absent; bleaching extends all over the spectrum. Different from **[FeL](BF$_4$)$_2$** the tren-derivative exhibits an additional intense MLCT-like transition in this spectral region. Bleaching of this intense transition through **ls** depopulation obviously outweighs the MLCT contribution of the meta-stable **hs**-state. Both the transient spectra and decay kinetics are only marginally affected by the solvent. I.e., the spectra recorded in MeOH and MeCN match almost ideally (squares and circles in Figure 4a). There is no spectral evolution during the course of transient decay. Decay profiles can be convincingly fitted to single-exponential decay functions, with time constants that are independent of the wavelength across the complete spectrum. The return to baseline absorption points to highly reversible photophysics and is in line with the implications of photo-induced SCO.

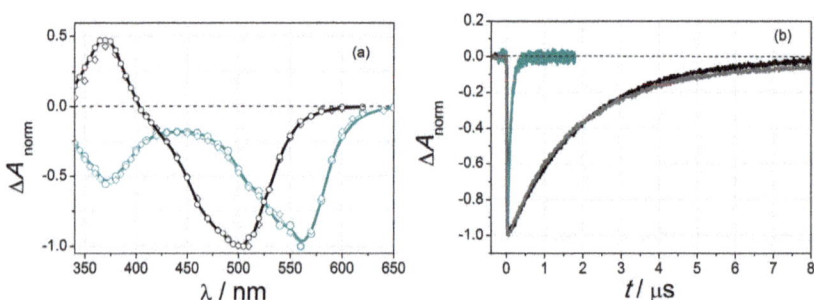

Figure 4. (**a**) Normalized transient absorption spectra recorded directly after laser excitation at 532 nm (MeOH; T = 298 K; $I_{exc} \approx$ 3 mJ; fwhm $\approx$ 9 ns): **[FeL]$^{2+}$** (OD$_{532}$ = 0.36; black) and **[Fe(tren)py$_3$]$^{2+}$** (OD$_{532}$ = 0.15; green); additional data recorded in MeCN are given as square symbols; (**b**) transient decay profiles recorded at 500 nm for **[FeL]$^{2+}$** (black: MeOH; grey: MeCN) and at 560 nm for **[Fe(tren)py$_3$]$^{2+}$** (green).

The lifetimes of transient decay, τ_{obs} ($\tau_{obs}^{-1} = k_{obs} = k_{H\to L} + k_{L\to H}$; $k_{H\to L}$ and $k_{L\to H}$ denote the rate constants of the **hs**$\to$**ls** and the **ls**$\to$**hs** transitions, respectively) are obtained from mono-exponential fits of the experimental data at variable temperature. As both compounds under study are **ls**-compounds with very small contributions of the **hs**-state at experimentally available temperatures, it becomes $k_{L\to H} \ll k_{H\to L}$, so that $k_{obs} \approx k_{H\to L}$. That is, the decay profiles in the transient absorption spectra are fully dominated by the relaxation of the *minority* component. It is noted at this point that intrinsically slower techniques may allow an ad-mixing of the decay of the majority component (see the NMR discussion below).

Table 5. Pertinent results from LFP of the iron(II) complexes in MeCN.

	[FeL]$^{2+}$	[Fe(tren)py$_3$]$^{2+}$
E_a [kJ·mol^{-1}]	22(1) [a]	7.8(3) [b]
$k_{H\to L}$ [10^6 s^{-1}] [b]	0.46	15.9
$k_{L\to H}$ [10^6 s^{-1}] [c]	<0.005	<0.2

[a] thermal barriers of 21.9 ± 0.3, 21.6 ± 0.3 and 20.9 ± 0.6 kJ·mol^{-1} have been recorded in MeOH, dichloromethane and water, respectively [25]; [b] measured in acetone, taken from [43]; decay rate constant at room temperature; [c] estimated with $K_{SCO} = \gamma_{hs}/\gamma_{ls} < 0.01$ at room temperature.

Notably the decay dynamics differ considerably among both compounds (Figure 4b). The decay rate constant $k_{obs} = 1.6 \times 10^7$ s^{-1} of **[Fe(tren)py$_3$](ClO$_4$)$_2$** is in qualitative agreement with previous measurements in acetone [43]. Decay of **[FeL](BF$_4$)$_2$** is substantially slower, yielding a decay rate constant of $k_{obs} = 4.6 \times 10^5$ s^{-1}. The sharp difference in the decay dynamics of both compounds is attributed to distinctly different activation barriers. Arrhenius treatment of the rate constants of **[FeL]**$^{2+}$ measured within a temperature range 250 K < T < 320 K gives a linear plot, ln(k_{obs}) vs. $1/T$ (Figure S3). A thermal barrier of $E_a = 22.2$ kJ·mol^{-1} is extracted in MeCN solution (Table 5), which well compares with previous experiments in other solvents, but deviates strongly from the reported value of $E_a = 7.8$ kJ·mol^{-1} of **[Fe(tren)py$_3$]**$^{2+}$. It is noted that our data on **[Fe(tren)py$_3$]**$^{2+}$, admittedly preliminary, indicate a somewhat stronger temperature dependence of its spin-state relaxation: While we obtain perfect agreement with the reported rate constants at $T = 260$ K, we record a lifetime of 65 ns at 293 K, which is significantly shorter than Hendrickson's 95 ns. With our preliminary data, we estimate $E_a \approx 12$ kJ·mol^{-1}, which appears to be in closer agreement with NMR-derived dynamics (see below). A more extended LFP study will of course be required to validate our result. Suffice to state here that the thermal barrier in **[Fe(tren)py$_3$]**$^{2+}$ will be around 10 kJ·mol^{-1}.

Although the solvent properties in the series MeCN, MeOH, water, dichloromethane vary widely (polarity, hydrogen bonding, internal pressure), the thermal SCO barriers of **[FeL]**$^{2+}$ are identical within experimental error. The indifference of the SCO kinetics to solvent properties is in line with an inherently small solvent dependence of the SCO phenomenon deduced by Navon, Bixon and Jortner from statistical arguments already in 1980 [61]. A major steering effect of the solvent on the SCO thermodynamics has been recently reported by Halcrow et al. and Petzold et al., though [62,63]. In both cases, strong effects of solvent-complex hydrogen bonding were invoked to rationalize the results. Respective effects on the SCO kinetics of these systems have not been reported yet. In this context, it is interesting to note that the SCO kinetics of a hydroxyl-substituted congener of **[FeL]**$^{2+}$ have been previously found to vary significantly with solvent [26].

2.3.2. VT-NMR Spectroscopy

The above results of optical spectroscopy in concert with the large preference for the **ls**-state in DFT are manifest also in the room temperature ^{1}H-NMR spectra in [D$_3$]MeCN of **[FeL]**$^{2+}$ and **[Fe(tren)py$_3$]**$^{2+}$. The resonances cover the range typical of diamagnetic compounds (9 ppm > δ > 1 ppm) and they largely stay in this region irrespective of temperature. Figure 5a highlights the temperature dependence of the "aromatic proton" region of **[FeL]**$^{2+}$ in [D$_3$]MeCN between 245 K < T < 340 K, measured at a frequency of 400 MHz. Evidently temperature has little effect on the peak positions, shifting the resonances by, at most, 0.6 ppm across the temperature range (Figure 5b). The sign and the relative size of the temperature induced shifts correlate with the residual spin density on the respective proton sites (from DFT studies). This points to a slight growing-in of **hs**-species at elevated temperatures. Measurements at even higher temperatures in [D$_5$]nitrobenzene at 500 MHz corroborate the picture of a beginning SCO with still minor contributions of the **hs**-species (Figure S4); by contrast, in pure **hs** complexes of polypyridine ligands, ortho-proton resonances are typically low-field shifted beyond δ > 50 ppm. For instance, the ortho-proton resonance of **[FeL′]**$^{2+}$, which is **hs** in [D$_3$]MeCN at all accessible temperatures, was recorded at δ= 82 ppm at T = 295 K [27]. A maximum shift of the H^1 resonance in **[FeL]**$^{2+}$ by $\Delta\delta$ = 6 ppm at close to 400 K therefore refutes contributions of the **hs**-state exceeding γ_{hs} > 0.10. Largely the same observations are made for solutions of **[Fe(tren)py$_3$]**$^{2+}$ in [D$_5$]nitrobenzene (500 MHz), although the thermal shift of some protons is somewhat larger (H^5; $\Delta\delta$ = 9 ppm; Figure S5). Taken together, the observations fully support the prediction of high-lying SCO transition points at T$_{1/2}$ > 450 K in both compounds. We note further that the datasets of **[FeL]**$^{2+}$ in [D$_3$]MeCN and [D$_5$]nitrobenzene nicely overlap (H^1; filled and open symbols in Figure 5b; a Curie plot of the chemical shifts is given in Figure S6), ruling out significant ligand-structure effects of the SCO thermodynamics.

(a)

(b)

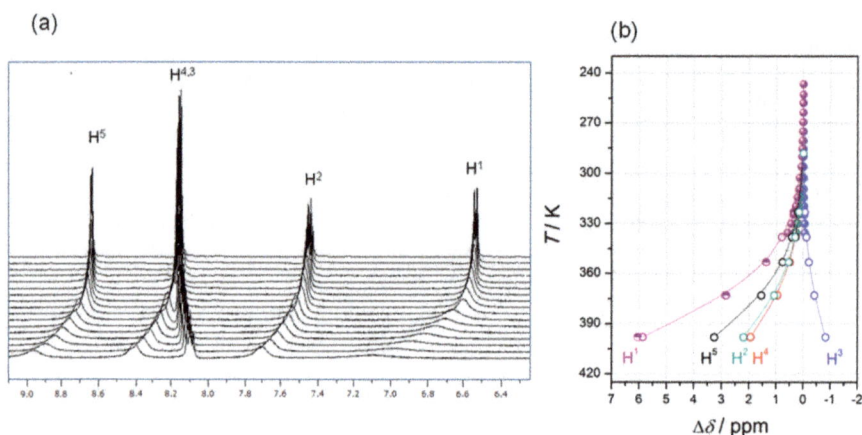

Figure 5. (a) VT-^{1}H-NMR spectra of [**FeL**](BF$_4$)$_2$ (400 MHz; [D$_3$]MeCN; zoom on "aromatic protons"; intensities are not to scale; (b) temperature-dependent drift of resonances due to H$^{1-5}$ in [**FeL**](BF$_4$)$_2$ (for proton assignment, see Scheme 1); open symbols: [**FeL**](BF$_4$)$_2$ (500 MHz; [D$_5$]nitrobenzene); half-filled symbols: [**Fe(tren)py$_3$**](ClO$_4$)$_2$ (H^1; 500 MHz; in [D$_5$]nitrobenzene).

Only at temperatures close to the freezing point of MeCN the spin–spin coupling in [**FeL**]$^{2+}$ is resolved. By contrast, the resonance of H^1 in [**Fe(tren)py$_3$**]$^{2+}$ is already well resolved at 288 K (500 MHz). The spectral information is in both cases progressively lost upon increasing the temperature: The moderate temperature-induced drift of the resonances is accompanied by substantial broadening of the resonances in both complexes, however, with significantly sharper lines in [**Fe(tren)py$_3$**]$^{2+}$ at all temperatures. Notably, the extent of line broadening of [**FeL**]$^{2+}$ at a given temperature increases with the Larmor frequency. At T = 298 K, the linewidth of the H^1 resonance in [**FeL**]$^{2+}$ (*LW*, full-width at half maximum) amounts to 19 Hz, 34 Hz and 42 Hz (the latter value at T = 288 K), when measured at 200 MHz, 400 MHz and 500 MHz, respectively. The field dependence points to a dynamic origin of the line broadening. In previous work on bis-meridional coordinated iron(II) complexes [63–65], we have recorded field-dependent NMR line broadening, which could be quantitatively analyzed in terms of a two-sites chemical exchange model. In particular, the linewidths could be correlated with the kinetics of SCO ($k_{obs} = k_{L \to H} + k_{H \to L}$). In the fast-exchange limit of a two-sites model, chemical exchange distinctly contributes to the transverse relaxation time R_2 and, thus, to the linewidth *LW* ($R_2 = \pi \times LW$). The R_2 values can be approximated from Equation (2) with $\Delta\omega_H = (C^0/T + C^1) \times \omega$ ($\Delta\omega_H$ denotes the difference in proton Larmor frequency of **ls** and **hs** state; C^0 and C^1 denote the Curie constant and its first-order correction).

$$R_2 = \{(1 - \gamma_{hs})R_{2(ls)} + \gamma_{hs} \cdot R_{2(hs)}\} + \{\gamma_{hs}(1 - \gamma_{hs})\frac{\Delta\omega^2}{k_{obs}}\} \quad (2)$$

As such the NMR method requires knowledge of the chemical shifts of the (elusive) **ls** and **hs** state (C^0 and C^1) and the thermodynamic parameters of SCO ($\Delta_{SCO}S$ and $\Delta_{SCO}H$). In the current case of [**FeL**]$^{2+}$ and [**Fe(tren)py$_3$**]$^{2+}$, however, some simplifying assumptions are possible. Both compounds are (close to) entirely **ls** at room temperature, so that the second term in Equation (2) can be approximately neglected, as long as the temperature remains well below $T_{1/2}$. In a second step, the first term is taken to be independent of temperature, so that all temperature dependence now resides in the third term of Equation (2). With these assumptions and simplifications, we derive Equation (3), which quantifies the change of R_2 relative to the pure **ls** state ($\Delta R_2 \approx R_2 - R_{2(ls)}$).

$$\Delta R_2 \approx \gamma_{hs}(1 - \gamma_{hs})\frac{\Delta\omega^2}{k_{obs}} \tag{3}$$

With this simplified equation in hands, two aspects deserve attention. Firstly, the temperature dependence of the transverse relaxation and therefore of *LW*, must be expected to have a bell-like shape ($\gamma_{hs}(1 - \gamma_{hs})$, maximum at $T = T_{1/2}$). As the factor ($\Delta\omega^2/k_{obs}$) decreases continuously with increasing temperature, the maximum linewidth LW_{max} is located at $T_{max} < T_{1/2}$. Such bell-shaped *LW* versus *T* plots in the NMR spectra of bis-meridional coordinated iron (II) complexes have been recently reported by us together with a detailed analysis in terms of admixed SCO dynamics [63,64]. While, in principle, each of the parameters γ_{hs}, $\Delta\omega$ and k_{obs} in Equations (2) and (3) must be considered structure-dependent, the coinciding SCO energies $\Delta_{SCO}E$ from DFT (Table 4) and the close similarity of the temperature drift in the NMR spectra of **[FeL]**$^{2+}$ and **[Fe(tren)py$_3$]**$^{2+}$ (Figure 5b), rule out significant ligand-structure imposed differences in γ_{hs} and $\Delta\omega$. That is, within this couple of complexes, effects of the SCO dynamics on the NMR linewidths are expected to be effectively isolated from other structural effects.

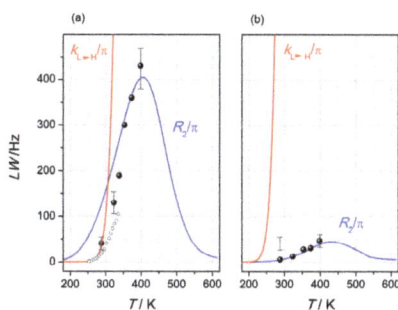

Figure 6. Temperature-dependent ^{1}H-NMR linewidths of proton H^1; filled symbols: 500 MHz; [D$_5$]nitrobenzene; open symbols: 400 MHz; [D$_3$]MeCN; blue lines: Model calculations of ^{1}H-NMR linewidths ($LW = R_2/\pi$) in terms of Equation (2) ($\Delta_{SCO}H = 40$ kJ·mol^{-1}; $\Delta_{SCO}S = 80$ J·(K·mol)$^{-1}$; C^0 = 40000 ppm·K; Arrhenius frequency factor: $A_{0(H\rightarrow L)} = 4 \times 10^9$ s^{-1}; 500 MHz); red line: temperature dependence of the **ls** relaxation rate constants $k_{L\rightarrow H}/\pi$ (Arrhenius frequency factor: $A_{0(L\rightarrow H)} = 6 \times 10^{13}$ s^{-1}); (a) **[FeL](BF$_4$)$_2$** with $E_a = 22$ kJ·mol^{-1}; (b) **[Fe(tren)py$_3$](ClO$_4$)$_2$**: $E_a = 12$ kJ·mol^{-1}.

Model calculations of the linewidths of H^1 in **[FeL]**$^{2+}$ and **[Fe(tren)py$_3$]**$^{2+}$ in terms of Equation (2), with reasonable estimates of $\Delta\omega$ and k_{obs}, show that this is actually the case and corroborate the validity of the simplifications leading to Equation (3). Thereby $\Delta_{SCO}H$ was fixed at 40 kJ·mol^{-1} (assuming $\Delta_{SCO}S = 80$ J·(K·mol)$^{-1}$, this corresponds to $T_{1/2} = 500$ K), while the Arrhenius barrier of the **hs** relaxation was allowed to vary between 22.0 kJ·mol^{-1} (as measured for **[FeL]**$^{2+}$) and 12.0 kJ·mol^{-1} (as estimated for **[Fe(tren)py$_3$]**$^{2+}$). A comparison of the measured linewidths and the modeling results in Figure 6 illustrates the impact of the SCO dynamics on the NMR linewidth. In particular, we find very close agreement between experiment and model for **[Fe(tren)py$_3$]**$^{2+}$ (Figure 6b). Fair qualitative agreement also prevails for **[FeL]**$^{2+}$ (Figure 6a) at higher temperatures. This overall agreement corroborates the underlying two-site model. Furthermore it gives indication of how strongly the differences in SCO dynamics can translate into NMR parameters. In turn, the SCO kinetics may be qualitatively estimated from NMR linewidths, provided that the inherent assumptions of Equation (2) are valid across the complete temperature range. In this respect we note an obviously sharper decrease of the linewidths in **[FeL]**$^{2+}$ below 340 K. I.e., at lower temperatures, the two-site model systematically overestimates the linewidths. Such deviations have been observed previously [64,65]. They have been attributed to the breakdown of the fast-exchange approximation inherent to the two-site model, rendering Equation (2) invalid at lower temperatures. In agreement with this notion, the experimental linewidths (400 MHz; open symbols in Figure 6a) well match the relaxation kinetics of the **ls** state at $T < 300$ K (red curve in Figure 6a). Both the observation of large maximum linewidths and the deviation from the two-site model at low temperature are fully in line with inherently slow SCO dynamics in **[FeL]²⁺**. In other words, the linewidths progressively reflect the dynamics of the majority spin state as soon as the SCO kinetics approaches the upper time limit of the NMR method; that is, time averaging among the spin states is lifted.

3. Discussion

Functional materials built from switchable molecular units are currently intensely sought after. Among the viable building blocks for such materials, SCO compounds based on iron(II) complexes hold high potential for the following reasons. Firstly, iron(II) complexes provide the largest magnetic change due to SCO; that is, $S_{hs} - S_{ls} = 2$. Secondly, in many cases the magnetic switching is accompanied by sharply varied optical response in the visible spectral regime, typically shuttling between deeply colored **ls** and faint or colorless **hs** states. Thus, thirdly, the systems are thermochromic and exhibit (after Vis excitation) negative photochromism; that is, iron(II) SCO complexes are photonically addressable. Finally, the wealth of experimentally accessible systems based on iron(II) provides a stable phenomenological background for structure prediction and search. One of the most significant drawbacks of iron(II)-based SCO compounds is their inherent kinetic lability towards thermal spin-state scrambling. In other words, an individual molecule in the **ls** state that can in principle be selectively addressed and driven to the **hs** state will lose its individuality spontaneously, and it will do so rapidly. In consequence, discrete spin states stable towards thermal randomization require either supramolecular concatenation via cooperative effects, cryogenic temperatures (LIESST effect) or sufficiently high thermal barriers.

While the first two alternatives have been studied with high intensity, the latter point attracted significantly less attention, although the concept of tuning of the SCO reaction coordinate points back to the late 1960s and has been recurrently discussed ever since [22]. The SCO reaction coordinate can be, in many cases, approximated as a (fully symmetric, isotropic) breathing of the coordination sphere (single-configurational coordinate model). This situation is represented by the parabolas given in black and red in Figure 7a. The **hs** parabola is shifted rightward with respect to the **ls** parabola along the reaction coordinate by a certain factor q, which mainly reflects the spin-state dependence of the metal-donor bond lengths. The horizontal shift of the parabolas establishes a thermal barrier E_a that has to be overcome during SCO. Notably for FeN₆ coordination compounds, if the isotropic model were strictly valid (which is not the case), both the shift factor q and the thermal barrier E_a become largely independent of ligand nature [11]. The only remaining option to tune and optimize the thermal barrier then is the variation of the driving force of SCO [61]. This option, favorably applied within complex families is represented by a green parabola in Figure 7a. Tuning of the reaction coordinate in this context therefore necessarily means to add an anisotropic component to the SCO induced molecular motion. Besides the coupled radial and angular motion in the hallmark system [Fe(tpy)₂]²⁺ [11,66–69], which has profound consequences for the SCO dynamics, only very few other systems have been reported to couple the breathing motion to other vibration or torsion modes. Based on a XRD-calibrated DFT study of iron(II) complexes, we have recently suggested that the coordinate of SCO is heavily affected by the nature of the ligand. That is, ligands of the extended-tripod type may impose a structural bias that distorts the **hs**-coordination sphere towards the trigonal prism. As shown in Figure 7a in terms of the emblematic picture of intersecting parabolas, such additional structural changes will directly affect the reaction coordinate and, in consequence, will massively affect the thermal barrier (blue in Figure 7a).

This conclusion is fully corroborated by the results of the present study, as outlined above. The iron(II) complex of ligand **L** is clearly defined as being **ls**, having only slightly distorted octahedral N₆ environment. DFT optimization gives consistent results. The very same holds for the well-established **ls**-complex **[Fe(tren)py₃]²⁺**, which shows even smaller distortion both in XRD [53] and in DFT. In agreement with the strongly favored **ls** character of both compounds (^{1}H-NMR, UV-Vis, and DFT), direct structural information on the elusive **hs** states was not available in both cases. The analogous zinc(II) complexes **[ZnL]$^{2+}$** (this work) and **[Zn(tren)py$_3$]$^{2+}$** (from [53]), taken as real-world surrogates of the elusive **hs**-species, reveal substantial structural differences that indicate a major structure steering effect of the ligand. In particular, besides elongated metal–nitrogen bond lengths, **[Zn(tren)py$_3$]$^{2+}$** mostly echoes the metrics of **[Fe(tren)py$_3$]$^{2+}$**. This view is fully confirmed by computed structures of **hs-[Fe(tren)py$_3$]$^{2+}$**, optimized with DFT methods on several levels of theory. In consequence, SCO in **[Fe(tren)py$_3$]$^{2+}$** may be termed "structure-conservative", in that an overall octahedral coordination prevails in both spin states. Strikingly, this is not the case in **[FeL]$^{2+}$**, where SCO induces quite severe molecular rearrangements in addition to Fe–N breathing. This conclusion inevitably arises both from the experimental structure of **[ZnL]$^{2+}$** and from the DFT-derived structural data of **hs-[FeL]$^{2+}$**, which give consistent results. In particular, the spin transition induces significant torsional motion within the coordination sphere along the Fe–P–S direction, so that both the zinc(II) complex and the optimized **hs-[FeL]$^{2+}$** structures are no longer adequately described as being octahedral. Overall, SCO in **[FeL]$^{2+}$** cannot be termed "structure-conservative", as the **hs** structure rather approaches a trigonal prismatic structure. As a matter of fact, essentially prismatic coordination of **hs**-iron(II) has been recently reported by us to hold for **[FeL']**(BF$_4$)$_2$ and a closely related complex. We thus conclude that SCO in **[FeL]$^{2+}$** likewise shuttles between an octahedral **ls** state and a trigonal prismatic **hs** state.

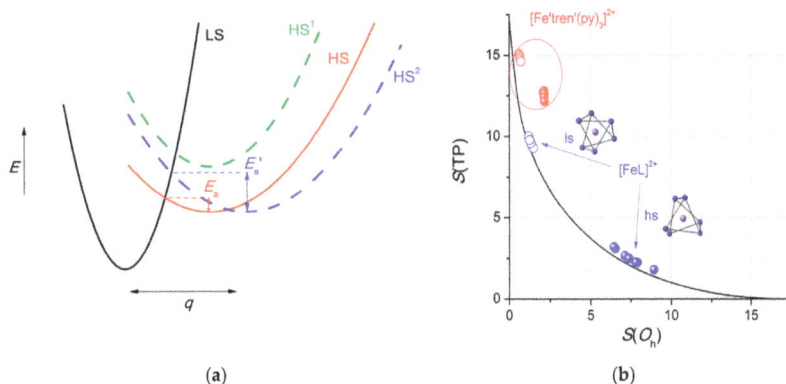

(a)   (b)

Figure 7. (**a**) Parabolic representation of the potential surface of an SCO complex; effects of vertical (green; variation of $\Delta_{SCO}E$) and horizontal displacements (blue; variation of q) are highlighted; arrows denote the activation barriers.; (**b**) shape map of the O_h–TP structure transition in **[FeL]$^{2+}$** (blue) and **[Zn(tren)py$_3$]$^{2+}$** (red); line: ideal Bailar-type trigonal twist.

Analysis of the structural changes of SCO in terms of the continuous shape measures of Avnir and Alvarez further accentuates the distinctly different influence of the ligand. While the (computed) **ls** and the **hs** state structures of **[Fe(tren)py$_3$]$^{2+}$** give close-lying features on the O_h–TP shape map (red in Figure 7b), a much larger area of the map is covered by the **ls** and the **hs** state structures of **[FeL]$^{2+}$** (blue in Figure 7b). Furthermore, both the **ls** and the **hs** state structures of **[FeL]$^{2+}$** match the implications of Bailar's trigonal twist pathway (line in Figure 7b; [28]). SCO in **[FeL]$^{2+}$** along the Bailar twist is thus highly probable. Additional XRD and DFT work is currently under way to validate this conclusion.

Once having the structural details of SCO in both compounds clarified, the question naturally arises, whether or not the modulated reaction coordinate in **[FeL]$^{2+}$** gives rise to decelerated SCO dynamics. Corresponding results from time-resolved laser spectroscopy and VT-NMR spectroscopy leave no doubt that this is actually the case. The bleach recovery curves of transient absorption recorded after Vis excitation reveal decay rate constants for [Fe(tren)py₃]²⁺ and [FeL]²⁺, which differ by more than one order of magnitude at room temperature, but tend to deviate even more strongly as the temperature is reduced. In consequence, the thermal barrier of SCO in [FeL]²⁺ is almost doubled with respect to [Fe(tren)py₃]²⁺. Consistent with the LFP results, dynamic line broadening of the NMR resonances is much more expressed for [FeL]²⁺. In agreement with the structural argumentation outlined above, we associate the qualitatively different thermal barriers with qualitatively modulated reaction coordinates. In contrast with this conclusion, we rule out the trivial effects of SCO driving force on the reaction barriers (green in Figure 7a). As DFT-derived SCO energies and thermal drifts of the resonances in the NMR spectra have shown, [Fe(tren)py₃]²⁺ and [FeL]²⁺ share very similar SCO energies.

4. Materials and Methods

4.1. Computational Methods

All DFT calculations were performed using ORCA2.9.1 [70]. Large TZVPP basis sets [71] were used throughout. The functional scan covers GGA type functionals (PBE [72], BP86 [73]), meta-GGA type functionals (TPSS [74]), hybrid functional (TPSSh [75], B3LYP [76,77]) and four derivatives of B3LYP (20% exact exchange a_0) with reduced exact exchange ($0.00 < a_0 < 0.15$). The structures of [FeL]²⁺ and of the well-established [Fe(tren)py₃]²⁺ were optimized in their **ls** and **hs** states using each functional. The structures of the respective zinc(II) complexes [ZnL]²⁺ and [Zn(tren)py₃]²⁺ were optimized with BP86 and B3LYP* ($a_0 = 0.15$). Cartesian coordinates of the B3LYP*-optimized structures of all complexes are given in Tables S4–S9 in the ESI. Apparent SCO energies ($\Delta_{SCO}E = E(\textbf{hs}) - E(\textbf{ls})$; negative and positive values of $\Delta_{SCO}E$ denote **hs** and **ls** states, respectively) were derived within the B3LYP family of functionals through application of Jakubikova's concept [57] and referenced to the experimentally known systems [FeL'](BF₄)₂ [27] and [Fe(tren)imid₃](BF₄)₂ [44–48]. The SCF energies were converged to 10^{-7} Hartree in energy. Dispersion contributions were approximated using Grimme's DFT-D6 atom-pairwise dispersion corrections [78]. Within the B3LYP derivatives dispersion was approximated with the parent B3LYP functional's parameters. Numerical frequency calculations of the BP86-derived structures revealed the stationary points to be minima on the potential surface.

4.2. Materials and General Techniques

All manipulations were performed under an argon atmosphere using standard Schlenk techniques. Diethyl ether was distilled from sodium–potassium alloy/benzophenone and acetonitrile from CaH₂ prior to use. CD₃CN was vacuum transferred from CaH into thoroughly dried glassware equipped with Young Teflon valves. (S)P(N^MeNCHPy)₃, was synthesized according to literature procedures [30]. All other chemicals were purchased from Aldrich (Schnelldorf, Germany,) or ABCR (Karlsruhe, Germany,) and were used as received.

4.3. Instrumentation

Elemental analyses and mass spectrometric investigations were carried out in the institutional technical laboratories. UV/Vis spectra in solution were measured using a Varian Cary 50 spectrometer (Varian, Darmstadt, Germany) equipped with a UV/Vis quartz immersion probe (light path 1 mm, Hellma), in a home-built measuring cell. LFP experiments were performed with the 532 nm output of a Nd:YAG laser system [79]. Transient decays were recorded at individual wavelengths by the step-scan method in the range of 320 to 700 nm and obtained as the mean signal of eight pulses. Spectral resolution was in the range of ±5 nm. The duration of the pulses (*fwhm* ca. 8 ns; 2–3 mJ per pulse) was generally much shorter than the decay lifetimes of the transient signals, so that deconvolution was not required for kinetic analysis. Solutions of [FeL](BF₄)₂ and [Fe(tren)py₃](ClO₄)₂ in high-purity methanol or MeCN (conc.(Fe) $\approx 1 \times 10^{-3}$ M) were rigorously deoxygenated by flushing with analytical grade argon for 20 min prior to and kept under argon during measurement in sealed quartz cuvettes. For data acquisition at variable temperatures we used a temperature-controlled cell holder (Quantum Northwest, model TC 125). IR spectra were measured using the ATR technique (attenuated total reflection) on a Thermo Nicolet iS5 FT-IR spectrometer (Thermo Fisher Scientific, Berlin, Germany) in the range from 4000 cm^{-1} to 400 cm^{-1}. Solution NMR spectra were recorded with Bruker Avance instruments (Bruker, Rheinstetten, Germany) operating at ^{1}H Larmor frequencies of 200, 400 and 500 MHz and are referenced according to IUPAC recommendations [80]. Chemical shifts are given relative to TMS for ^{13}C and ^{1}H, and H_3PO_4 for ^{31}P. Proton spin–spin coupling patterns in the ^{1}H-NMR spectra of **[FeL](BF$_4$)$_2$** and **[ZnL](ClO$_4$)$_2$** were simulated in MestReC. Linewidths were measured by fitting to Lorenzian shaped curve in Topshim2.1 package. No corrections for possible H–H coupling were applied. The linewidths are generally susceptible to temperature variations and should be seen as good estimations with ca. 10% uncertainty.

4.4. X-ray Crystal Structure Determination

Crystals suitable for X-ray diffraction were obtained from acetonitrile layered with diethyl ether or by diethyl-ether diffusion into acetonitrile solutions. Data were collected at 150.0(1) K using an "Agilent Xcalibur" diffractometer (Agilent Technologies, Yarnton, Oxfordshire, United Kingdom) equipped with a goniometer in κ geometry, a "Sapphire3" CCD-detector, and a graphite-monochromated "Enhance" Mo $K\alpha$ source (λ = 0.71073 Å). Diffraction images were integrated with CRYSALISPRO [81]. An analytical absorption correction using a multifaceted crystal model was performed [82]. The structure was solved with SHELXT-2014 [83] using a dual-space method and refined with SHELXL-2017 [84] against F_o^2 data using the full-matrix least-squares algorithm. OLEX2 was employed as a front end [85]. All non-hydrogen atoms were refined anisotropically. Hydrogen atoms were located on difference Fourier maps (for the complex cations) or their positions were inferred from neighboring sites (for acetonitrile molecules). They were refined with standard riding models. Structure graphics were produced using MERCURY [86].

In **[ZnL](ClO$_4$)$_2$**, the three separate acetonitrile molecules are disordered over the same center of inversion. They have been modelled as rigid bodies (fragment imported from FragmentDB [87]) using enhanced rigid-bond and ADP-similarity restraints. A tightened isotropy restraint was applied to the otherwise physically meaningless anisotropic displacement parameters (ADPs) of the terminal carbon atoms C42 and C52. The occupancy of all acetonitrile molecules was fixed at 0.5 after checking the number of electrons contained in their volume segment using PLATON/SQUEEZE [88]. In **[FeL](BF$_4$)$_2$**, the tetrafluoroborate ion containing B40 exhibits disorder and was modeled in two discrete positions as regularly tetrahedral using tight 1,2-distance, 1,3-distance, rigid-bond and ADP-similarity restraints. Occupancies refined to 0.649(9)/0.351(9); the average B–F bond length refined to 1.355(6) Å. Additionally, the acetonitrile molecule was found rotationally disordered and was modelled in two discrete positions using tight rigid-bond and ADP-similarity restraints. The occupancies refined to 0.74(2)/0.26(2). Due to problems with the beam stop and reflection overlap, seven and ten ill-measured low-angle reflections had to be excluded at the final stage of the refinement of **[ZnL](ClO$_4$)$_2$** and **[FeL](BF$_4$)$_2$**, respectively. The relevant details of the crystals, data collection, and structure refinement are found in Table 1.

4.5. [FeL](BF$_4$)$_2$: [{(S)P(N^{Me}NPy)$_3$}Fe](BF$_4$)$_2$

The synthesis of the iron(II) complex from a tetrafluoroborate precursor has been described in detail recently [25]. Single crystals suitable for X ray diffraction were obtained from acetonitrile solutions at 4 °C by isothermal diffusion of diethylether within a few days. Yield: 80%. Elemental analysis: Calc. (%) for C$_{21}$H$_{24}$FeN$_9$SPB$_2$F$_8$·0.5 EtOH: C 36.80, N 17.56, H 3.79, S 4.47, found: C 37.70, N 17.58, H 3.48, S 4.02. ^{1}H-NMR (400 MHz, [D$_3$]MeCN, −32 °C): δ [ppm] = 8.63 (d, ^4J = 2.8 Hz, 3H, H5), 8.15 (m, 6H, H3,4), 7.43 (m, 3H, H2), 6.52 (d, ^3J = 5.6 Hz, 3H, H1) 3.47 (d, ^3J = 12.8 Hz, 9H, N–Me). {^{1}H}^{13}C-NMR (50.32 MHz, [D$_3$]MeCN, 25 °C): δ [ppm] = 156.2 (s, 3C, pyC), 154.7 (s, 3C, pyC), 151.3 (d, ^3J = 8 Hz, 3C, C=N$_{ald}$), 140.5 (s, 3C, C$_{Ar}$), 129.2 (s, 3C, C$_{Ar}$), 128.1 (s, 3C, C$_{Ar}$), 38.4 (d, 9C,

^2J = 3 Hz, N–CH$_3$). {^{1}H}^{31}P-NMR (80.95 MHz, [D$_3$]MeCN, 25 °C): δ [ppm] = 65.5. HR-MS (ESI) Calc. for C$_{21}$H$_{24}$FeN$_9$PS: $[M]^{2+}$ = 260.5473; found: $[M]^{2+}$ 260.5475. IR (KBr): $\tilde{v}$ [cm^{-1}] = 3432bs, 3102w, 3056w, 2923w, 1607s, 1576m, 1466s, 1444m, 1243m, 1161s, 1052s, 952m, 897w, 802s, 572w, 519w.

4.6. [ZnL](ClO$_4$)$_2$: [{(S)P(N^{Me}NPy)$_3$}Zn](ClO$_4$)$_2$

To a stirred solution of (S)P(N^{Me}NCHPy)$_3$ (104 mg, 0.22 mmol) in 4 mL degassed ethanol a solution of [Zn(ClO$_4$)$_2$ × 6 H$_2$O] (83 mg, 0.22 mmol) in 2 mL ethanol was added dropwise at room temperature. After stirring the obtained colorless suspension over night the reaction mixture was filtered and the precipitate washed with 2 × 3 mL diethylether. Colorless crystals suitable for X-ray diffraction were obtained from acetonitrile solutions of the complex salt by isothermal diffusion of diethylether at 4 °C within a few days. Yield: 123 mg (77%). elemental analysis (%) calcd. for C$_{21}$H$_{24}$N$_9$O$_6$PSZnCl$_2$O$_8$: C 34.56, H 3.31, N 17.27, S 4.39; found: C 34.36, H 3.28, N 17.23, S 4.06; ^{1}H-NMR (200 MHz, 298 K, [D$_3$]MeCN): δ = 3.53 (d, $^3J_{HP}$ = 10.8 Hz, 9H, N–CH$_3$), 7.72 (ddd, $^3J_{HH}$ = 7.8 Hz, $^3J_{HH}$ = 5.0 Hz, $^4J_{HH}$ = 1.5 Hz, 3H, H^2), 7.87 (ddd, $^3J_{HH}$ = 7.8 Hz, $^4J_{HH}$ = 1.5 Hz, $^5J_{HH}$ = 0.8 Hz, 3H, H^4), 8.12 (ddd, $^3J_{HH}$ = 5.0 Hz, $^4J_{HH}$ = 1.6 Hz, $^5J_{HH}$ = 0.8 Hz, 3H, H^1), 8.18 (s (br), 3H, H^5), 8.26 (dt, $^3J_{HH}$ = 7.8 Hz, $^4J_{HH}$ = 1.6 Hz, 3H, H^3) ppm; ^{13}C{^{1}H}-NMR (50.32 MHz, 298 K, [D$_3$]MeCN): δ = 35.3 (d, $^2J_{CP}$ = 10.0 Hz, N–CH$_3$), 127.1 (s, 3C, C$_{Ar}$), 127.8 (s, 3C, C$_{Ar}$), 139.8 (d, $^3J_{CP}$ = 9.0 Hz, C=N$_{ald}$), 142.7 (s, 3C, C$_{Ar}$), 148.1 (s, 3C, C$_{Ar}$), 148.8 (s, 3C, C$_{Ar}$) ppm; ^{31}P{^{1}H}-NMR (80.95 MHz, 298 K, [D$_3$]MeCN): δ = 56.2 (s) ppm.

Supplementary Materials: The following items are available online at www.mdpi.com/2304-6740/5/3/60/s1, CIFs and checkCIF reports. Table S1: Pertinent geometric parameters of DFT-optimized **ls-[FeL]**$^{2+}$, Table S2: Pertinent geometric parameters of DFT-optimized **ls-[Fe(tren)py$_3$]**$^{2+}$, Table S3. Pertinent geometric parameters of DFT-optimized **hs-[Fe(tren)py$_3$]**$^{2+}$, Figure S1: Dependence of DFT-derived apparent SCO energies $\Delta_{SCO}E$ on exact exchange a$_0$, Figure S2: VT-UV-Vis spectra of **[FeL](BF$_4$)$_2$**, Figure S3: Arrhenius plot of transient decay rate constants of **[FeL]**$^{2+}$, Figure S4: VT-^{1}H-NMR spectra of **[FeL](BF$_4$)$_2$**, Figure S5: VT-^{1}H-NMR spectra of **[Fe(tren)py$_3$](ClO$_4$)$_2$**, Figure S6: Curie plot of the proton chemical shifts, Table S4: Cartesian coordinates of DFT-optimized **ls-[FeL]**$^{2+}$, Table S5: Cartesian coordinates of DFT-optimized **hs-[FeL]**$^{2+}$, Table S6: Cartesian coordinates of DFT-optimized **[ZnL]**$^{2+}$, Table S7: Cartesian coordinates of DFT-optimized **ls-[Fe(tren)py$_3$]**$^{2+}$, Table S8: Cartesian coordinates of DFT-optimized **hs-[Fe(tren)py$_3$]**$^{2+}$, Table S9: Cartesian coordinates of DFT-optimized **[Zn(tren)py$_3$]**$^{2+}$.

Acknowledgments: Gerald Hörner acknowledges financial support of Deutsche Forschungsgemeinschaft (SFB 658; Elementary Processes in Molecular Switches at Surfaces), Andreas Grohmann (Technische Universität Berlin, Germany) for continuous support and Tomasz Pedzinski (Adam-Mickiewicz University Poznan, Poland) for assistance with laser photolysis.

Author Contributions: Gerald Hörner conceived and designed the experiments; Philipp Stock, Gerald Hörner and Holm Petzold performed the experiments; Holm Petzold and Gerald Hörner analyzed the data; Dennis Wiedemann contributed crystallographic work; Gerald Hörner wrote the paper.

Conflicts of Interest: The authors declare no conflict of interest.

References

1. Halcrow, M.A. (Ed.) *Spin Crossover Materials*; Wiley: Chichester, UK, 2013.

2. Kahn, O.; Martinez, C.J. Spin-Transition Polymers: From Molecular Materials toward Memory Devices. *Science* **1998**, *279*, 44–48. [CrossRef]

3. Létard, J.-F.; Guionneau, P.; Gouz-Capes, L. Towards Spin Crossover Applications. *Top. Curr. Chem.* **2004**, *235*, 221–249. [CrossRef]

4. Gütlich, P.; Gaspar, A.B.; Garcia, Y. Spin state switching in iron coordination compounds. *Beilstein J. Org. Chem.* **2013**, *9*, 342–391. [CrossRef] [PubMed]

5. Bousseksou, A.; Molnar, G.; Salmon, L.; Nicolazzi, W. Molecular spin crossover phenomenon: Recent achievements and prospects. *Chem. Soc. Rev.* **2011**, *40*, 3313–3335. [CrossRef] [PubMed]

6. Sato, O.; Tao, J.; Zhang, Y.-Z. Controlling Magnetic Properties through External Stimuli. *Angew. Chem. Int. Ed.* **2006**, *46*, 2152. [CrossRef] [PubMed]

7. Kumar, K.S.; Ruben, M. Emerging trends in spin crossover (SCO) based functional materials and devices. *Coord. Chem. Rev.* **2017**, *346*, 176–205. [CrossRef]

8. Decurtins, S.; Gütlich, P.; Köhler, C.P.; Spiering, H.; Hauser, A. Light-induced excited spin state trapping in a transition-metal complex: The hexa-1-propyltetrazole-iron (II) tetrafluoroborate spin-crossover system. *Chem. Phys. Lett.* **1984**, *105*, 1–4. [CrossRef]

9. Letard, J.-F.; Asthana, S.; Shepherd, H.J.; Guionneau, P.; Goeta, A.E.; Suemura, N.; Ishikawa, R.; Kaizaki, S. Photomagnetism of a sym-*cis*-Dithiocyanato Iron(II) Complex with a Tetradentate *N,N'*-Bis(2-pyridylmethyl)1,2-ethanediamine Ligand. *Chem. Eur. J.* **2012**, *19*, 5924–5934. [CrossRef] [PubMed]

10. Letard, J.-F.; Capes, L.; Chastanet, G.; Molinar, N.; Letard, S.; Real, J.-A.; Kahn, O. Critical temperature of the LIESST effect in iron(II) spin crossover compounds. *Chem. Phys. Lett.* **1999**, *313*, 115–120. [CrossRef]

11. Hauser, A.; Enachescu, C.; Daku, M.L.; Vargas, A.; Amstutz, N. Low-temperature lifetimes of metastable high-spin states in spin-crossover and in low-spin iron(II) compounds: The rule and exceptions to the rule. *Coord. Chem. Rev.* **2006**, *250*, 1642–1652. [CrossRef]

12. Klingele, J.; Kaase, D.; Schmucker, M.; Lan, Y.H.; Chastanet, G.; Letard, J.-F. Thermal Spin Crossover and LIESST Effect Observed in Complexes [Fe(LCh)$_2$(NCX)$_2$] [LCh = 2,5-Di(2-Pyridyl)-1,3,4-Chalcadiazole; Ch = O, S, Se; X = S, Se, BH$_3$]. *Inorg. Chem.* **2013**, *52*, 6000–6010. [CrossRef] [PubMed]

13. Dose, E.V.; Hoselton, M.A.; Sutin, N.; Tweedle, M.F.; Wilson, L.J. Dynamics of intersystem crossing processes in solution for six-coordinate d5, d6, and d7 spin-equilibrium metal complexes of iron(III), iron(II), and cobalt(II). *J. Am. Chem. Soc.* **1978**, *100*, 1141–1147. [CrossRef]

14. Garvey, J.J.; Lawthers, I. Photochemically-induced perturbation of the 1A $\rightleftharpoons$ 5T equilibrium in FeII complexes by pulsed laser irradiation in the metal-to-ligand charge-transfer absorption band. *J. Chem. Soc. Chem. Commun.* **1982**, 906–907. [CrossRef]

15. Beattie, J.K. Dynamics of Spin Equilibria in Metal Complexes. *Adv. Inorg. Chem.* **1988**, *32*, 1–53. [CrossRef]

16. Toftlund, H. Spin equilibria in iron(II) complexes. *Coord. Chem. Rev.* **1989**, *94*, 67–108. [CrossRef]

17. Xie, C.-L.; Hendrickson, D.N. Mechanism of spin-state interconversion in ferrous spin-crossover complexes: Direct evidence for quantum mechanical tunneling. *J. Am. Chem. Soc.* **1987**, *109*, 6981–6988. [CrossRef]

18. Al-Obaidi, A.H.R.; Jensen, K.B.; McGarvey, J.J.; Toftlund, H.; Jensen, B.; Bell, S.E.J.; Carroll, J.G. Structural and Kinetic Studies of Spin Crossover in an Iron(II) Complex with a Novel Tripodal Ligand. *Inorg. Chem.* **1996**, *35*, 5055–5060. [CrossRef] [PubMed]

19. Chang, H.-R.; McCusker, J.K.; Toftlund, H.; Wilson, S.R.; Trautwein, A.X.; Winkler, H.; Hendrickson, D.N. [Tetrakis(2-pyridylmethyl)ethylenediamine]iron(II) perchlorate, the first rapidly interconverting ferrous spin-crossover complex. *J. Am. Chem. Soc.* **1990**, *112*, 6814–6827. [CrossRef]

20. McCusker, J.K.; Toftlund, H.; Rheingold, A.L.; Hendrickson, D.N. Ligand conformational changes affecting 5T2 → 1A1 intersystem crossing in a ferrous complex. *J. Am. Chem. Soc.* **1993**, *115*, 1797–1804. [CrossRef]

21. Brady, C.; McGarvey, J.J.; McCusker, J.K.; Toftlund, H.; Hendrickson, D.N. Time-Resolved Relaxation Studies of Spin Crossover Systems in Solution. *Top. Curr. Chem.* **2004**, *235*, 1–22. [CrossRef]

22. McCusker, J.K.; Rheingold, A.L.; Hendrickson, D.N. Variable-Temperature Studies of Laser-Initiated 5T2 → 1A1 Intersystem Crossing in Spin-Crossover Complexes: Empirical Correlations between Activation Parameters and Ligand Structure in a Series of Polypyridyl Ferrous Complexes. *Inorg. Chem.* **1996**, *35*, 2100–2112. [CrossRef]

23. Hain, S.K.; Heinemann, F.W.; Gieb, K.; Müller, P.; Hörner, G.; Grohmann, A. On the Spin Behaviour of Iron(II)–Dipyridyltriazine Complexes and Their Performance as Thermal and Photonic Spin Switches. *Eur. J. Inorg. Chem.* **2010**, 221–232. [CrossRef]

24. Fatur, S.M.; Shepard, S.G.; Higgins, R.F.; Shores, M.P.; Damrauer, N.H. A Synthetically Tunable System To Control MLCT Excited-State Lifetimes and Spin States in Iron(II) Polypyridines. *J. Am. Chem. Soc.* **2017**, *139*, 4493–4505. [CrossRef] [PubMed]

25. Stock, P.; Pędziński, T.; Spintig, N.; Grohmann, A.; Hörner, G. High Intrinsic Barriers against Spin-State Relaxation in Iron(II)-Complex Solutions. *Chem. Eur. J.* **2013**, *19*, 839–842. [CrossRef] [PubMed]

26. Stock, P.; Spintig, N.; Scholz, J.; Epping, J.D.; Oelsner, C.; Wiedemann, D.; Grohmann, A.; Hörner, G. Spin-State Dynamics of a Photochromic Iron(II) Complex and its Immobilization on Oxide Surfaces Via Phenol Anchors. *J. Coord. Chem.* **2015**, *68*, 3099–3115. [CrossRef]

27. Stock, P.; Deck, E.; Hohnstein, S.; Korzekwa, J.; Meyer, K.; Heinemann, F.W.; Breher, F.; Hörner, G. Molecular Spin Crossover in Slow Motion: Light-Induced Spin-State Transitions in Trigonal Prismatic Iron(II) Complexes. *Inorg. Chem.* **2016**, *55*, 5254–5265. [CrossRef] [PubMed]

28. Bailar, J.C., Jr. Some problems in the stereochemistry of coordination compounds: Introductory lecture. *J. Inorg. Nucl. Chem.* **1958**, *8*, 165–175. [CrossRef]

29. Trapp, I.; Löble, M.W.; Meyer, J.; Breher, F. Copper complexes of tripodal κ⁶*N*-donor ligands: A structural, EPR spectroscopic and electrochemical study. *Inorg. Chim. Acta* **2011**, *374*, 373–384. [CrossRef]
30. Löble, M.W.; Casimiro, M.; Thielemann, D.T.; Oña-Burgos, P.; Fernandez, I.; Roesky, P.W.; Breher, F. 1H,89Y HMQC and Further NMR Spectroscopic and X-ray Diffraction Investigations on Yttrium-Containing Complexes Exhibiting Various Nuclearities. *Chem. Eur. J.* **2012**, *18*, 5325–5334. [CrossRef] [PubMed]
31. Chandrasekhar, V.; Azhakar, R.; Pandian, B.M.; Boomishankar, R.; Steiner, A. A phosphorus-supported multisite coordination ligand containing three imidazolyl arms and its metalation behaviour. An unprecedented co-existence of mononuclear and macrocyclic dinuclear Zn(II) complexes in the same unit cell of a crystalline lattice. *Dalton Trans.* **2008**, *37*, 5962–5969. [CrossRef] [PubMed]
32. Marchivie, M.; Guionneau, P.; Letard, J.-F.; Chasseau, D. Photo-induced spin-transition: The role of the iron(II) environment distortion. *Acta Crystallogr. Sect. B Struct. Sci.* **2005**, *61*, 25–28. [CrossRef] [PubMed]
33. Fleischer, E.B.; Gebala, A.E.; Swift, D.R.; Tasker, P.A. Trigonal prismatic-octahedral coordination. Complexes of intermediate geometry. *Inorg. Chem.* **1972**, *11*, 2775–2784. [CrossRef]
34. Wentworth, R.A.D. Trigonal prismatic vs. octahedral stereochemistry in complexes derived from innocent ligands. *Coord. Chem. Rev.* **1972**, *9*, 171–187. [CrossRef]
35. Purcell, K.F. Pseudorotational intersystem crossing in d6 complexes. *J. Am. Chem. Soc.* **1979**, *101*, 5147–5152. [CrossRef]
36. Vanquickenborne, L.G.; Pierloot, K. Role of spin change in the stereomobile reactions of strong-field d6 transition-metal complexes. *Inorg. Chem.* **1981**, *20*, 3673–3677. [CrossRef]
37. Comba, P. Coordination geometries of hexaamine cage complexes. *Inorg. Chem.* **1989**, *28*, 426–431. [CrossRef]
38. Larsen, E.; La Mar, G.N.; Wagner, B.E.; Parks, J.E.; Holm, R.H. Three-dimensional macrocyclic encapsulation reactions. III. Geometrical and electronic features of tris(diimine) complexes of trigonal-prismatic, antiprismatic, and intermediate stereochemistry. *Inorg. Chem.* **1972**, *11*, 2652–2668. [CrossRef]
39. Comba, P.; Sargeson, A.M.; Engelhardt, L.M.; Harrowfield, J.M.B.; White, A.H.; Horn, E.; Snow, M.R. Analysis of trigonal-prismatic and octahedral preferences in hexaamine cage complexes. *Inorg. Chem.* **1985**, *24*, 2325–2327. [CrossRef]
40. Alvarez, S.; Avnir, D.; Llunell, M.; Pinsky, M. Continuous symmetry maps and shape classification. The case of six-coordinated metal compounds. *New J. Chem.* **2002**, *26*, 996–1009. [CrossRef]
41. Knight, J.C.; Alvarez, S.; Amoroso, A.J.; Edwards, P.G.; Singh, N. A novel bipyridine-based hexadentate tripodal framework with a strong preference for trigonal prismatic co-ordination geometries. *Dalton Trans.* **2010**, *39*, 3870–3883. [CrossRef] [PubMed]
42. Alvarez, S. Distortion Pathways of Transition Metal Coordination Polyhedra Induced by Chelating Topology. *Chem. Rev.* **2015**, *115*, 13447–13483. [CrossRef] [PubMed]
43. Conti, A.J.; Xie, C.L.; Hendrickson, D.N. Tunneling in spin-state interconversion of ferrous spin-crossover complexes. Concentration dependence of apparent activation energy determined in solution by laser-flash photolysis. *J. Am. Chem. Soc.* **1989**, *111*, 1171–1180. [CrossRef]
44. Brewer, G.; Olida, M.J.; Schmiedekamp, A.M.; Viragh, C.; Zavalij, P.Y. A DFT computational study of spin crossover in iron(III) and iron(II) tripodal imidazole complexes. A comparison of experiment with calculations. *Dalton Trans.* **2006**, *35*, 5617. [CrossRef] [PubMed]
45. Sunatsuki, Y.; Ikuta, Y.; Matsumoto, N.; Ohta, H.; Kojima, M.; Iijima, S.; Hayami, S.; Maeda, Y.; Kaizaki, S.; Dahan, F.; Tuchagues, J.-P. An Unprecedented Homochiral Mixed-Valence Spin-Crossover Compound. *Angew. Chem. Int. Ed.* **2003**, *42*, 1614. [CrossRef] [PubMed]
46. Ohta, H.; Sunatsuki, Y.; Kojima, M.; Iijima, S.; Akashi, H.; Matsumoto, N. A Tripodal Ligand Containing Three Imidazole Groups Inducing Spin Crossover in Both Fe(II) and Fe(III) Complexes; Structures and Spin Crossover Behaviors of the Complexes. *Chem. Lett.* **2004**, *33*, 350. [CrossRef]
47. Sunatsuki, Y.; Ohta, H.; Kojima, M.; Ikuta, Y.; Goto, Y.; Matsumoto, N.; Iijima, S.; Akashi, H.; Kaizaki, S.; Dahan, F.; et al. Supramolecular Spin-Crossover Iron Complexes Based on Imidazole-Imidazolate Hydrogen Bonds. *Inorg. Chem.* **2004**, *43*, 4154. [CrossRef] [PubMed]
48. Brewer, C.; Brewer, G.; Luckett, C.; Marbury, G.S.; Viragh, C.; Beatty, A.M.; Scheidt, W.R. Proton Control of Oxidation and Spin State in a Series of Iron Tripodal Imidazole Complexes. *Inorg. Chem.* **2004**, *43*, 2402. [CrossRef] [PubMed]
49. Halcrow, M.A. Structure: Function relationships in molecular spin-crossover complexes. *Chem. Soc. Rev.* **2011**, *40*, 4119–4142. [CrossRef] [PubMed]

50. Weber, B.; Bauer, W.; Obel, J. An Iron(II) Spin-Crossover Complex with a 70 K Wide Thermal Hysteresis Loop. *Angew. Chem Int. Ed.* **2008**, *47*, 10098–10101. [CrossRef] [PubMed]

51. Stock, P.; Erbe, A.; Buck, M.; Wiedemann, D.; Menard, H.; Hörner, G.; Grohmann, A. Thiocyanate Anchors for Salt-like Iron(II) Complexes on Au(111): Promises and Caveats. *Z. Naturforsch. B A J. Chem. Sci.* **2014**, *69*, 1164–1180. [CrossRef]

52. Hostettler, M.; Törnroos, K.W.; Chernyshow, D.; Vangdal, B.; Bürgi, H.-B. Challenges in Engineering Spin Crossover: Structures and Magnetic Properties of Six Alcohol Solvates of Iron(II) Tris(2-picolylamine) Dichloride. *Angew. Chem. Int. Ed.* **2004**, *43*, 4589–4594. [CrossRef] [PubMed]

53. Kirchner, R.M.; Mealli, C.; Bailey, M.; Howe, N.; Torre, L.P.; Wilson, L.J.; Andrews, L.C.; Rose, N.J.; Lingafelter, E.C. The Variable Coordination Chemistry of a Potentially Heptadentate Ligand with a Series of 3d Transition Metal Ions. The Chemistry and the Strcutures of $[M(py3tren)]^{2+}$, where M(II) = Mn, Fe, Co, Ni, Cu and Zn and (py3tren) = $N\{CH_2CH_2N=C(H)(C_5H_4N)\}_3$. *Coord. Chem. Rev.* **1987**, *77*, 89–163. [CrossRef]

54. Lazar, H.Z.; Forestier, T.; Barrett, S.A.; Kilner, C.A.; Letard, J.-F.; Halcrow, M.A. Thermal and light-induced spin-crossover in salts of the heptadentate complex [tris(4-{pyrazol-3-yl}-3-aza-3-butenyl)amine]iron(II). *Dalton Trans.* **2007**, *36*, 4276–4285. [CrossRef] [PubMed]

55. Reiher, M.; Salomon, O.; Hess, B.A. Reparameterization of hybrid functionals based on energy differences of states of different multiplicity. *Theor. Chem. Acc.* **2001**, *107*, 48–55. [CrossRef]

56. Lawson Daku, M.L.; Vargas, A.; Hauser, A.; Fouqueau, A.; Casida, M.E. Assessment of Density Functionals for the High-Spin/Low-Spin Energy Difference in the Low-Spin Iron(II) Tris(2,2'-bipyridine) Complex. *ChemPhysChem* **2005**, *6*, 1393–1410. [CrossRef] [PubMed]

57. Bowman, D.N.; Jakubikova, E. Low-Spin versus High-Spin Ground State in Pseudo-Octahedral Iron Complexes. *Inorg. Chem.* **2012**, *51*, 6011–6019. [CrossRef] [PubMed]

58. Bowman, D.N.; Bondarev, A.; Mukherjee, S.; Jakubikova, E. Tuning the Electronic Structure of Fe(II) Polypyridines via Donor Atom and Ligand Scaffold Modifications: A Computational Study. *Inorg. Chem.* **2015**, *54*, 8786–8793. [CrossRef] [PubMed]

59. Mengel, A.K.C.; Förster, C.; Breivogel, A.; Mack, K.; Ochsmann, J.R.; Laquai, F.; Ksenofontov, V.; Heinze, K. Heteroleptic Push-Pull Substituted Iron(II) Bis(tridentate) Complex with Low Energy Charge Transfer States. *Chem. Eur. J.* **2015**, *21*, 704–714. [CrossRef] [PubMed]

60. Jamula, L.L.; Brown, A.M.; Guo, D.; McCusker, J.K. Synthesis and characterization of a high-symmetry ferrous polypyridyl complex: Approaching the $^5T_2/^3T_1$ crossing point for Fe^{II}. *Inorg. Chem.* **2014**, *53*, 15–17. [CrossRef] [PubMed]

61. Buhks, E.; Navon, G.; Bixon, M.; Jortner, J. Spin Conversion Processes in Solutions. *J. Am. Chem. Soc.* **1980**, *102*, 2918–2923. [CrossRef]

62. Barrett, S.A.; Kilner, C.A.; Halcrow, M.A. Spin-crossover in [Fe(3-bpp)$_2$][BF$_4$]$_2$ in different solvents—A dramatic stabilisation of the low-spin state in water. *Dalton Trans.* **2011**, *40*, 12021–12024. [CrossRef] [PubMed]

63. Petzold, H.; Djomgoue, P.; Hörner, G.; Speck, J.M.; Rüffer, T.; Schaarschmidt, D. 1H-NMR spectroscopic elucidation in solution of the kinetics and thermodynamics of spin crossover for an exceptionally robust Fe^{2+} complex. *Dalton Trans.* **2016**, *45*, 13798–13809. [CrossRef] [PubMed]

64. Petzold, H.; Djomgoue, P.; Hörner, G.; Heider, S.; Lochenie, C.; Weber, B.; Rüffer, T.; Schaarschmidt, D. Spin state variability in Fe^{2+} complexes of substituted (2-(pyridin-2-yl)-1,10-phenanthroline) ligands as versatile terpyridine analogues. *Dalton Trans.* **2017**, *46*, 6218–6229. [CrossRef] [PubMed]

65. Petzold, H.; Djomgoue, P.; Hörner, G.; Lochenie, C.; Weber, B.; Rüffer, T. Bis-meridional Fe^{2+} SCO complexes of phenyl and pyridyl substituted 2-(pyridin-2-yl)-1,10-phenanthrolines. *Dalton Trans.* **2017**. (under Review).

66. Renz, F.; Oshio, H.; Ksenofontov, V.; Waldeck, M.; Spiering, H.; Gütlich, P. Strong Field Iron(II) Complex Converted by Light into a Long-Lived High-Spin State. *Angew. Chem. Int. Ed.* **2000**, *39*, 3699–3700. [CrossRef]

67. Vanko, G.; Bordage, A.; Papai, M.; Haldrup, K.; Glatzel, P.; March, A.M.; Doumy, G.; Britz, A.; Galler, A.; Assefa, T.; et al. Detailed Characterization of a Nanosecond-Lived Excited State: X-ray and Theoretical Investigation of the Quintet State in Photoexcited [Fe(terpy)$_2$]$^{2+}$. *J. Phys. Chem. C* **2015**, *119*, 5888–5902. [CrossRef] [PubMed]

68. Zhang, X.; Lawson Daku, M.L.; Zhang, J.; Suarez-Alcantara, K.; Jennings, G.; Kurtz, C.A.; Canton, S.E. Dynamic Jahn–Teller Effect in the Metastable High-Spin State of Solvated [Fe(terpy)$_2$]$^{2+}$. *J. Phys. Chem. C* **2015**, *119*, 3312–3321. [CrossRef]

69. Canton, S.E.; Zhang, X.; Lawson Daku, M.L.; Liu, Y.; Zhang, J.; Alvarez, S. Mapping the Ultrafast Changes of Continuous Shape Measures in Photoexcited Spin Crossover Complexes without Long-Range Order. *J. Phys. Chem. C* **2015**, *119*, 3322–3330. [CrossRef]

70. Neese, F. The ORCA program system. *WIREs Comput. Mol. Sci.* **2012**, *2*, 73–78. [CrossRef]

71. Schäfer, A.; Horn, H.; Ahlrichs, R. Fully optimized contracted Gaussian basis sets for atoms Li to Kr. *J. Chem. Phys.* **1992**, *97*, 2571–2577. [CrossRef]

72. Perdew, J.P.; Burke, K.; Ernzerhof, M. Generalized Gradient Approximation Made Simple. *Phys. Rev. Lett.* **1996**, *77*, 3865–3868. [CrossRef] [PubMed]

73. Becke, A.D. Density-functional exchange-energy approximation with correct asymptotic behavior. *Phys. Rev. A* **1988**, *38*, 3098–3100. [CrossRef]

74. Kanai, Y.; Wang, X.; Selloni, A.; Car, R. Testing the TPSS meta-generalized-gradient-approximation exchange-correlation functional in calculations of transition states and reaction barriers. *J. Chem. Phys.* **2006**, *125*, 234104. [CrossRef] [PubMed]

75. Staroverov, V.N.; Scuseria, G.E.; Tao, J.; Perdew, J.P. Comparative assessment of a new nonempirical density functional: Molecules and hydrogen-bonded complexes. *J. Chem. Phys.* **2003**, *119*, 12129. [CrossRef]

76. Becke, A.D. Density-functional thermochemistry. III. The role of exact exchange. *J. Chem. Phys.* **1993**, *98*, 5648–5652. [CrossRef]

77. Lee, C.; Yang, W.; Parr, R.G. Development of the Colle-Salvetti correlation-energy formula into a functional of the electron density. *Phys. Rev. B* **1988**, *37*, 785–789. [CrossRef]

78. Grimme, S. Semiempirical GGA-type density functional constructed with a long-range dispersion correction. *J. Comput. Chem.* **2006**, *27*, 1787–1799. [CrossRef] [PubMed]

79. Pędziński, T.; Markiewicz, A.; Marciniak, B. Photosensitized oxidation of methionine derivatives. Laser flash photolysis studies. *Res. Chem. Intermed.* **2009**, *35*, 497–506. [CrossRef]

80. Harris, R.K.; Becker, E.D.; Cabral de Menezes, S.M.; Granger, P.; Hoffman, R.E.; Zilm, K.W. Further conventions for NMR shielding and chemical shifts. *Pure Appl. Chem.* **2008**, *80*, 59–84. [CrossRef]

81. *CrysAlisPro Software System*, version 1.171.37. Intelligent Data Collection and Processing Software for Small Molecule and Protein Crystallography. Agilent Technologies: Oxford, UK, 2014.

82. Clark, R.C.; Reid, J.S. The analytical calculation of absorption in multifaceted crystals. *Acta Crystallogr. Sect. A Found. Crystallogr.* **1995**, *51*, 887–897. [CrossRef]

83. Sheldrick, G. *SHELXT*—Integrated space-group and crystal-structure determination. *Acta Crystallogr. Sect. A Found. Adv.* **2015**, *71*, 3–8. [CrossRef] [PubMed]

84. Sheldrick, G. Crystal structure refinement with *SHELXL*. *Acta Crystallogr. Sect. C Struct. Chem.* **2015**, *71*, 3–8. [CrossRef] [PubMed]

85. Dolomanov, O.V.; Bourhis, L.J.; Gildea, R.J.; Howard, J.A.K.; Puschmann, H. OLEX2: A complete structure solution, refinement and analysis program. *J. Appl. Crystallogr.* **2009**, *42*, 339–341. [CrossRef]

86. Macrae, C.F.; Bruno, I.J.; Chisholm, J.A.; Edgington, P.R.; McCabe, P.; Pidcock, E.; Rodriguez-Monge, L.; Taylor, R.; van de Streek, J.; Wood, P.A. Mercury CSD 2.0—New features for the visualization and investigation of crystal structures. *J. Appl. Crystallogr.* **2008**, *41*, 466–470. [CrossRef]

87. Kratzert, D.; Holstein, J.J.; Krossing, I. DSR: Enhanced modelling and refinement of disordered structures with *SHELXL*. *J. Appl. Crystallogr.* **2015**, *48*, 933–938. [CrossRef] [PubMed]

88. Spek, A.L. PLATON SQUEEZE: A tool for the calculation of the disordered solvent contribution to the calculated structure factors. *Acta Crystallogr. Sect. C Struct. Chem.* **2015**, *71*, 9–18. [CrossRef] [PubMed]

inorganics

MDPI

Article

Structure and Spin State of Iron(II) Assembled Complexes Using 9,10-Bis(4-pyridyl)anthracene as Bridging Ligand

Saki Iwai [1], Keisuke Yoshinami [1] and Satoru Nakashima [1,2,*]

[1]   Graduate School of Science, Hiroshima University, Higashi-Hiroshima 739-8526, Japan;
    m173497@hiroshima-u.ac.jp (S.I.); m165715@hiroshima-u.ac.jp (K.Y.)
[2]   Natural Science Center for Basic Research and Development, Hiroshima University,
    Higashi-Hiroshima 739-8526, Japan
*   Correspondence: snaka@hiroshima-u.ac.jp; Tel.: +81-82-424-6291

Received: 31 July 2017; Accepted: 7 September 2017; Published: 12 September 2017

Abstract: Assembled complexes, $[Fe(NCX)_2(bpanth)_2]_n$ (X = S, Se, BH_3; bpanth = 9,10-bis (4-pyridyl)anthracene), were synthesized. The iron for the three complexes was in temperature-independent high spin state by ^{57}Fe Mössbauer spectroscopy and magnetic susceptibility measurement. X-ray structural analysis revealed the interpenetrated structure of $[Fe(NCS)_2(bpanth)_2]_n$. In the local structure around the iron atom, the coordinated pyridine planes were shown to be a parallel type, which is in accordance with the results investigated by density functional theory (DFT) calculation. This complex (X = S) has CH–π interactions between the H atom of coordinated pyridine and the neighboring anthracene of the other 2D grid. It was suggested that the interpenetrated structure was supported by the stabilization of CH–π interaction, and this intermolecular interaction forced the relatively unstable parallel structure. That is, the unstable local structure is compensated by the stabilization due to intermolecular interaction, which controlled the spin state as high spin state.

Keywords: assembled complex; Mössbauer spectroscopy; spin crossover; intermolecular interaction controlling local structure

1. Introduction

Iron(II) octahedral assembled complexes can take two spin states (high spin (HS); S = 2 and low spin (LS); S = 0) in intermediate ligand field [1]. They may show a spin-crossover (SCO) phenomenon between HS state and LS state by several external field stimuli, such as temperature, pressure, and light-illumination. Assembled complexes have an important position in the spin-crossover chemistry [2,3], because they have a variety of structures and the spin state is sometimes controlled by the adsorption/desorption of guest molecules. Reference [4] provides a theoretical review. We have studied iron(II) assembled complexes using various bipyridine-type ligands [5,6]. The change of complex structure and the intermolecular interaction were predicted to influence the SCO phenomenon. The SCO behavior of iron(II) assembled complexes is related to the dihedral angle of pyridine ligands to Fe–NCX. The SCO appears when the local structure is propeller-type; on the other hand, SCO does not occur when the local structure is parallel or distorted propeller-type, which was demonstrated by density functional theory (DFT) calculation [7,8]. There is also a possibility that the weak π-accepter property of pyridine affects the spin state depending on the local structure around iron atom.

The iron(II) complex $[Fe(NCX)_2(bpb)_2]_n$ (X = BH_3; bpb = 1,4-bis(4-pyridyl)benzene) [9] shows SCO phenomenon. We introduced a methyl substituent to the benzene ring of bpb

(DMBPB = 1,4-dimethyl-2,5-bis(4-pyridyl)benzene) to control the dihedral angle between benzene and pyridine in the bridging ligand [10]. We expected that such change would affect the coordination angle of pyridine to Fe–NCX. Complexes of Ni and Cu using 9,10-bis(4-pyridyl)anthracene (bpanth) including guest molecules have some intermolecular interactions through π electrons [11,12]. In the present study, iron(II) assembled complexes were synthesized using bpanth which has many π-electrons on the bridging ligand. We expected to reveal the relation between intermolecular interactions (π–π or CH–π) and SCO phenomenon.

2. Results

2.1. ^{57}Fe Mössbauer Spectroscopy

The ^{57}Fe Mössbauer spectra of complexes [Fe(NCX)$_2$(bpanth)$_2$] (**1** (X = S), **2** (X = Se) and **3** (X = BH$_3$)) at 78 and 298 K are shown in Figure 1. All spectra showed doublet peaks. The values of isomer shift (I.S.) and quadrupole splitting (Q.S.) are summarized in Table 1. All values of I.S. are close to 1.00 mm·s^{-1}. This is the typical value for Fe(II)-HS state. This indicates that the SCO phenomenon does not occur from 78 K to room temperature. Q.S. value decreased in the order of NCS > NCSe > NCBH$_3$, as shown in Table 1.

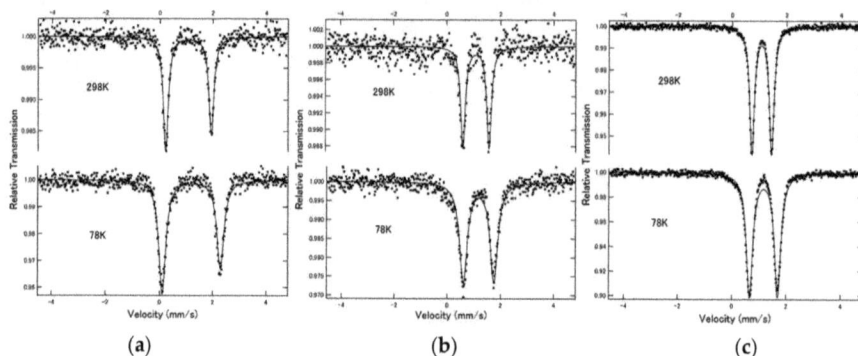

Figure 1. ^{57}Fe Mössbauer spectra of (a) **1**; (b) **2**; and (c) **3**.

Table 1. Mössbauer parameters for complexes **1**, **2**, and **3**.

Complex	T/K	I.S./mm·s^{-1}	Q.S./mm·s^{-1}	Γ/mm·s^{-1}	Relative Area/%
1	298	0.95	1.72	0.23	100
	78	1.12	2.18	0.32	100
2	298	0.95	0.97	0.21	100
	78	1.04	1.14	0.34	100
3	298	0.95	0.72	0.21	100
	78	1.04	1.03	0.28	100

2.2. Magnetic Suscesptibility

Figure 2 shows the magnetic susceptibilities of all complexes obtained on a SQUID magnetometer. The values of $\chi_M T$ of complexes **1**, **2**, and **3** are nearly constant (ca. 3.5 cm^3·K·mol^{-1}). These values reveal that irons are in Fe(II)-HS state. This suggests that all complexes do not show the SCO phenomenon, although slight possibility of HS form quenching remains.

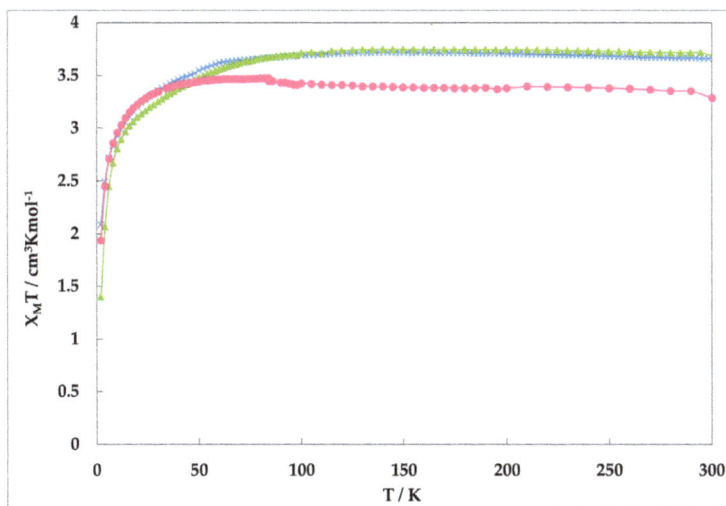

Figure 2. Magnetic susceptibility of **1** (red circle), **2** (green triangle), and **3** (blue cross).

2.3. Crystal Structure

The single crystal of $[Fe(NCS)_2(bpanth)_2]_n$ was obtained by diffusion method. Table 2 shows crystallographic data for complex **1** at 173 K. Figure 3a shows the Oak Ridge Thermal-Ellipsoid Plot Program (ORTEP) drawing and Figure 3b depicts the packing view of complex **1**. Guest molecule was not observed.

Table 2. Crystallographic data for complex **1**.

Complex	1
Formula	$FeC_{50}H_{32}N_6S_2$
$M/g \cdot mol^{-1}$	836.78
Crystal system	orthorhombic
Space group	Ibam
$a/Å$	9.6020(7)
$b/Å$	18.9796(14)
$c/Å$	23.8123(18)
$\alpha/°$	90
$\beta/°$	90
$\gamma/°$	90
$V/Å^3$	4339.6(6)
Z	4
T/K	173.0(2)
μ/mm^{-1}	0.485
ρ_{calcd} $(g \cdot cm^{-3})$	1.286
R	0.0711
wR	0.1786
Goodness of fit	0.907

The iron of complex **1** is coordinated by bpanth in the equatorial plane, and by NCS in the axial position. The bond length of Fe–N (NCS) is 2.110(3) Å, and that of Fe–N(bpanth) is 2.246(3) Å. In general, the distance of Fe–N for octahedral complex in Fe(II)-HS is ca. 2.2 Å. These values suggest that the iron of this complex takes the HS state at 173 K. This result corresponds to the results of

Mössbauer spectroscopy and magnetic measurement. Figure 3b shows the packing view of the interpenetrated structure. 2D grids are interpenetrated to each other.

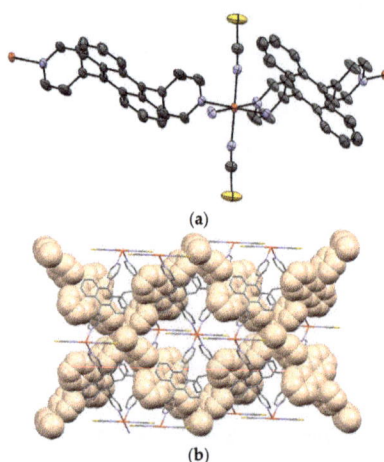

(a)

(b)

Figure 3. The ORTEP drawing (**a**) and the packing view (**b**) of complex **1**.

2.4. Powder X-ray Diffraction (PXRD) Pattern

Figure 4 shows the PXRD patterns of complexes, **1**, **2**, and **3**. The complexes are obtained from EtOH–dichloromethane solution as well as water–dichloromethane solution. The pattern of the sample obtained from EtOH–dichloromethane solution is similar to that from water–dichloromethane solution in each complex, suggesting no effect of solvent in the crystal structure. The simulation pattern using single crystal data is almost the same as the experimental pattern of **1**, suggesting the same structure between single crystal and the powder sample. However, a small amount of the other crystal was also suggested.

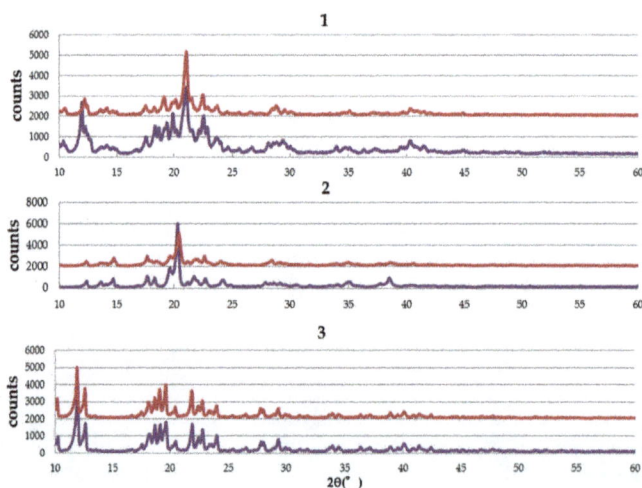

Figure 4. The powder X-ray diffraction (PXRD) patterns of **1**, **2**, and **3**. Red and purple show the pattern of the crystals from EtOH–dichloromethane solution and water–dichloromethane solution, respectively.

3. Discussion

The measurements of the ^{57}Fe Mössbauer spectroscopy and magnetic susceptibility suggest that all complexes **1**, **2**, and **3** do not show the SCO phenomenon and take Fe(II)-HS state from 5 to 300 K. The obtained single crystal of complex **1** was analyzed by XRD, and the structure was revealed. Figure 5 illustrates the local structure around the iron atom. It can be easily seen that the planes of opposite pyridine rings are parallel to each other. As mentioned in the introduction, an earlier report [7,8] stated that complexes of this type do not undergo SCO. The present result corresponds to this report. Figure 5 shows why: In order for the iron atom to become LS, the ligands must approach the iron. In this conformation, this would cause hydrogen atoms of neighboring rings to come unreasonably close to each other.

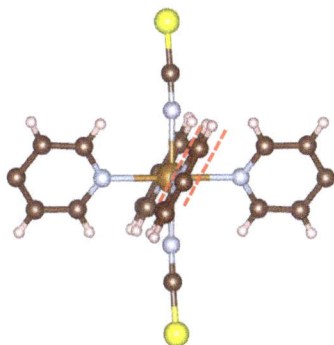

Figure 5. The local structure around the iron of complex **1**. Coordinated pyridine planes are parallel.

Figure 3b demonstrates the packing view of interpenetrated structure by 2D grids. It is suggested that this complex has intermolecular interactions. Figure 6 displays the focus picture of coordinated bridging ligand and neighboring bpanth of the other 2D grid. The pyridine of the upper ligand and the neighboring anthracene are orthogonal to each other. The distance between carbon atom and hydrogen atom (shown using a pink line) is 2.760(6) Å. This value is shorter than the sum (2.900 Å) of van der Waals radius for carbon and hydrogen atoms, and the C–H⋯C angle is 164.8(3) deg. This is the evidence that complex **1** has the CH–π interactions in the assembled structure. In this manner, this complex has the interpenetrated structure supported by the stabilization of CH–π interaction. Consequently, the coordinated pyridines facing each other are forced to become parallel, which is an unstable local structure.

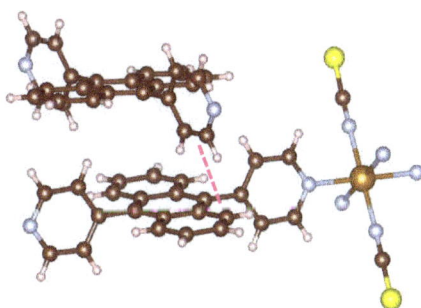

Figure 6. CH–π interaction between 2D grids.

Therefore, intermolecular interactions such as CH–π may prevent SCO phenomenon in the assembled complex, because such interactions affect the assembled structure and local structure around the iron atom. When an assembled complex has more flexible ligand or structure, it will have a possibility of showing the SCO phenomenon, because both the intermolecular interaction and the stability of the local structure around the iron atom do not need to compete.

4. Materials and Methods

4.1. Synthesis of Ligand and Complexes

4.1.1. Ligand

9,10-bis(4-pyridyl)anthracene (bpanth) was obtained by Suzuki–Miyaura cross-coupling reaction using 4-pyridyl boronic acid (18 mmol) and 9,10-dibromoanthracene (4 mmol) in toluene (60 mL) and water (60 mL). 1,1′-bis[(diphenylphosphino)ferrocene]dichloropalladium (1.2 mmol) as catalyst and Na_2CO_3 (20 mmol) as base were used in this reaction. The mixture was refluxed for 96 h under inactive gas atmosphere. The product was purified by column chromatography using alumina as the stationary phase and toluene as eluent. The alumina was deactivated by adding water (10 volume % to alumina), and then it was recrystallized from toluene. The bpanth was obtained as yellow crystal (yield: 72%) and identified by ^{1}H-NMR and elemental analysis. Anal. calcd. for $C_{24}H_{16}N_2$: C, 86.64%; H, 4.81%; N, 8.42%. Found for bpanth: C, 86.82%; H, 4.86%; N, 8.32%. ^{1}H-NMR (CDCl$_3$): δ 7.39 (q, 2H anthH$_{1,4}$), 7.44 (d, 2H, anthH$_{2,3}$), 7.60 (q, 2H PyH$_\beta$), 8.87 (d, 2H PyH$_\alpha$) ppm.

4.1.2. Complexes

Powder Complex of $[Fe(NCX)_2(bpanth)_2]_n$ (X = S **(1)**, Se **(2)**, BH$_3$ **(3)**)

$FeSO_4\cdot7H_2O$ (0.2 mmol), RNCX salt (0.4 mmol: R = K, X = S; R = K, X = Se; R = Na, X = BH$_3$) and ascorbic acid were dissolved in degassed water (15 mL). The bpanth (0.4 mmol) in dichloromethane (15 mL) and the water solution were mixed directly and were stirred for 30 min. Yellow powder in the X = S case (yield: 92%) and in the X = Se case (yield: 89%) and yellowish white powder in the X = BH$_3$ case (yield: 66%) were obtained. Three products were identified by elemental analysis. Anal. calcd. for $C_{50}H_{32}N_6S_2Fe\cdot0.8(C_2H_5OH)$ (for single crystals below using EtOH as solvent): C, 70.93%; H, 4.25%; N, 9.62%. Found for $[Fe(bpanth)_2(NCS)_2\cdot0.8(C_2H_5OH)]_n$: C, 70.64%; H, 3.96%; N, 10.05%. Anal. calcd. for $C_{50}H_{32}N_6Se_2Fe$: C, 64.53%; H, 3.47%; N, 9.03%. Found for $[Fe(bpanth)_2(NCSe)_2]_n$: C, 64.71%; H, 3.16%; N, 8.92%. Anal. calcd. for $C_{50}H_{38}N_6B_2Fe$: C, 75.03%; H, 4.79%; N, 10.50%. Found for $[Fe(bpanth)_2(NCBH_3)_2]_n$: C, 73.10%; H, 4.53%; N, 10.07%.

Single Crystal of $[Fe(NCS)_2(bpanth)_2]_n$

$FeSO_4\cdot7H_2O$ (0.2 mmol) and ascorbic acid were dissolved in ethanol (7 mL) and stirred for 10 min. This solution was added to a solution of KNCS salt (0.4 mmol) in ethanol (7 mL) and was stirred for 10 min. This turbid solution was filtered, and clear solution was obtained. The clear solution was put in the bottom of a Schlenk flask. Mixed solvent of ethanol (15 mL) and dichloromethane (15 mL) was carefully added in the flask as middle layer. The bpanth (0.4 mmol) in dichloromethane (15 mL) was added as upper layer. The yellow single crystal was obtained and analyzed by single-crystal X-ray diffraction. CCDC 1573711 contains the supplementary crystallographic data for this paper (see Supplementary Materials). These data can be obtained free of charge via http://www.ccdc.cam.ac.uk/conts/retrieving.html.

4.2. Instrumental Analysis

4.2.1. ^{57}Fe Mössbauer Spectroscopy

^{57}Fe Mössbauer spectra were obtained for both samples from water–dichloromethane solution and EtOH–dichloromethane solution at 78, 298 K with a ^{57}Co (Rh) radiation source moving in constant acceleration mode on Wissel MB-500 (Wissenschaftliche Electronik GmbH, Starnberg, Germany). The samples were cooled from 298 to 78 K for about 1 h. The 78 K was controlled during the measurement. The Mössbauer parameters were obtained by least-squares fitting to Lorentzian peaks. The spectra were calibrated by the six lines of α-Fe, the center of which was taken as zero isomer shift. The samples from water–dichloromethane solution and EtOH–dichloromethane solution did not show a significant difference in each complex.

4.2.2. Magnetic Susceptibility Measurement

Magnetic susceptibility measurement was performed for the samples obtained from water–dichloromethane solution on a Quantum Design MPMS-5S SQUID apparatus (Quantum Design, Inc., San Diego, CA, USA). Magnetic field was 1000 Oe. The data of **1** were obtained for cooling mode at scan rate of 10 K/min from 300 to 200 K, 5 K/min from 200 to 100 K, and 2 K/min from 100 to 2 K. The data of **2** were obtained for heating mode at scan rate of 2 K/min from 300 to 2 K after the cooling from 300 to 2 K at the rate of 10 K/min. The data of **3** were obtained for cooling mode at scan rate of 5 K/min from 300 to 100 K, and 2 K/min from 100 to 2 K. The solvate molecules are not included in the calculation of molecular weight.

4.2.3. Powder X-ray Diffraction Pattern

Powder X-ray diffraction (PXRD) patterns were measured on RIGAKU RINT-2000 using Cu Kα radiation (Rigaku Corp., Akishima, Japan). Scan rate was 2 degrees/min, step was 0.02 degrees, and scan region of 2θ was from 10 to 60 degrees for complexes **1**, **2**, and **3**. All measurements were performed at room temperature under air condition.

5. Conclusions

We synthesized three assembled complexes [Fe(NCX)$_2$(bpanth)$_2$]$_n$ (X = S, Se, BH$_3$; bpanth = 9,10-bis(4-pyridyl)anthracene). ^{57}Fe Mössbauer spectroscopic measurement and magnetic susceptibility measurement revealed that the spin crossover phenomenon does not occur, and it remains in the Fe(II)-HS state. X-ray structural analysis suggested that complex **1** has the parallel-type local structure around the iron atom and there is CH–π interactions between 2D grids in interpenetrated structure. This overall structure is supported by the stabilization of CH–π interaction. However, it causes the formation of unstable parallel-type local structure which takes Fe(II)-HS state. Intermolecular interactions such as CH–π can prevent SCO phenomenon in assembled complex, because such interactions affect the assembled structure and local structure around the iron atom. When an assembled complex has more flexible ligand or structure, it has a possibility of showing the SCO phenomenon. We need to reveal the structures for other complexes and the effect to SCO from the point of interaction.

Supplementary Materials: The following are available online at www.mdpi.com/2304-6740/5/3/61/s1. Cif and cif checked files.

Acknowledgments: We thank Inoue and members of Solid Material Chemistry Research Group, Hiroshima University for supporting magnetic susceptibility measurement and PXRD analysis. We also thank Mouri, the Natural Science Center for Basic Research and Development, Hiroshima University for the measurement of elemental analysis. We would like to thank Wiseman Chisale Bekelesi (MSc), Hiroshima University, for proofreading of English.

Author Contributions: Saki Iwai was involved in all stages of the work, including planning, conducting experiments, analyzing the data; Keisuke Yoshinami was involved in the discussion of the results and work planning; Satoru Nakashima acted as the supervisor and helped the data analysis and work planning; and all the authors wrote the paper and contributed to the revision of the paper.

Conflicts of Interest: The authors declare no conflict of interest.

References

1. Gütlich, P.; Bill, E.; Trautwein, A.L. *Mössbauer Spectroscopy and Transition Metal Chemistry*; Springer: Berlin/Heidelberg, Germany, 2011; pp. 391–476, ISBN 978-3-540-88427-9.
2. Real, J.A.; André, E.; Muñoz, M.C.; Julve, M.; Granier, T.; Bousseksou, A.; Varret, F. Spin crossover in a catenane supramolecular system. *Science* **1995**, *268*, 265–267. [CrossRef] [PubMed]
3. Halder, G.J.; Kepert, C.J.; Moubaraki, B.; Murray, K.S.; Cashion, J.D. Guest-Dependent Spin Crossover in a Nanoporous Molecular Framework Material. *Science* **2002**, *298*, 1762–1765. [CrossRef] [PubMed]
4. Banerjee, H.; Chakraborty, S.; Saha-Dasgupta, T. Design and Control of Cooperativity in Spin-Crossover in Metal-Organic Complexes: A Theoretical Overview. *Inorganics* **2017**, *5*, 47. [CrossRef]
5. Morita, S.; Nakashima, S.; Yamada, K.; Inoue, K. Occurrence of the Spin-crossover Phenomenon of Assembled complexes, $Fe(NCX)_2(bpa)_2$ (X = S, BH_3; bpa = 1,2-bis(4-pyridyl)ethane) by Enclathrating Organic Guest Molecule. *Chem. Lett.* **2006**, *35*, 1042–1043. [CrossRef]
6. Atsuchi, M.; Higashikawa, H.; Yoshida, Y.; Nakashima, S.; Inoue, K. Novel 2D Interpenetrated Structure and Occurrence of the Spin-crossover Phenomena of Assembled complexes, $Fe(NCX)_2(bpp)_2$ (X = S, Se, BH_3; bpp = 1,3-bis(4-pyridyl)propane). *Chem. Lett.* **2007**, *36*, 1064–1065. [CrossRef]
7. Kaneko, M.; Tokinobu, S.; Nakashima, S. Density Functional Study on Spin-crossover Phenomena of Assembled Complexes, $[Fe(NCX)_2(bpa)_2]_n$ (X = S, Se, BH_3; bpa: 1,2-bis(4-pyridyl)ethane). *Chem. Lett.* **2013**, *42*, 1432–1434. [CrossRef]
8. Kaneko, M.; Nakashima, S. Computational Study on Thermal Spin-Crossover Behavior for Coordination Polymers Possessing *trans*-$[Fe(NCS)_2(pyridine)_4]$ Unit. *Bull. Chem. Soc. Jpn.* **2015**, *88*, 1164–1170. [CrossRef]
9. Wu, X.-R.; Shi, H.-Y.; Wei, R.-J.; Li, J.; Zheng, L.-S.; Tao, J. Coligand and Solvent Effects on the Architectures and Spin-Crossover Properties of (4,4)-Connected Iron(II) Coordination Polymers. *Inorg. Chem.* **2015**, *54*, 3773–3780. [CrossRef] [PubMed]
10. Yoshinami, K.; Kaneko, M.; Yasuhara, H.; Nakashima, S. Effect of methyl substituent on the spin state of iron(II) assembled complex using 1,4-bis(4-pyridyl)benzene. *Radioisotopes* **2017**, in press.
11. Biradha, K.; Fujita, N. 2D and 1D Coordination Polymers with the Ability for Inclusion of Guest Molecules: Nitrobenzene, Benzene, Alkoxysilanes. *J. Incl. Phenom. Macrocycl. Chem.* **2001**, *49*, 201–208. [CrossRef]
12. Marin, G.; Andruh, M.; Madalan, A.M.; Blake, A.J.; Wilson, C.; Champness, N.R.; Schröder, M. Structural Diversity in Metal–Organic Frameworks Derived from Binuclear Alkoxo-Bridged Copper(II) Nodes and Pyridyl Linkers. *Cryst. Growth Des.* **2009**, *8*, 964–975. [CrossRef]

![inorganics logo] *inorganics*

MDPI

Article

Investigation of the Spin Crossover Properties of Three Dinulear Fe(II) Triple Helicates by Variation of the Steric Nature of the Ligand Type

Alexander R. Craze [1], Natasha F. Sciortino [2], Mohan M. Badbhade [3], Cameron J. Kepert [2], Christopher E. Marjo [3] and Feng Li [1,*]

[1] School of Science and Health, Western Sydney University, Locked Bag 1797, Penrith, Sydney, NSW 2751, Australia; 17717986@student.westernsydney.edu.au

[2] School of Chemistry, University of Sydney, Sydney, NSW 2006, Australia; natasha.sciortino@sydney.edu.au (N.F.S.); cameron.kepert@sydney.edu.au (C.J.K.)

[3] Mark Wainwright Analytical Centre, University of New South Wales, Sydney, NSW 2052, Australia; m.bhadbhade@unsw.edu.au (M.M.B.); c.marjo@unsw.edu.au (C.E.M.)

* Correspondence: Feng.Li@westernsydney.edu.au; Tel.: +61-2-9685-9987

Received: 14 August 2017; Accepted: 13 September 2017; Published: 21 September 2017

Abstract: The investigation of new spin-crossover (SCO) compounds plays an important role in understanding the key design factors involved, informing the synthesis of materials for future applications in electronic and sensing devices. In this report, three bis-bidentate ligands were synthesized by Schiff base condensation of imidazole-4-carbaldehyde with 4,4-diaminodiphenylmethane (L^1), 4,4'-diaminodiphenyl sulfide (L^2) and 4,4'-diaminodiphenyl ether (L^3) respectively. Their dinuclear Fe(II) triple helicates were obtained by complexation with $Fe(BF_4)_2 \cdot 6H_2O$ in acetonitrile. The aim of this study was to examine the influence of the steric nature of the ligand central atom (–X–, where X = CH_2, S or O) on the spin-crossover profile of the compound. The magnetic behaviours of these compounds were investigated and subsequently correlated to the structural information from single-crystal X-ray crystallographic experiments. All compounds $[Fe_2(L^1)_3](BF_4)_2$ (**1**), $[Fe_2(L^2)_3](BF_4)_2$ (**2**) and $[Fe_2(L^3)_3](BF_4)_2$ (**3**), demonstrated approximately half-spin transitions, with $T_{1/2}\downarrow$ values of 155, 115 and 150 K respectively, corresponding to one high-spin (HS) and one low-spin (LS) Fe(II) centre in a [LS–HS] state at 50 K. This was also confirmed by crystallographic studies, for example, bond lengths and the octahedral distortion parameter ($\sum$) at 100 K. The three-dimensional arrangement of the HS and LS Fe(II) centres throughout the crystal lattice was different for the three compounds, and differing extents of intermolecular interactions between BF_4^- counter ions and imidazole N–H were present. The three compounds displayed similar spin-transition profiles, with **2** (–S–) possessing the steepest nature. The shape of the spin transition can be altered in this manner, and this is likely due to the subtle effects that the steric nature of the central atom has on the crystal packing (and thus inter-helical Fe–Fe separation), intermolecular interactions and Fe–Fe intra-helical separations.

Keywords: spin-crossover; triple helicate; dinuclear; Fe(II) complex

---

1. Introduction

The design of new spin-crossover (SCO) coordination complexes is a challenge at the forefront of the field of magnetic molecular materials [1–6]. Maintained interest in these compounds stems from their potential applications in data storage, molecular switching and sensing devices [7,8]. The ability to understand and control the SCO properties of molecular materials has been of ongoing interest and research. The intricate effect of intermolecular interactions and magnetic coupling on the cooperativity of SCO systems has greatly complicated this understanding. Spin transition is generally observed in first-row transition metal coordination complexes with electronic configurations in the range d^4–d^7 [9,10]. These transitions produce a change in the magnetic, optical and structural properties of the material. Most commonly, Fe(II) (d^6) is implemented in an approximately octahedral coordination environment, which, when in an appropriate ligand field, can undergo a transition from the paramagnetic (S = 2) high-spin (HS) state ($^5T_2$) to a diamagnetic (S = 0) low-spin (LS) electronic configuration ($^1A_1$) [11–14].

Advances in the design of large metallo-supramolecular complexes, utilising metal–ligand interactions to synthesise nanoscale architectures such as universal ravels, [15] molecular knots, [16,17] cages [18] and helices [19–21]—not so easily accessible through the use of covalent bonds—are providing SCO complexes with increasing complexity of design [22–24]. The development of synthetic pathways for the construction of metallo-supramolecular SCO architectures that do not require extensive ligand synthesis, and that utilise commercially available materials, provides greater accessibility of these materials for future applications of SCO materials [25].

Homoleptic tris-diimine moieties have been extensively explored as ligand donors in SCO complexes [26], with the 2,4-imidazole-imine functional groups commonly displaying SCO behaviour in Fe(II) complexes of this manner [24,27]. Imidazole-imine groups create N–H hydrogen-bond donor sites towards the exterior of the chemical structure, providing an increased potential for the organisation of the induvial subunits into a larger network within the crystal structure, leading to magnetic cooperativity between molecular subunits [25].

To the best of our knowledge, there are only a few examples of Fe(II) dinuclear triple helicates that are capable of undergoing a reversible spin transition [25,28–32], all of which utilise the imidazole-imine moiety. Such compounds are composed of three ligands bridging two metal ion centres in a helical architecture. Hannon and co-workers explored the structural and magnetic differences of various counter ions and metal ion identities in a helicate series using carbon as the central atom of the ligand (–CH$_2$–) [25]. Two years later, the Mössbauer and magnetic properties of such complexes with the ClO$_4^-$ counter ion were examined by Garcia and Gütlich [28]. More recently, further studies of such helicates were conducted by Kruger [29,30], employing an oxygen central atom (–O–), identifying the importance of the degree of solvation on the spin transition of these compounds, as well as investigating the light-induced trapping of the excited state. Alternatively, the design and synthesis of coordination cages reported by Fujita [33], showed that the steric nature of the central atom of the ligand (C, S or O) could be manipulated to give subtly different ligand angles, which produced profound changes in the overall supramolecular architecture.

After investigations of such dincluear Fe(II) SCO triple helicates, such as counter ions and solvents, this paper sought to synthesise three Fe(II) dinuclear triple helicates, ([Fe$_2$(**L^1**)$_3$](BF$_4$)$_2$ (**1**), [Fe$_2$(**L^2**)$_3$](BF$_4$)$_2$ (**2**) and [Fe$_2$(**L^3**)$_3$](BF$_4$)$_2$ (**3**)), and investigate the effects of the steric nature of three different central ligand atom identities (–X–, where X = CH$_2$, S or O, Figure 1) on the structure and spin transition of their respective helicate architectures. Changing the identity of this central atom of the ligand may have several effects on the compounds. First, manipulation of the angle with which the ligand coordinates to two Fe(II) centres could influence the intra-helical separation (Fe···Fe separation). The investigation of the –CH$_2$– compound, **1**, by Hannon and Gütlich [25,28], found this separation to be 11.56 Å, while in the –O– compounds of Kruger and co-workers the distances of 11.35 and 11.45 Å for compounds with MeCN and H$_2$O solvent inclusions were respectively exhibited. Second, slight changes in this angle could affect the geometry of the coordination environment (octahedral distortion), and therefore affect the temperature of transition [34]. Such structural manipulation could also cause a change in the preferred three-dimensional packing arrangement of the helicates in the crystal lattice, imposing changings on the intermolecular interactions within the crystal lattice and inter-helical Fe–Fe separation. These factors could all affect the cooperativity between the Fe(II) centres in the lattice, and as a result impact the spin-transition profile.

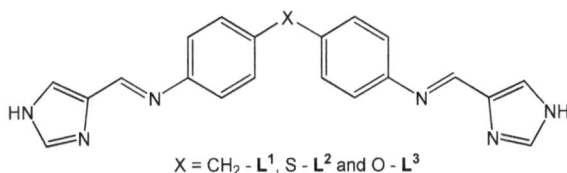

X = CH$_2$ - **L^1**, S - **L^2** and O - **L^3**

Figure 1. Schematic representation of the ligands **L^1**, **L^2** and **L^3**.

2. Results and Discussion

2.1. Syntheses

Three Schiff base bis-bidentate ligands **L^1**, **L^2** and **L^3** were prepared in 88%, 91% and 93% yield, respectively, from the condensation of imidazole-4-carbaldehyde with 4,4-diaminodiphenylmethane, 4,4′-diaminobiphenyl sulfide or 4,4′-diaminodiphenyl ether. ^{1}H NMR spectra (Figures S1–S3) and high-resolution electrospray ionization (HR-ESI) mass spectrometry results were consistent with the proposed structures of **L^1** **L^2** and **L^3**, with the NMR spectra, confirming the formation of bis-bidentate linking species. In the HR-mass spectra (Figures S4–S6), peaks for [Ligand+H]$^+$ were observed at m/z 355.1606 (for [**L^1**+H]$^+$), m/z 357.1367 (for [**L^2**+H]$^+$) and m/z 357.1325 (for [**L^3**+H]$^+$); the appropriate isotope patterns for [Ligand+H]$^+$ were evident (inserts in Figures S4–S6) with the isotopic distributions in excellent agreement with their simulated patterns. The further reaction of **L^1**, **L^2** and **L^3** with iron(II) tetrafluoroborate in acetonitrile, respectively, followed by the slow diffusion of diethyl ether into the reaction mixture, produced large light orange crystals of complexes **1**, **2** and **3** of suitable quality for X-ray diffraction studies. HR-ESI mass spectrometry results (Figures S7–S9) revealed a series of peaks of various charges corresponding to $\{[Fe_2(\mathbf{L}^m)_3][BF_4]_{(4-n)}\}^{n+}$ ($m = 1$–3 and $n = 2$–4), which are consistent with the successive loss of $[BF_4]^-$ anions, with the isotopic distributions for all the above species being in good agreement with their simulated patterns (inserts in Figures S7–S9). However, a singly charged species of $\{[Fe_2(\mathbf{L}^m)_3][BF_4]_3\}^+$ ($m = 1$–3), resulting from the loss of one $[BF_4]^-$ anion was not observed, even at low intensity for all three compounds.

2.2. Magnetic Susceptibility Measurements

Magnetic susceptibility data was collected over the temperature range of 50–350 K. The measurements were performed on polycrystalline samples with a field of 0.5 T; a scan rate of 4 K/min. The $\chi_M T$ versus T plots for the compounds **1**, **2** and **3** are shown in Figure 2. The samples were pre-dried for two weeks in a vacuum oven and heated to 400 K for 30 min in an initial heating cycle to remove acetonitrile solvent molecules included in the lattice (see DSC-TGA measurements Figures S10–S12). The samples were then cooled to 50 K and subsequently heated to 350 K. The heating and cooling modes demonstrated a small hysteresis for all three compounds **1**–**3** (Figures 2 and S19–S21).

The $\chi_M T$ versus T plot of **1**, in the cooling mode, is characteristic of a half SCO, with broad transition taking place between 60 and 250 K. The room temperature $\chi_M T$ of 7.71 cm^3·K·mol^{-1} corresponds to two magnetically uncoupled octahedral Fe(II) centres with an $^5T_2$ (S = 2) ground state. As the temperature was lowered, the $\chi_M T$ stayed relatively constant until 240 K, after which it dropped in a single step manner, with a $T_{1/2}\downarrow$ value of 155 K, towards a value of 4.41 cm^3·K·mol^{-1} at 50 K. This represents a slightly incomplete half spin transition, with approximately 40% of the Fe(II) centres transitioning to a LS^{1}A$_1$ (S = 0) state. Alternatively, the heating mode demonstrated a $T_{1/2}\uparrow$ value of 170 K, indicating a thermal hysteresis of 15 K.

The spin transition of **2** demonstrated a more abrupt character. For the cooling mode, the room temperature $\chi_M T$ value of 7.65 cm^3·K·mol^{-1} was consistent with a full [HS–HS] state. Upon cooling, the $\chi_M T$ also decreased in a single step manner, with a $T_{1/2}\downarrow$ value of 115 K. At 50 K, the $\chi_M T$ value dropping to 3.15 cm^3·K·mol^{-1}, which corresponded to c.a. 60% of the Fe(II) metal ions having undergone a thermally-induced spin conversion to the low-spin $^1A_1$ (S = 0) from the high-spin $^5T_2$ (S = 2) electronic state. A similar thermal hysteresis of 15 K between the cooling and heating mode was observed ($T_{1/2}\uparrow$ 130 K for heating mode).

Finally, **3** behaved in a similar manner, exhibiting a more gradual spin transition, with a $T_{1/2}$ of 150 K and 165 K for the cooling and heating cycles respectively, with a 15 K thermal hysteresis. The room temperature $\chi_M T$ value of 7.67 cm^3·K·mol^{-1} again showed complete [HS–HS] occupation of the Fe(II) centres. This decreased in a single step manner to a value of 3.92 cm^3·K·mol^{-1} at 50 K, at which point approximately 50% of Fe(II) had transitioned to the low spin state.

Of the three compounds, the sulphur derivative demonstrated the most complete half spin transition and a more abrupt nature. The carbon and oxygen derivatives exhibited very similar $\chi_M T$ versus T plots, although the oxygen derivative possessed a slightly more complete half transition.

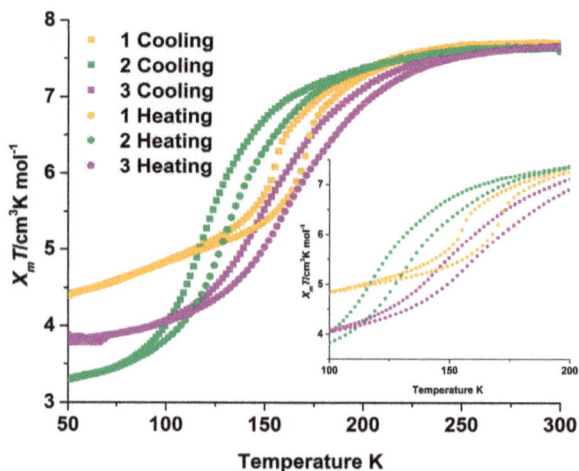

Figure 2. Magnetic susceptibility $\chi_M T$ versus T plots for **1–3**, at a scan rate of 4 K/min over the temperature range of 50–300 K, in both the cooling (square) and heating (circle) modes. For clarity, the inset shows the spin transition between 100 and 200 K.

2.3. Magneto-Structural Correlations

Single crystal X-ray diffraction experiments were performed at 100 K for **1**, **2** and **3** respectively (Figure 3). The complex **1**, which was previously reported by Hannon et al. [25] at 100 K, crystallised in the monoclinic space group *C2/c*, with half of the helicate in the asymmetric unit. The Fe(II) atom is coordinated by three N_{imine} and three $N_{imidazole}$ donors in a distorted octahedral environment, with octahedral distortion parameters (Σ) of 76.30 (Fe01) degrees (Table 1). The N–Fe(II) coordinate bond lengths were 2.12 Å (Fe01). These intermediate values suggest a mixed population of HS and LS-state Fe(II) centres within the helicates of the crystal lattice ([LS–HS]) at 100 K [29,30,34,35]. The two Fe(II) centres are separated by an inter-helical distance of 11.72 Å (Table 2). Hydrogen bonding interactions are present between the BF$_4^-$ counter ions and the imidazole N–H (F···N 2.71 and 2.80 Å), as well as between acetonitrile C–H and counter ion F (C···F 2.71 Å).

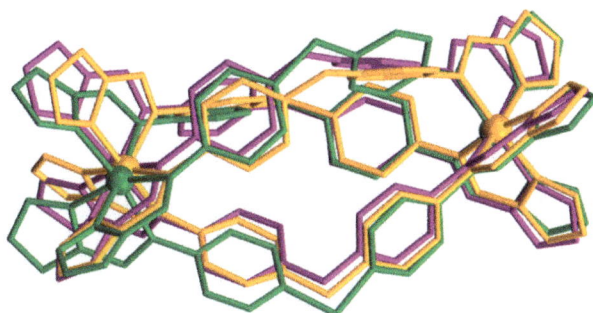

Figure 3. Schematic overlay representation of the X-ray structures of **1** (orange), **2** (green) and **3** (purple) at 100 K. Counter Ions, solvent molecules and hydrogen atoms have been excluded for clarity.

The complex **1** packs in infinite 2D sheets along the *c* and *a*-crystallographic axes in rows of helicates orientated in a diagonal manner with respect to the adjacent helicates along these axes (Figure 4). These 2D sheets stack upon one another along the *b*-axis in an offset manner. The observed differences in magnetic susceptibility results between **1** and those reported by Hannon et al. may be a result of the different packing arrangements these helicates demonstrate and the absence of the formation of a two-dimensional hydrogen-bonding network in **1**. As such, differences in molecular arrangement and non-covalent interactions leads to separate degrees of cooperativity between Fe(II) centres throughout the crystal lattice. The lesser extent to which counter ions provide intermolecular interactions between individual helicates may explain why the spin-crossover observed in this study is less abrupt in nature [36]. The asymmetric unit obtained by Hannon et al. contained two waters, two acetonitriles and one diisopropyl ether molecule, whereas the crystal structure of **1** contained only two partially occupied acetonitrile molecules in the asymmetric unit. The packing differences, along with the alternate solvent molecules, could have led to different degrees of cooperativity between Fe(II) centres of the crystal lattice. It is also to be noted that **1** was crystallised by means of a diethylether vapour diffusion, while Hannon et al. employed a diisopropyl ether diffusion.

The magnetic susceptibility measurement at 100 K (4.84 K·mol^{-1}) suggests approximately 40% of Fe(II) centres have transitioned to a low spin state. This is confirmed by crystal data measurements, which show Fe–N bond lengths and distortion parameters representative of a mixed spin-state population, [LS–HS], of Fe(II) centres. At this stage, single crystal diffraction experiments were unable to be performed at high temperatures.

Table 1. Comparison of important spin cross-over related crystallographic values.

Compound	Σ (Degrees)	Average Fe(II)–N Bond Lengths (Å)	High Spin or Low Spin Fe(II) in the Asymmetric Unit
1 100 K	Fe01—76.30	Fe01—2.13	1 mixed HS/LS-state population
2 100 K	Fe01—59.4 Fe02—90.3	Fe01—2.00 Fe02—2.18	1 HS, 1 LS
3 100 K	Fe01—77.2 Fe02—85.2	Fe01—2.10 Fe02—2.18	1 HS, 1 mixed HS/LS state population

Table 2. Comparison of helical geometric parameters in **1**, **2** and **3**.

Complex	–X– Angle (Degrees)	Intra-Helical Separation (Å)	Closest Inter-Helical Separation (Å)
1	113.6	11.72	8.71
2	104.9	11.78	8.31
3	115.8	11.62	7.96

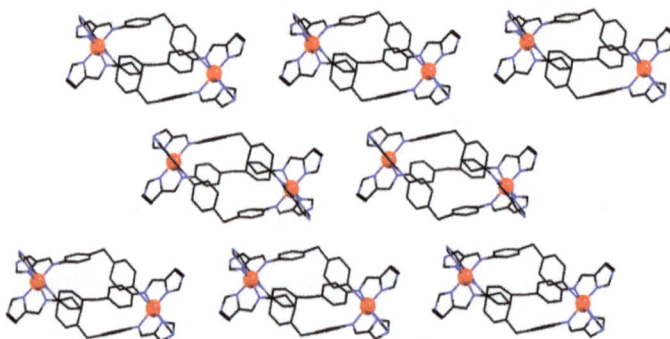

Figure 4. Schematic representation of the crystal packing structure of **1** at 100 K, demonstrating the formation of 2D sheets of interdigitated helicates. Solvent molecules and counter ions have been excluded for clarity. Mixed spin state populated Fe(II) metal centres are represented by red spheres. The crystallographic *c*-axis runs along the length of the page, and the *a*-axis down the page.

At 100 K, complex **2** crystallised in the triclinic space group $P\bar{1}$. A single unique helicate was present in the asymmetric unit, encompassing one high-spin and one low-spin Fe(II) centre (Figure 5). This is confirmed by average Σ values of 90.3 and 59.4 degrees and the average Fe(II)–N distances of 2.18 and 2.00 Å for the HS and LS Fe(II) centres respectively [34,35,37], which is consistent with magnetic measurements that suggest a 53% transition to the low spin state at 100 K ($\chi_M T$ 7.67 cm^3·K·mol^{-1}). The intra-helical distance in this case is 11.78 Å. These helicates are linked by hydrogen bonding between the uncoordinated $N_{Imidazole}$ N–H and BF$_4^-$ counter ions (F···N 2.78 and 2.90 Å) along the *a*-axis. Along this axis, the adjacent centres linked by hydrogen-bonding are of the same spin-state. Packing along the *c*-axis lengthwise, in a slightly interdigitated manner, positions the closest Fe(II) centres of adjacent helicates so that they are of the opposite spin state (inter-helical distance 12.31 Å) (Figure 5). Similarly, along the *b*-axis, inversion of the complexes produces alternating HS and LS Fe(II) centres, so that centres of opposite spin states are closest to one another (inter-helical distance 8.31 Å) (Figure 6). This is considerably closer than the intra-helical distance between Fe(II) metal ions (Table 2). In this manner, the high and low spin states alternate throughout the crystal lattice at 100 K. The helicates are not linked by hydrogen bonding along the *b* or *c*-axis. Hydrogen-bonding interactions linking complexes of like spin-states along the *a*-axis suggest some degree of cooperativity between Fe(II) centres along this crystallographic axis. Three of the six aryl substituents participate in intramolecular edge-to-face CH···π interactions, stabilising larger torsion angles between planes of the two phenyl groups of the same ligand.

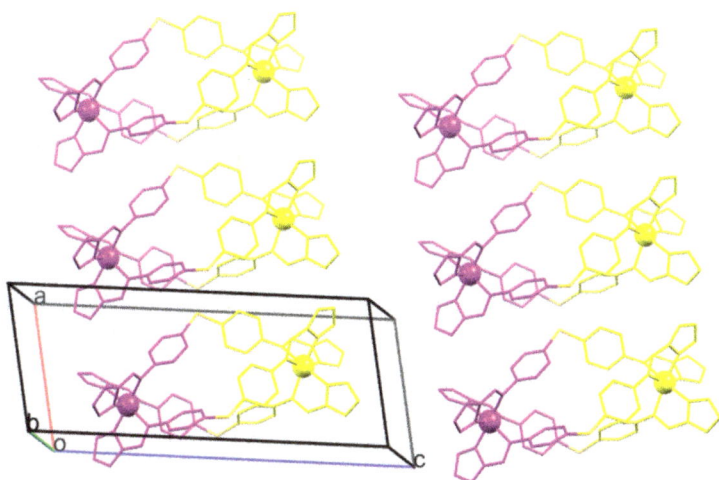

Figure 5. Schematic representation of the X-ray crystal structure of **2** at 100 K showing the distribution of high (yellow) and low spin (purple) Fe(II) centres along the crystallographic *c*- and *a*-axis. Hydrogens, solvent and counter ions have been excluded for clarity.

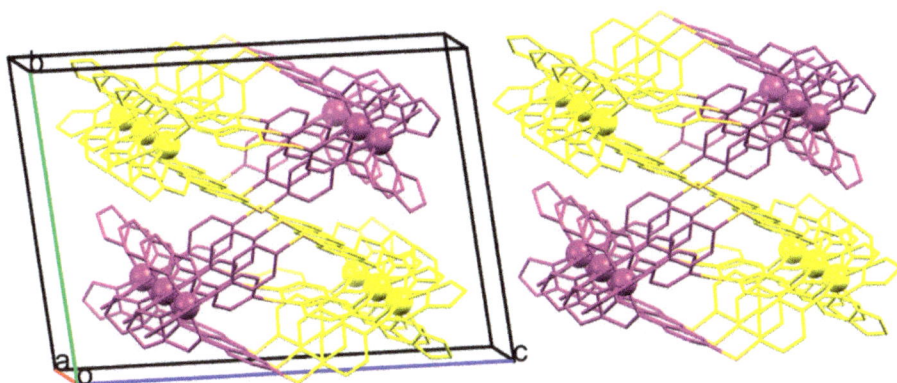

Figure 6. Schematic representation of the X-ray crystal structure of **2** at 100 K showing the distribution of high (yellow) and low spin (purple) Fe(II) centres in alternating HS and LS centres along the *b*- and *c*-axis, and in chains of the same spin state along the *a*-axis. Hydrogens, solvent and counter ions have been excluded for clarity.

Finally, complex **3** at 100 K also crystallised in the triclinic space group $P\overline{1}$, with one molecule in the asymmetric unit. The Σ octahedral distortion and Fe–N bond lengths in this case suggest one HS and one mixed spin state populated Fe(II) centre (2.18 and 2.10 Å and Σ values of 85.2 and 77.2 degrees). Again, this agrees with the magnetic susceptibility data at 100 K (4.07 $cm^3 \cdot K \cdot mol^{-1}$), which indicates that around 53% of Fe(II) centres are in the HS state. At 100 K, the complex packs in undulating layers of complexes along the crystallographic *a*-axis. Along the *a* and *b*-axis, the Fe(II) centres in neighbouring helicates pack so that they are adjacent to Fe(II) ions of the same spin state (Figure 7). These form infinite row-like domains of Fe(II) atoms throughout the lattice with the same spin state. Hydrogen bonding interactions between non-bonding imidazole nitrogens and BF_4^- counter ions (F···N 2.87, 2.74, 2.89 and 2.83 Å) link HS centres to mixed-spin state centres of adjacent complexes along the *c*-axis. The intra-helical Fe⋯Fe separation is 11.62 Å, while the shortest inter-helical distance is 7.96 Å. This intra-helical distance is very similar to that found by Archer et al. [30] in an oxygen helicate derivative with methylated imidazole moieties. Two of the six aryl rings of the dinuclear triple helicate participate in edge-to-face CH⋯π interactions of 2.54 and 2.83 Å.

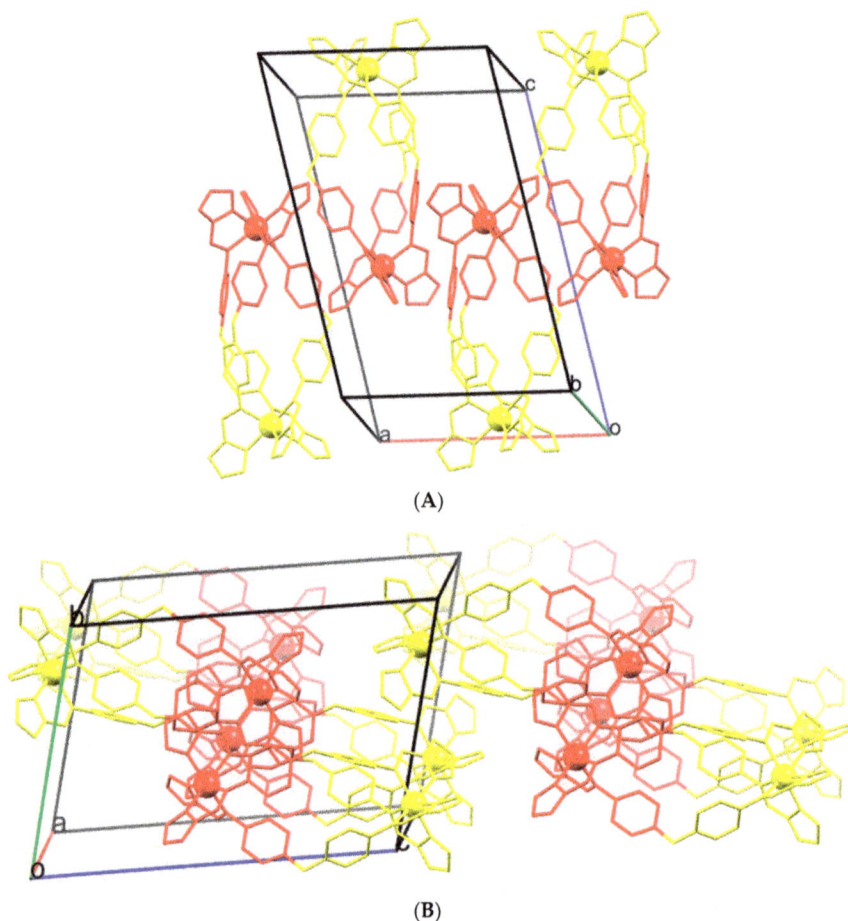

(A)

(B)

Figure 7. Schematic representations of the crystal packing structure of **3** at 100 K showing the packing of HS (yellow) and mixed spin state (red) Fe(II) centres in undulating rows of metal ions with similar spin states throughout the crystal lattice. Image (**A**) shows the packing in two-dimensions and image (**B**) shows three-dimensional arrangement of Fe(II) centres. Hydrogens, solvent and counter ions have been omitted for clarity.

This study was interested in how the angle formed between the linking atom (C, S or O) and the two phenyl rings of the ligand (–X–, where X = CH$_2$, S or O) affects the nature of the spin transition. Differences in this angle could influence the spin transition by altering the intra-helical distance, or by altering the orientation with which the complexes pack, therefore changing the degree of intermolecular interactions and cooperativity between metal centres. The –X– ligand angle was largest for the –O– in **3** (115.760 degrees), while –CH$_2$– (**1**) and –S– (**2**) were 113.6 and 104.9 degrees respectively. The packing orientations of the three structures at 100 K differ significantly. This could be a factor of the intra-helical distance and other geometric properties of the helicates, as well as the different intermolecular interactions with solvent and counter ion molecules. Interestingly, the magnetic susceptibility curves of **1** and **3** are very similar in shape ($T_{1/2}\downarrow$ 155 and 150 K respectively), as are their –X– angles. On the other hand, **2** displayed a much steeper curve ($T_{1/2}\downarrow$ 115 K), and in turn a larger intra-helical separation. Complex **2** also displayed a greater degree of intermolecular interactions mediated by the BF_4^- anions, linking the helicates into chains along the *a*-axis. In this structure, these "linked" Fe(II) centres were of the same spin-state (see above), suggesting a greater degree of cooperativity between Fe(II) centres in this structure. As has been shown previously, the intermolecular interactions, particularly hydrogen bonding, are a crucial influence on the cooperativity of the spin transition [36,38,39]. This cooperativity may be a factor influencing the more abrupt nature of the spin transition in **2**. To surmise, altering the steric nature of the central ligand atom (C, O or S) shows a slight influence on the spin transition, allowing the spin transition to be altered in this manner. Although the direct cause of this difference is not clear, it is more likely to be a consequence of subtle changes in the packing arrangement, the inter and intra-helical separations, as well as the degree of intermolecular interactions within the crystal structure. Important crystallographic information for compounds **1**, **2** and **3** is included in Table 3.

Table 3. Crystallographic data for the compounds measured in this experiment.

	1·(CH₃CN)	2·2 (CH₃CN)	3·2 (CH₃CN)
Empirical formula	$C_{65}H_{57}B_4F_{16}Fe_2N_{19}$	$C_{64}H_{54}B_4F_{16}Fe_2N_{20}S_3$	$C_{64}H_{54}B_4F_{16}Fe_2N_{20}O_3$
Formula weight	1563.23	1658.39	1610.21
Temperature/K	100	100	100
Crystal system	monoclinic	triclinic	triclinic
Space group	$C2/c$	$P\bar{1}$	$P\bar{1}$
a/Å	21.131(4)	9.5210(19)	13.837(3)
b/Å	17.391(4)	16.723(3)	13.855(3)
c/Å	21.069(4)	23.621(5)	20.684(4)
α/°	90	95.29(3)	77.43(3)
β/°	107.00(3)	100.29(3)	77.58(3)
γ/°	90	92.93(3)	86.73(3)
Volume/Å3	7404(3)	3675.8(13)	3779.5(15)
Z	4	2	2
ρcalc.g/cm³	1.402	1.498	1.415
μ/mm 1	0.485	0.576	0.481
F(000)	3184.0	1684.0	1636.0
2Θ range for data collection/°	0.02 × 0.01 × 0.01	0.015 × 0.0075 × 0.0075	0.02 × 0.01 × 0.01
Index ranges	Synchrotron (λ = 0.7108)	Synchrotron (λ = 0.7108)	Synchrotron (λ = 0.7108)
Reflections collected	3.09 to 52.998	1.762 to 52.742	2.062 to 53.998
Independent reflections	$-26 \leq h \leq 26, -21 \leq k \leq 21,$ $-26 \leq l \leq 26$	$-11 \leq h \leq 11, -20 \leq k \leq 20,$ $-29 \leq l \leq 29$	$-17 \leq h \leq 17, -17 \leq k \leq 17,$ $-26 \leq l \leq 26$
Data/restraints/parameters	50,460	45,853	57,096
Goodness-of-fit on F^2	6995 [R_{int} = 0.0697, R_{sigma} = 0.0341]	13,521 [R_{int} = 0.0971, R_{sigma} = 0.0898]	14,860 [R_{int} = 0.0446, R_{sigma} = 0.0360]
Final R indexes [$I \geq = 2\sigma$ (*I*)]	6995/218/630	13,521/202/1047	14,860/33/1095
Final R indexes [all data]	1.119	1.157	1.072
Largest diff. peak/hole/e Å-3	$R_1 - 0.1050$, w$R_2 = 0.2997$	$R_1 - 0.1103$, w$R_2 = 0.2958$	$R_1 = 0.0691$, w$R_2 = 0.1967$

3. Materials and Methods

All reagents and solvents were purchased from commercial sources, with no further purification being undertaken. ^{1}H NMR spectra were recorded on a Bruker 400 MHz spectrometer (Bruker AXS GmbH, Karlsruhe, Germany). High resolution ESI-MS data were acquired using a Waters Xevo QToF mass spectrometer (Waters, Milford, MA, USA), operating in positive ion mode with a desolvation temperature of 120, desolvation gas flow of 450 and varying sample and extraction cone temperatures. A waters lock spray system was used to calibrate the high-resolution masses. FT-IR measurements were undertaken on a Bruker Vertex 70 (Bruker AXS GmbH, Karlsruhe, Germany) with a diamond ATR stage. DSC and TGA measurements were performed using a simultaneous thermal analysis (STA) 449 C Jupiter instrument (Netzsch Australia Pty Ltd., Sydney, Australia). The STA measurements were performed using an aluminium crucible; nitrogen was used as both the protective and purge gases, and the temperature range of 30–200 °C was cycled at a rate of $10 \cdot K \cdot min^{-1}$.

3.1. Preparation of (L^1, L^2 and L^3)

The methanol (15 mL) solution of 0.75 mmol of 4,4′methylenedianiline, 4,4-oxydianiline or 4,4-thiodianiline was added dropwise to 1.50 mmol of imidazole-4-carbaldehyde in 15 mL methanol under stirring. Three drops of glacial acetic acid were added and the reaction mixture was refluxed overnight. The mixture was allowed to cool to room temperature, and subsequently cooled in a refrigerator. The white precipitate was filtered, washed with MeOH and air dried.

L^1: Yield 88%. ^{1}H NMR (DMSO, 400 MHz) δ (ppm) 8.44 (s, 2H), 7.83 (s, 2H), 7.65 (br, 2H), 7.26 (d, 4H), 7.14 (d, 4H), 3.95 (s, 2H); ESI-MS (positive-ion detection, CH_3CN, m/z): cald. for $[L^1+H]^+$, 355.1671; found 355.1606; FT-IR (ATR ν_{max}/cm^{-1}): 3023, 2821, 1626, 1501, 1221, 1093, 873, 841, 620, 537 cm^{-1}.

L^2: Yield 91%. ^{1}H NMR (DMSO, 400 MHz) δ (ppm) 8.46 (s. 2H), 7.83 (d, 3H), 7.56 (br, 1H), 7.36 (d, 4H), 7.23 (d, 4H); ESI-MS (positive-ion detection, CH_3CN, m/z): cald. for $[L^2+H]^+$, 373.1235; found 373.1167; FT-IR (ATR ν_{max}/cm^{-1}): 2579, 1622, 1491, 1243, 1111, 834, 622, 535 cm^{-1}.

L^3: Yield 93%. ^{1}H NMR (DMSO, 400 MHz) δ (ppm) 8.45 (s. 2H), 7.84 (s, 2H), 7.26 (d, 4H), 7.04 (d, 4H), 3.38 (s), 2.51 (p); ESI-MS (positive-ion detection, CH_3CN, m/z): cald. for $[L^3+H]^+$, 357.1464; found 357.1325; FT-IR (ATR ν_{max}/cm^{-1}): 2583, 1630, 1578, 1364, 1110, 1087, 867, 821, 777, 620, 552 cm^{-1}.

3.2. Preparation of **1**, **2** and **3**

$Fe(BF_4)_2 \cdot 6H_2O$ (0.2 mmol) in 10 mL of acetonitrile was slowly added to a suspension of L^1, L^2 or L^3 (0.3 mmol) in 20 mL of acetonitrile. The reaction mixture was heated at 70 °C and stirred for 3 h, leading to a clear orange solution. The solution was filtered, with a slow diffusion of diethyl ether into the acetonitrile solution, resulting in the formation of large orange crystals. These were then allowed to dry under vacuum. Single crystals were taken from the same sample and used for the X-ray study.

1: Yield 81%. ESI-HRMS (positive ion detection, CH_3CN, m/z): cald. for $[Fe_2(L^1)_2]^{4+}$, 293.5881; found 293.5724; FT-IR (ATR ν_{max}/cm^{-1}): 3311, 1619, 1491, 1225, 1006, 613 cm^{-1}.

2: Yield 78%. ESI-HRMS (positive-ion detection, CH_3CN, m/z): cald. for $[Fe_2(L^2)_2]^{4+}$, 307.0554; found 307.0481; FT-IR (ATR ν_{max}/cm^{-1}): 3310, 2981, 1615, 1004, 613, 562 cm^{-1}.

3: Yield 85%. ESI-HRMS (positive-ion detection, CH_3CN, m/z): cald. for $[Fe_2(L^3)_2]^{4+}$, 295.0726; found 295.0609; FT-IR (ATR ν_{max}/cm^{-1}): 3310, 3150, 1618, 1490, 1225, 1200, 1007, 861, 838, 613, 520 cm^{-1}.

3.3. X-ray Crystallography

The X-ray crystallography experiments were performed at the Australian Synchrotron, using silicon double crystal monochromated radiation at 100 K [40,41]. The crystal was rotated through Phi angle of 1–360 degrees. Data was collected at 100 K for each structure. Data integration and reduction was undertaken with XDS [42]. An empirical absorption correction was then applied using *SADABS* at the Australian Synchrotron [43]. The structures were solved by direct methods and the full-matrix least-squares refinements were carried out using a suite of *SHELX* programs [44,45] via the *OLEX2* graphical interface [46]. Non-hydrogen atoms were refined anisotropically. Carbon-bound hydrogen atoms were included in idealised positions and refined using a riding model. The crystallographic data in CIF format has been deposited at the Cambridge Crystallographic Data Centre with CCDC 1568781–1568783. It is available free of charge from the Cambridge Crystallographic Data Centre, 12 Union Road, Cambridge CB2 1 EZ, UK; fax: (+44) 1223-336-033; or e-mail: deposit@ccdc.cam.ac.uk. Specific refinement details and crystallographic data for each structure are present above and in the supporting information. The asymmetric unit of **1** contains one half of a helicate, two BF_4^- counter ions and two acetonitrile molecules, with occupancies of 0.25. One counter ion was modelled in two parts, one of 0.8 and one of 0.2 occupancies, the latter of which was modelled isotropically. Both solvent molecules were modelled isotropically. Compound **2** possessed one helicate, four BF_4^- counter ions and two acetonitrile solvent molecules. Disorder in two BF_4^- counter ions were modelled in two parts.

Compound **3** possessed an asymmetric unit with one full helicate, four BF_4^- counter ions and four acetonitrile molecules with an occupancy of 0.5 each, one of which was modelled in two parts of 0.25 occupancy. These later two were modelled isotropically.

3.4. Magnetic Susceptibility Measurements

Samples of crystalline material were dried for two weeks in a vacuum oven at 340 K prior to magnetic measurement. Data for magnetic susceptibility measurements were collected on a Quantum Design Versalab Measurement System (Quautum Design, San Diego, CA, USA) with a Vibrating Sample Magnetometer (VSM) attachment. Measurements were taken continuously under an applied field of 0.5 T. For each experiment, samples were first heated in situ to 400 K for 30 min to ensure complete solvent loss, and subsequently cycled over the temperature range 50–350 K at a heating rate of $4\ K \cdot min^{-1}$.

4. Conclusions

In conclusion, a series of two novel and one previously studied Fe(II) dinuclear triple helicates was presented, utilising a homoleptic tris-2,4-diimine coordination environment to induce a low temperature spin transition. The identity of the central atom of the ligand was changed to deduce the effects of the different steric nature of these ligands on the structure and spin-transition profile of these compounds (**1**, **2** and **3**). All three compounds displayed a transition of approximately half of the Fe(II) centres at 50 K. Compound **1** (–CH$_2$–), **2** (–S–) and **3** (–O–) completed a transition of 40%, 60% and 50% of Fe(II) ions to the S = 0 LS state respectively at 50 K, with a $T_{1/2}\downarrow$ value of 150, 115 and 155 K in the cooling mode, and thermal hysteresis of 15 K for **1**, **2** and **3**. Compound **2** showed the most complete and abrupt spin transition. The three compounds crystallised in the same space group, although displayed different packing arrangements, intermolecular interactions and arrangements of HS and LS centres throughout the lattice. The ligand (–X–) angle was found to be 113.6, 104.9 and 115.8 degrees for compounds **1**, **2** and **3** respectively, corresponding to intra-helical Fe···Fe separations of 11.72 11.79 and 11.62 Å. Changing the steric nature of the central ligand atom (C, S or O) produced helicate complexes with slight differences in their magnetic behaviours, which allows the magnetic properties of these materials to be altered in this way. Although the direct mechanism of this change cannot be confirmed and is likely to be the result of changes in crystal packing, the degree of intermolecular interactions, intra-helical distances and the electronic effects of the substituted ligand central atoms could cause differing degrees of cooperativity between Fe(II) metal centres.

Supplementary Materials: The following are available online at www.mdpi.com/2304-6740/5/4/62/s1. Experimental data for HRESI-MS, FT-IR, NMR and TGA-DSC, cif and cif-checked files.

Acknowledgments: The authors would like to thank the Western Sydney University (WSU) for research funding, the Advanced Characterization Facility at WSU and Sydney University at which the magnetic measurements were performed. Alexander R. Craze would like to thank WSU and The Australian nuclear Science and Technology Organization (ANSTO) for Masters Scholarships. The crystallographic data were undertaken on the MX1 beamline of the Australian Synchrotron, Clayton, Victoria, Australia. We also thank Australian Synchrotron for travel support and their staff for beamline assistance.

Author Contributions: Alexander R. Craze performed the synthetic, experimental, single crystal X-ray and magnetic studies. Natasha F. Sciortino and Cameron J. Kepert assisted in the magnetic studies. Mohan M. Badbhade and Christopher E. Marjo assisted with crystallographic studies. Alexander R. Craze and Feng Li prepared the manuscript, and Feng Li directed the work.

Conflicts of Interest: The authors declare no conflict of interest.

References

1. Gütlich, P.; Gaspar, A.B.; Garcia, Y. Spin state switching in iron coordination compounds. *Beilstein J. Org. Chem.* **2013**, *9*, 342–391. [CrossRef] [PubMed]
2. Brooker, S. Spin crossover with thermal hysteresis: Practicalities and lessons learnt. *Chem. Soc. Rev.* **2015**, *44*, 2880–2892. [CrossRef] [PubMed]

3.   Halcrow, M.A. Structure: Function relationships in molecular spin-crossover complexes. *Chem. Soc. Rev.* **2011**, *40*, 4119–4142. [CrossRef] [PubMed]

4.   Bousseksou, A.; Molnár, G.; Salmon, L.; Nicolazzi, W. Molecular spin crossover phenomenon: Recent achievements and prospects. *Chem. Soc. Rev.* **2011**, *40*, 3313–3335. [CrossRef] [PubMed]

5.   Leita, B.A.; Moubaraki, B.; Murray, K.S.; Smith, J.P.; Cashion, J.D. Structure and magnetism of a new pyrazolate bridged iron(II) spin crossover complex displaying a single HS–HS to LS–LS transition. *Chem. Commun.* **2004**, 156–157. [CrossRef] [PubMed]

6.   Toftlund, H. Spin equilibrium in solutions. *Monatshefte Chem.* **2001**, *132*, 1269–1277. [CrossRef]

7.   Kahn, O.; Martinez, C.J. Spin-Transition Polymers: From Molecular Materials toward Memory Devices. *Science* **1998**, *279*, 44–48. [CrossRef]

8.   Halcrow, M.A. *Spin-Crossover Materials: Properties and Applications*; John Wiley & Sons Ltd.: Oxford, UK, 2013. [CrossRef]

9.   Hayami, S.; Shigeyoshi, Y.; Akita, M.; Inoue, K.; Kato, K.; Osaka, K.; Takata, M.; Kawajiri, R.; Mitani, T.; Maeda, Y. Reverse Spin Transition Triggered by a Structural Phase Transition. *Angew. Chem. Int. Ed. Engl.* **2005**, *44*, 4899–4903. [CrossRef] [PubMed]

10. Hayami, S.; Moriyama, R.; Shigeyoshi, Y.; Kawajiri, R.; Mitani, T.; Akita, M.; Inoue, K.; Maeda, Y. Spin-Crossover Cobalt(II) Compound with Banana-Shaped Structure. *Inorg. Chem.* **2005**, *44*, 7295–7297. [CrossRef] [PubMed]

11. Zhao, X.-H.; Zhang, S.-L.; Shao, D.; Wang, X.-Y. Spin Crossover in $[Fe(2\text{-Picolylamine})_3]^{(2+)}$ Adjusted by Organosulfonate Anions. *Inorg. Chem.* **2015**, *54*, 7857–7867. [CrossRef] [PubMed]

12. Dupouy, G.; Marchivie, M.; Triki, S.; Sala-Pala, J.; Gómez-García, C.J.; Pillet, S.; Lecomte, C.; Letard, J.-F. Photoinduced HS state in the first spin-crossover chain containing a cyanocarbanion as bridging ligand. *Chem. Commun.* **2009**, 3404–3406. [CrossRef] [PubMed]

13. Garcia, Y.; Niel, V.; Muñoz, M.C.; Real, J.A. Spin Crossover in 1D, 2D and 3D Polymeric Fe(II) Networks. In *Spin Crossover in Transition Metal Compounds I*; Gütlich, P., Goodwin, H.A., Eds.; Springer: Heidelberg/Berlin, Germany, 2004; pp. 229–257.

14. Gütlich, P.; Garcia, Y.; Goodwin, H.A. Spin crossover phenomena in Fe(II) complexes. *Chem. Soc. Rev.* **2000**, *29*, 419–427. [CrossRef]

15. Li, F.; Clegg, J.K.; Lindoy, L.F.; Macquart, R.B.; Meehan, G.V. Metallosupramolecular self-assembly of a universal 3-ravel. *Nat. Commun.* **2011**, *2*, 1208. [CrossRef] [PubMed]

16. Baxter, P.N.W.; Lehn, J.-M.; Sleiman, H.; Rissanen, K. Multicomponent Self-Assembly: Generation and Crystal Structure of a Trimetallic[4]Pseudorotaxane. *Angew. Chem. Int. Ed.* **1997**, *36*, 1294–1296. [CrossRef]

17. Sleiman, H.; Baxter, P.N.W.; Lehn, J.-M.; Airola, K.; Rissanen, K. Multicomponent Self-Assembly: Generation of Rigid-Rack Multimetallic Pseudorotaxanes. *Inorg. Chem.* **1997**, *36*, 4734–4742. [CrossRef] [PubMed]

18. Ward, M.D.; Raithby, P.R. Functional behaviour from controlled self-assembly: Challenges and prospects. *Chem. Soc. Rev.* **2013**, *42*, 1619–1636. [CrossRef] [PubMed]

19. Hannon, M.J.; Childs, L.J. Helices and Helicates: Beautiful Supramolecular Motifs with Emerging Applications. *Supramol. Chem.* **2004**, *16*, 7–22. [CrossRef]

20. Albrecht, M. "Let's Twist Again" Double-Stranded, Triple-Stranded, and Circular Helicates. *Chem. Rev.* **2001**, *101*, 3457–3498. [CrossRef] [PubMed]

21. Hora, S.; Hagiwara, H. High-Temperature Wide Thermal Hysteresis of an Iron(II) Dinuclear Double Helicate. *Inorganics* **2017**, *5*, 49. [CrossRef]

22. Breuning, E.; Ruben, M.; Lehn, J.-M.; Renz, F.; Garcia, Y.; Ksenofontov, V.; Gutlich, P.; Wegelius, E.; Rissanen, K. Spin Crossover in a Supramolecular $Fe_4{}^{II}$ [2×2] Grid Triggered by Temperature, Pressure, and Light. *Angew. Chem. Int. Ed.* **2000**, *39*, 2504–2507. [CrossRef]

23. Struch, N.; Bannwarth, C.; Ronson, T.K.; Lorenz, Y.; Mienert, B.; Wagner, N.; Engeser, M.; Bill, E.; Puttready, R.; Rissanen, K.; et al. An Octanuclear Metallosupramolecular Cage Designed To Exhibit Spin-Crossover Behavior. *Angew. Chem. Int. Ed.* **2017**, *56*, 4930–4935. [CrossRef] [PubMed]

24. Li, L.; Saigo, N.; Zhang, Y.; Fanna, D.J.; Shepherd, N.D.; Clegg, J.K.; Zheng, R.; Hayami, S.; Lindoy, L.F.; Aldrich-Wright, J.R.; et al. A large spin-crossover $[Fe_4L_4]^{8+}$ tetrahedral cage. *J. Mater. Chem. C* **2015**, *3*, 7878–7882. [CrossRef]

25. Tuna, F.; Lees, M.R.; Clarkson, G.J.; Hannon, M.J. Readily prepared metallo-supramolecular triple helicates designed to exhibit spin-crossover behaviour. *Chem. Eur. J.* **2004**, *10*, 5737–5750. [CrossRef] [PubMed]

26. Phan, H.; Hrudka, J.J.; Igimbayeva, D.; Lawson Daku, L.M.; Shatruk, M. A Simple Approach for Predicting the Spin State of Homoleptic Fe(II) Tris-diimine Complexes. *J. Am. Chem. Soc.* **2017**, *139*, 6437–6447. [CrossRef] [PubMed]

27. Fujinami, T.; Nishi, K.; Matsumoto, N.; Iijima, S.; Halcrow, M.A.; Sunatsuki, Y.; Kojima, M. 1D and 2D assembly structures by imidazole···chloride hydrogen bonds of iron(II) complexes [Fe^{II}(HL$^{n\text{-Pr}}$)$_3$]Cl·Y (HL$^{n\text{-Pr}}$ = 2-methylimidazol-4-yl-methylideneamino-n-propyl; Y = AsF$_6$, BF$_4$) and their spin states. *Dalton Trans.* **2011**, *40*, 12301–12309. [CrossRef] [PubMed]

28. Garcia, Y.; Grunert, C.M.; Reiman, S.; van Campenhoudt, O.; Gütlich, P. The Two-Step Spin Conversion in a Supramolecular Triple Helicate Dinuclear Iron(II) Complex Studied by Mössbauer Spectroscopy. *Eur. J. Inorg. Chem.* **2006**, *2006*, 3333–3339. [CrossRef]

29. Pelleteret, D.; Clérac, R.; Mathonière, C.; Harté, E.; Schmitt, W.; Kruger, P.E. Asymmetric spin crossover behaviour and evidence of light-induced excited spin state trapping in a dinuclear iron(II) helicate. *Chem. Commun.* **2009**, *2*, 221–223. [CrossRef] [PubMed]

30. Archer, R.J.; Hawes, C.S.; Jameson, G.N.L.; McKee, V.; Moubaraki, B.; Chilton, N.F.; Murray, K.S.; Kruger, P.E. Partial spin crossover behaviour in a dinuclear iron(II) triple helicate. *Dalton Trans.* **2011**, *40*, 12368–12373. [CrossRef] [PubMed]

31. Telfer, S.G.; Bocquet, B.; Williams, A.F. Thermal Spin Crossover in Binuclear Iron(II) Helicates: Negative Cooperativity and a Mixed Spin State in Solution. *Inorg. Chem.* **2001**, *40*, 4818–4820. [CrossRef] [PubMed]

32. Fujita, K.; Kawamoto, R.; Tsubouchi, R.; Sunatsuki, Y.; Kojima, M.; Iijima, S.; Matsumoto, N. Spin States of Mono- and Dinuclear Iron(II) Complexes with Bis(imidazolylimine) Ligands. *Chem. Lett.* **2007**, *36*, 1284–1285. [CrossRef]

33. Sun, Q.-F.; Iwasa, J.; Ogawa, D.; Ishido, Y.; Sato, S.; Ozeki, T.; Sei, T.; Yamaguchi, K.; Fujita, M. Self-Assembled M$_{24}$L$_{48}$ Polyhedra and Their Sharp Structural Switch upon Subtle Ligand Variation. *Science* **2010**, *328*, 1144–1147. [CrossRef] [PubMed]

34. Marchivie, M.; Guionneau, P.; Létard, J.-F.; Chasseau, D. Photo-induced spin-transition: The role of the iron(II) environment distortion. *Acta Crystallogr. Sect. B* **2005**, *61*, 25–28. [CrossRef] [PubMed]

35. Dupouy, G.; Marchivie, M.; Triki, S.; Sala-Pala, J.; Salaün, J.-Y.; Gómez-García, C.J.; Guionneau, P. The Key Role of the Intermolecular π-π Interactions in the Presence of Spin Crossover in Neutral [Fe(abpt)$_2$A$_2$] Complexes (A = Terminal Monoanion N Ligand). *Inorg. Chem.* **2008**, *47*, 8921–8931. [CrossRef] [PubMed]

36. Milin, E.; Benaicha, B.; El Hajj, F.; Patinec, V.; Triki, S.; Marchivie, M.; Gomez-Garcia, C.J.; Pillet, S. Magnetic Bistability in Macrocycle-Based FeII Spin-Crossover Complexes: Counter Ion and Solvent Effects. *Eur. J. Inorg. Chem.* **2016**, *34*, 5305–5314. [CrossRef]

37. Guionneau, P.; Gac, F.L.; Lakhoufi, S.; Kaiba, A.; Chasseau, D.; Létard, J.-F.; Negrier, P.; Mondieig, D.; Howard, J.A.K.; Leger, J.-M. X-ray diffraction investigation of a spin crossover hysteresis loop. *J. Phys.* **2007**, *19*, 326211. [CrossRef]

38. Harding, D.J.; Phonsri, W.; Harding, P.; Gass, I.A.; Murray, K.S.; Moubaraki, B.; Cashion, J.D.; Liu, L.; Telfer, S.G. Abrupt spin crossover in an iron(III) quinolylsalicylaldimine complex: Structural insights and solvent effects. *Chem. Commun.* **2013**, *49*, 6340–6342. [CrossRef] [PubMed]

39. Dorbes, S.; Valade, L.; Real, J.A.; Faulmann, C. [Fe(sal$_2$-trien)][Ni(dmit)$_2$]: Towards switchable spin crossover molecular conductors. *Chem. Commun.* **2005**, 69–71. [CrossRef] [PubMed]

40. McPhillips, T.M.; McPhillips, S.E.; Chiu, H.J.; Cohen, A.E.; Deacon, A.M.; Ellis, P.J.; Garman, E.; Gonzalex, A.; Sauter, N.K.; Phizackerly, R.P.; et al. Blu-Ice and the distributed control system: Software for data acquisition and instrument control at macromolecular crystallography beamlines. *J. Synchrotron Radiat.* **2002**, *9*, 401–406. [CrossRef] [PubMed]

41. Cowieson, N.P.; Aragao, D.; Clift, M.; Ericsson, D.J.; Gee, C.; Harrop, S.J.; Mudie, N.; Panjikar, S.; Price, J.R.; Riboldi-Tunnicliffe, A.; et al. MX1: A bending-magnet crystallography beamline serving both chemical and macromolecular crystallography communities at the Australian Synchrotron. *J. Synchrotron Radiat.* **2015**, *22*, 187–190. [CrossRef] [PubMed]

42. Kabsch, W. XDS. *J. Appl. Crystallogr.* **1993**, *26*, 795–800. [CrossRef]

43. Bruker. *SADABS*, Version 2.11; Bruker AXS Inc.: Fitchburg, WI, USA, 2001.

44. Sheldrick, G.M. *SHELX-2014: Programs for Crystal Structure Analysis*; University of Göttingen: Göttingen, Germany, 2014.

45. Sheldrick, G.M. Crystal structure refinement with *SHELXL*. *Acta Crystallogr. Sect. C* **2015**, *71*, 3–8. [CrossRef] [PubMed]

46. Dolomanov, O.V.; Bourhis, L.J.; Gildea, R.J.; Howard, J.A.K.; Puschmann, H. *OLEX2*: A complete structure solution, refinement and analysis program. *J. Appl. Crystallogr.* **2009**, *42*, 339–341. [CrossRef]

inorganics

MDPI

Article

Metal Substitution Effect on a Three-Dimensional Cyanido-Bridged Fe Spin-Crossover Network

Kenta Imoto [1], Shinjiro Takano [1] and Shin-ichi Ohkoshi [1,2,*]

[1]   Department of Chemistry, School of Science, The University of Tokyo, 7-3-1 Hongo, Bunkyo-ku,
       Tokyo 113-0033, Japan; imoto@chem.s.u-tokyo.ac.jp (K.I.); stakano@chem.s.u-tokyo.ac.jp (S.T.)
[2]   Cryogenic Research Center, The University of Tokyo, 2-11-16 Yayoi, Bunkyo-ku, Tokyo 113-0032, Japan
*   Correspondence: ohkoshi@chem.s.u-tokyo.ac.jp; Tel.: +81-3-5841-4331

Received: 4 September 2017; Accepted: 21 September 2017; Published: 24 September 2017

Abstract: We report the Co^{II}-substitution effect on a cyanido-bridged three-dimensional Fe^{II} spin-crossover network, $Fe_2[Nb(CN)_8](4\text{-pyridinealdoxime})_8 \cdot 2H_2O$. A series of iron–cobalt octacyanidoniobate, $(Fe_xCo_{1-x})_2[Nb(CN)_8](4\text{-pyridinealdoxime})_8 \cdot zH_2O$, was prepared. In this series, the behavior of Fe^{II} spin-crossover changes with the Co^{II} concentration. As the Co^{II} concentration increases, the transition of the spin-crossover becomes gradual and the transition temperature of the spin-crossover shifts towards a lower temperature. Additionally, this series shows magnetic phase transition at a low temperature. In particular, $(Fe_{0.21}Co_{0.79})_2[Nb(CN)_8](4\text{-pyridinealdoxime})_8 \cdot zH_2O$ exhibits a Curie temperature of 12 K and a large coercive field of 3100 Oe.

Keywords: spin-crossover; magnetic phase transition; cyanido-bridged metal assembly; metal substitution

1. Introduction

The spin-crossover phenomenon between low-spin (LS) and high-spin (HS) states has been extensively studied in many fields [1–13]. This phenomenon can be modulated by various physical and chemical stimulations (e.g., light, pressure, temperature, vapor molecule, and metal substitution), and it has potential for sensor and memory applications [14,15]. To control the spin-crossover behavior, the metal substitution effect on the spin-crossover behavior for some Fe^{II} spin-crossover materials has been investigated [16–25].

In the field of molecule-based magnets [26–30], cyanido-bridged metal assemblies have drawn attention because they exhibit various magnetic functionalities such as a high Curie temperature (T_c) [31–34], a charge transfer transition [35–42], and an externally stimulated phase transition phenomena [43–47]. In the recent years, we have synthesized several kinds of magnetic cyanido-bridged bimetal assemblies possessing Fe spin-crossover sites. For example, $CsFe[Cr(CN)_6] \cdot 1.3H_2O$ exhibits a spin-crossover phenomenon at 211 K in the cooling process ($T_{1/2}\downarrow$) and 238 K in the heating process ($T_{1/2}\uparrow$), and a ferromagnetic phase transition at 9 K [48]. $Fe_2[Nb(CN)_8](3\text{-pyridylmethanol})_8 \cdot 4.6H_2O$ shows a gradual spin-crossover phenomenon at 250 K and a ferrimagnetic phase transition at 12 K [49]. However, the photoresponsivities were not reported for these compounds.

In 2011, we synthesized a cyanido-bridged metal assembly, $Fe_2[Nb(CN)_8](4\text{-pyridinealdoxime})_8 \cdot 2H_2O$, which shows a spin-crossover phenomenon at 130 K [50]. When this material is irradiated with 473-nm light, a spontaneous magnetization is observed. This photoinduced ferrimagnetic phase exhibits a T_C value of 20 K and a coercive field (H_c) value of 240 Oe. This is the first demonstration of light-induced spin crossover ferrimagnetism. In 2014, we prepared $Fe_2[Nb(CN)_8](4\text{-bromopyridine})_8 \cdot 2H_2O$, which is the first chiral photomagnet, and observed 90° optical switching of the polarization plane of second harmonic light [51].

From the viewpoint of controlling the magnetic performance of a photomagnetic material, metal replacement is effective. In particular, $Co_2[Nb(CN)_8](4\text{-pyridinealdoxime})_8 \cdot 2H_2O$, which is a metal-substituted compound of $Fe_2[Nb(CN)_8](4\text{-pyridinealdoxime})_8 \cdot 2H_2O$ described above as the first photoinduced spin-crossover magnet, shows a large coercive field of 15,000 Oe [52]. In this work, we synthesize cyanido-bridged metal assemblies containing both Fe and Co ions, $(Fe_{1-x}Co_x)_2[Nb(CN)_8](4\text{-pyridinealdoxime})_8 \cdot zH_2O$, and discuss the crystal structures, spectroscopic properties, and magnetic properties.

2. Results and Discussions

2.1. Syntheses

The preparation of $(Fe_{1-x}Co_x)_2[Nb(CN)_8](4\text{-pyridinealdoxime})_8 \cdot zH_2O$ was performed by reacting a mixed aqueous solution of $FeCl_2 \cdot 4H_2O$, $CoCl_2 \cdot 6H_2O$, L-(+)-ascorbic acid, and 4-pyridinealdoxime, with an aqueous solution of $K_4[Nb(CN)_8] \cdot 2H_2O$ with Fe/Co ratios [Fe]/([Fe] + [Co]) of 0, 0.1, 0.25, 0.5, 0.75, and 1, yielding a microcrystalline powder. Stirring was continued for 1 h. Then the solution was filtered and washed twice by water. Elemental analyses indicate that the chemical formulae of the obtained compounds are $Fe_2[Nb(CN)_8](4\text{-pyridinealdoxime})_8 \cdot 3H_2O$ ($x = 0$, compound **1**), $(Fe_{0.92}Co_{0.08})_2[Nb(CN)_8](4\text{-pyridinealdoxime})_8 \cdot 3H_2O$ ($x = 0.08$, compound **2**), $(Fe_{0.71}Co_{0.29})_2[Nb(CN)_8](4\text{-pyridinealdoxime})_8 \cdot 3H_2O$ ($x = 0.29$, compound **3**), $(Fe_{0.50}Co_{0.50})_2[Nb(CN)_8]$ $(4\text{-pyridinealdoxime})_8 \cdot 3H_2O$ ($x = 0.50$, compound **4**), $(Fe_{0.21}Co_{0.79})_2[Nb(CN)_8](4\text{-pyridinealdoxime})_8 \cdot 3H_2O$ ($x = 0.79$, compound **5**), and $Co_2[Nb(CN)_8](4\text{-pyridinealdoxime})_8 \cdot 3H_2O$ ($x = 1$, compound **6**). (See Section 3) The compounds of **1** and **6** in this work correspond well to the formulae in our previous reports [50,52].

2.2. Crystal Structures and Spectroscopic Properties

Table 1 and Figure 1 show the results of the Rietveld analyses of the powder X-ray diffraction (XRD) patterns for **1–6**. Structural analyses show that the crystal structures of **1** and **6** are isostructural to those reported in our previous papers [50,52]. Rietveld analyses of the XRD patterns of **2–5** were performed using the crystal structure of **1** as the initial structure with the occupancies of Fe and Co based on the chemical formula. The lattice constant versus *x* value (Co content) plot shows that the lattice constant of the *a*-axis decreases from 20.2893 Å ($x = 0$) to 20.2105 Å ($x = 1$) (0.4% decrease), while that of the *c*-axis slightly decreases from 15.0224 Å ($x = 0$) to 15.0066 Å ($x = 1$) (0.1% decrease), and the unit cell volume decreases from 6184.1 $Å^3$ ($x = 0$) to 6129.7 $Å^3$ ($x = 1$) (0.9% decrease) with increasing *x* value (Table 1, Figure 2). It is noteworthy that the XRD peaks become broader with an increasing *x* value. The SEM images indicate that this broadening is caused by the reduction of the crystallite size (Figure S1).

Table 1. Crystal system, space group, and lattice constants of **1–6**.

	1	2	3	4	5	6
Crystal system	Tetragonal	Tetragonal	Tetragonal	Tetragonal	Tetragonal	Tetragonal
Space group	$I4_1/a$	$I4_1/a$	$I4_1/a$	$I4_1/a$	$I4_1/a$	$I4_1/a$
$a(b)/Å$	20.2893(5)	20.2683(5)	20.2572(6)	20.2453(8)	20.2203(12)	20.2105(1)
$c/Å$	15.0224(5)	15.0156(5)	15.0154(6)	15.0151(8)	15.0047(13)	15.0066(13)
$V/Å^3$	6184.1(3)	6168.5(3)	6161.6(3)	6154.2(5)	6134.8(8)	6129.7(7)

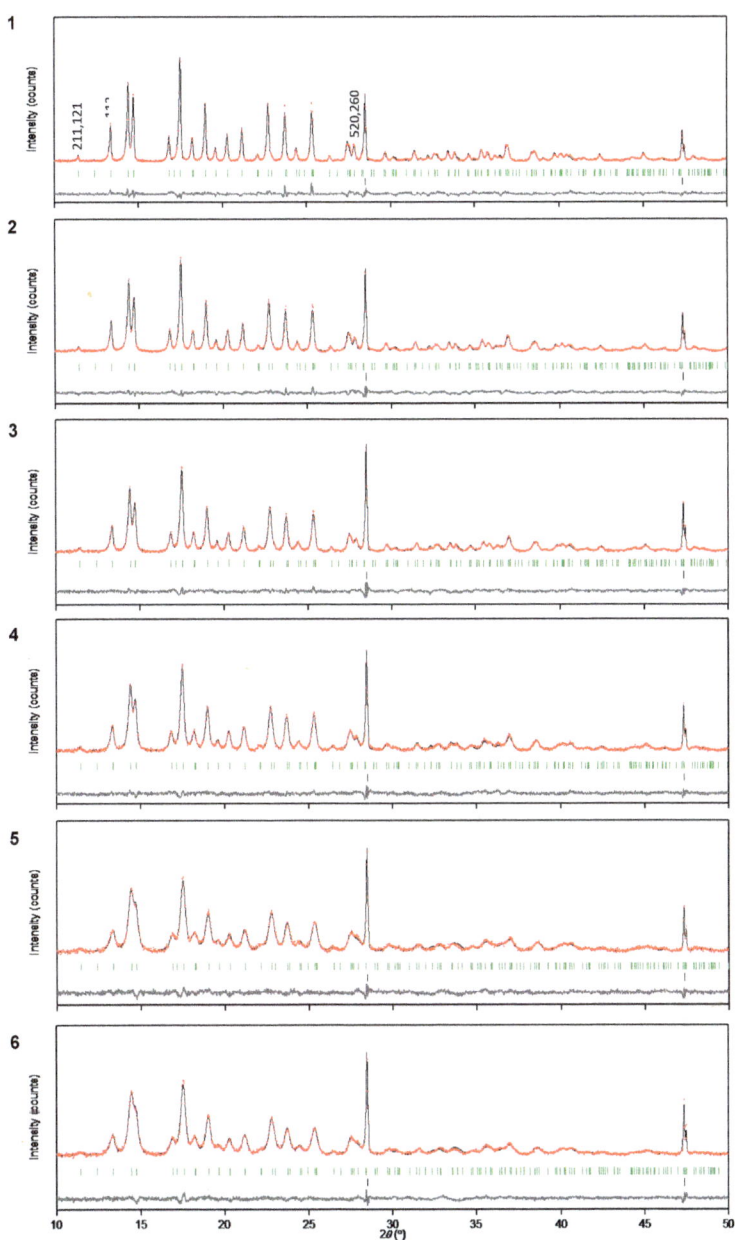

Figure 1. XRD patterns with Rietveld analyses of **1–6**. Red plots, black lines, blue lines, green bars, and black bars are the observed patterns, calculated patterns, residue between the calculated and observed patterns, calculated positions of the Bragg reflections in the sample, and those of the silicon (Si) standard, respectively. The XRD peaks due to Si are shown as black sticks. Representative reflection indexes are shown in the XRD pattern of **1**.

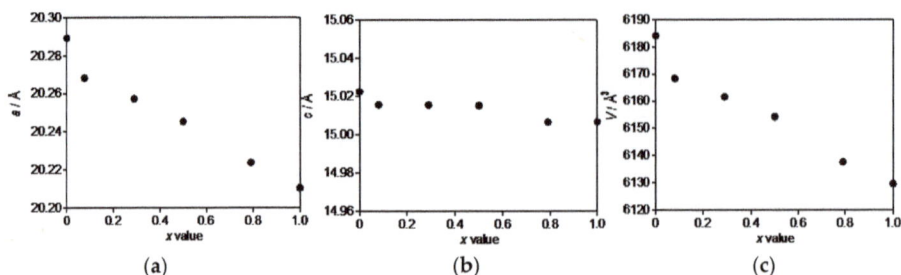

Figure 2. x dependence of (a) the a-axis; (b) the c-axis; and, (c) the unit cell volume (V).

The crystal structure and coordination geometries of this series are explained using **3** as an example. **3** has a tetragonal crystal structure in the $I4_1/a$ space group with $a = 20.2572(6)$ Å and $c = 15.0154(6)$ Å. The asymmetric unit is composed of a quarter of the $[Nb(CN)_8]$ anion, half of the $[M(\text{4-pyridinealdoxime})_4]$ (M = Fe or Co) cation, and a water molecule. Here, we assume that Fe and Co are randomly incorporated. The coordination geometries of the Nb and M sites are dodecahedron (D_{2d}) and pseudo-octahedron (D_{4h}), respectively. For the eight CN groups of $Nb(CN)_8$, four CN groups are bridged to the M ions, and the other four CN groups are not bridged. Two cyanide nitrogen atoms coordinate to the two axial positions of the M site and four pyridyl nitrogen atoms of 4-pyridinealdoxime are located at the other four equatorial positions. A cyanido-bridged three-dimensional (3D) network structure is formed by the M–NC–Nb component (Figure 3).

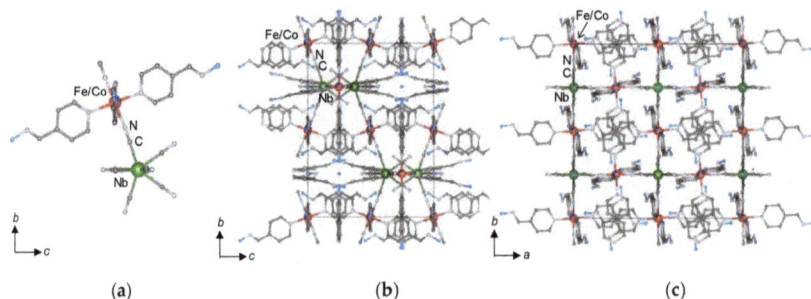

Figure 3. Crystal structure and coordination geometries around the metal centers for **3**. (a) Coordination geometry around the metal centers; (b) crystal structure viewed from the a-axis; and, (c) from the c-axis. Water molecules are omitted for clarity.

The infrared spectrum of **1** shows two CN stretching peaks at 2130 cm^{-1} and 2151 cm^{-1}, which are ascribed to the CN stretching peak due to non-bridged CN (Nb–CN) and bridged CN between Nb and Fe (Nb–CN–Fe), respectively. In **2–6**, a different peak is observed around 2160 cm^{-1}, and its intensity increases with an increasing Co^{II} concentration, while the peak around 2151 cm^{-1} decreases. This indicates that the peak around 2160 cm^{-1} is due to the bridged CN between Nb and Co (Nb–CN–Co) (Figure 4).

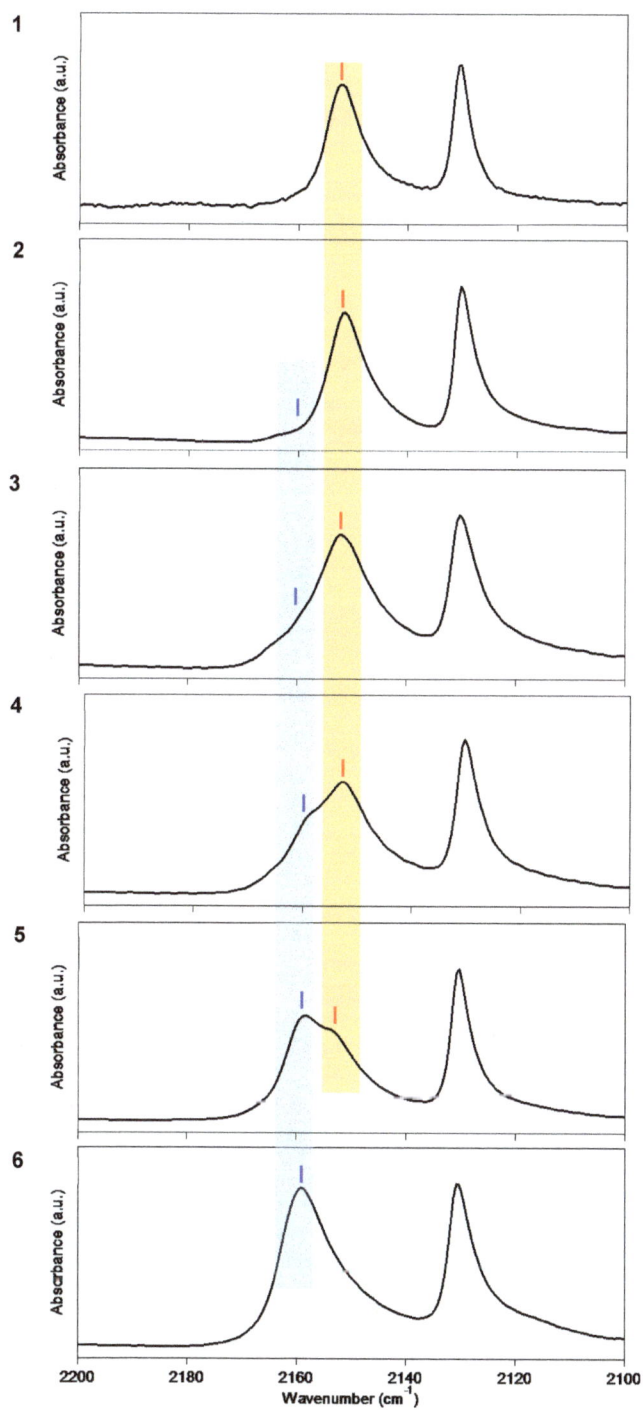

Figure 4. Infrared spectra of **1–6** at room temperature for the CN stretching region.

2.3. Magnetic Properties

Figure 5a shows the temperature dependence of the product of the magnetic susceptibility and the temperature ($\chi_M T$) of **1–6** under an external magnetic field of 5000 Oe. The $\chi_M T$ values of **1–6** at 300 K are 7.04, 7.08, 6.77, 6.11, 5.79, and 5.16 K·cm³·mol^{-1}, respectively. These values agree with the estimated values of 6.96, 6.83, 6.48, 6.14, 5.79, and 5.16 K·cm³·mol^{-1}, which are obtained by Equation (1)

$$\chi_M T = \frac{N_A \mu_B^2}{3 k_B} \{2 x g_{Co}^2 S_{Co}(S_{Co}+1) + 2(1-x) g_{Fe}^2 S_{Fe}(S_{Fe}+1) + g_{Nb}^2 S_{Nb}(S_{Nb}+1)\} \quad (1)$$

where N_A is Avogadro's constant, μ_B is the Bohr magneton, k_B is the Boltzmann constant, g_i is the g value of atom i, S_i is the spin quantum number of atom i, and x is the Co content [53,54], assuming $g_{Fe} = 2.1$, $g_{Co} = 2.4$, $g_{Nb} = 2.0$, $S_{Fe} = 2$, $S_{Co} = 3/2$, and $S_{Nb} = 1/2$.

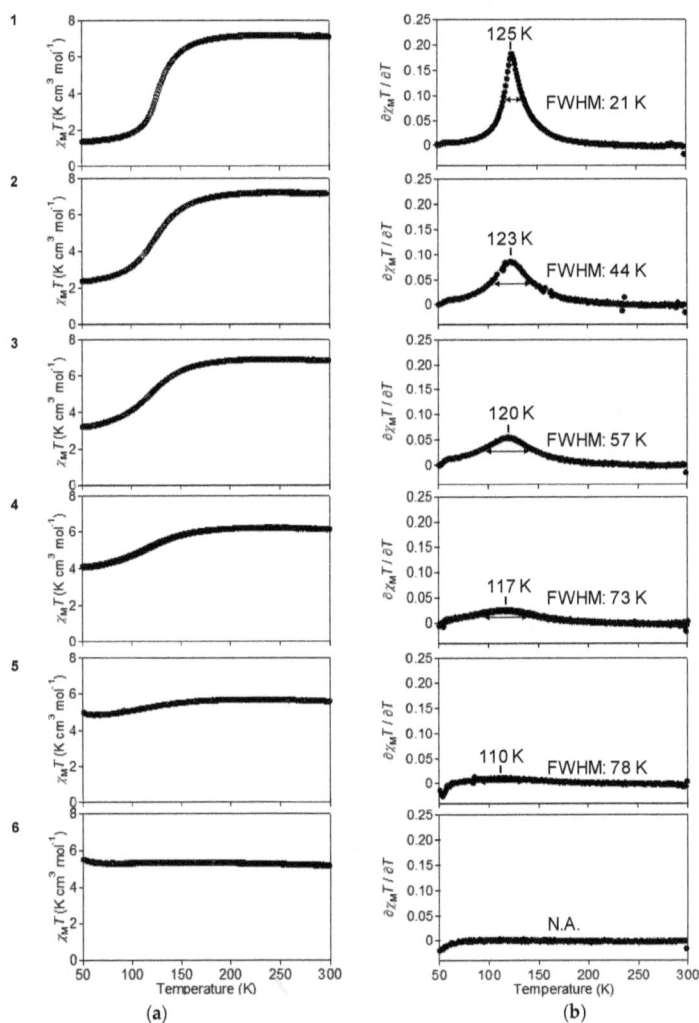

Figure 5. $\chi_M T$ vs. T plots (a) and $\partial(\chi_M T)/\partial T$ vs. T plots (b) of **1–6**.

As the temperature decreases, the $\chi_M T$ values decreases at intermediate temperatures in **1–5**, while the $\chi_M T$ value of **6** is almost constant between 50 K and 300 K. Thus, the occurrence of the Fe^{II} spin-crossover phenomenon for all of the Fe^{II} containing compounds is confirmed by the magnetic susceptibility measurements. The thermal spin-crossover temperature ($T_{1/2}$), which is estimated as the temperature where the temperature differential of $\chi_M T$ is maximized, shows that with an increasing x value, $T_{1/2}$ shifts to a lower temperature (Figure 5b and Figure S2). In addition, with an increasing x value, spin crossovers become more gradual. According to the reported observations [16–25], these results are explained as follows. Since the ionic radii of Co(II)$_{HS}$ (0.75 Å) is closer to Fe(II)$_{HS}$ (0.78 Å) than Fe(II)$_{LS}$ (0.61 Å) [55], the spin-crossover from Fe(II)$_{HS}$ to Fe(II)$_{LS}$ becomes unfavorable, leading to a decrease of the spin transition temperature. Additionally, because the distance between spin-crossover sites becomes longer by metal substitution, the cooperativity between spin-crossover sites decreases, resulting in a gradual spin-crossover behavior.

Next, we measured the magnetic properties in the low temperature region. Figure 6a shows the magnetization vs. temperature plots of **1–6** with cooling temperature at an external magnetic field of 10 Oe. The magnetization vs. temperature curves of **2** and **3** show a small shoulder below 15 K. **4**, **5**, and **6** clearly show spontaneous magnetization with critical temperatures (T_c) of 8 K, 12 K, and 18 K, respectively. The magnetization vs. external magnetic field plots at 2 K show that the magnetic coercive fields of **4**, **5**, and **6** are 1400 Oe, 3100 Oe, and 13,000 Oe, respectively (Figure 6b). The singleness of the T_c values and the shape of magnetic hysteresis loop indicate that Fe and Co are mixed with each other on the atomic level. The observation of such a large coercive field is attributed to the single-ion anisotropy of Co ion possessing an unquenched orbital angular momentum.

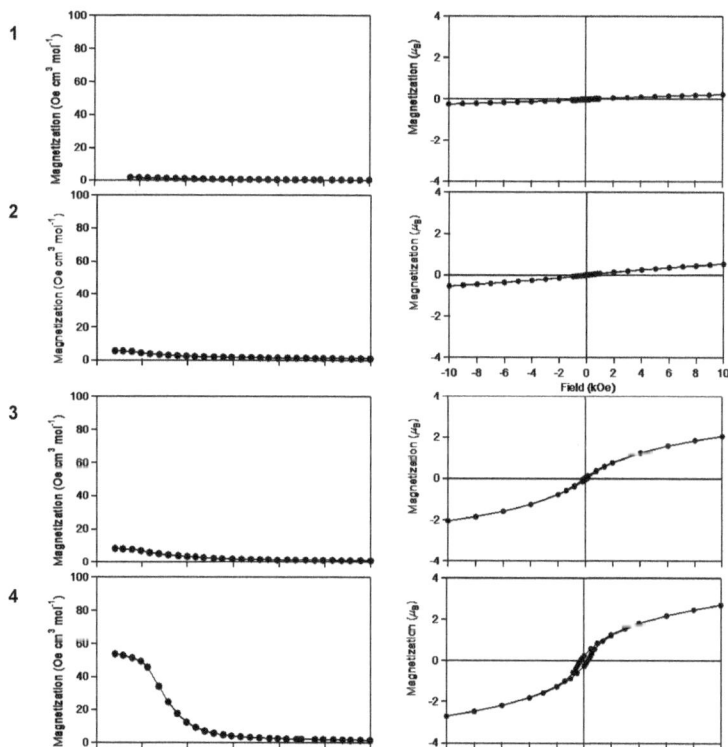

Figure 6. *Cont.*

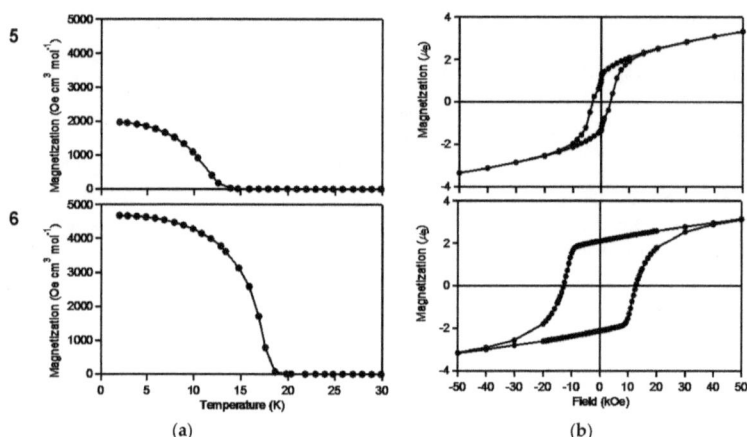

Figure 6. (a) Magnetization vs. temperature plots and (b) magnetization vs. external magnetic field plots of **1–6**.

3. Materials and Methods

3.1. Syntheses

$K_4[Nb(CN)_8]\cdot 2H_2O$ was synthesized according to the reported procedure [56]. Other reagents were purchased from commercial sources (Wako Pure Chemical Industries and Tokyo Chemical Industries) and were used without further purification. For the preparation of $(Fe_xCo_{1-x})_2[Nb(CN)_8](4\text{-pyridinealdoxime})_8\cdot zH_2O$, a 98-cm^3 aqueous solution containing $FeCl_2\cdot 4H_2O$ and $CoCl_2\cdot 6H_2O$ (0.2 mmol in total), L-(+)-ascorbic acid (0.4 mmol), and 4-pyridinealdoxime (4 mmol), was added to an aqueous solution (2 cm^3) of $K_4[Nb(CN)_8]\cdot 2H_2O$ (0.1 mmol) with Fe/Co ratios [Fe]/([Fe] + [Co]) of 0, 0.1, 0.25, 0.5, 0.75, and 1, yielding a microcrystalline powder. Stirring was continued for 1 h. Then the solution was filtered and washed twice by water. Elemental analyses indicate that the chemical formulae of the obtained compounds were $Fe_2[Nb(CN)_8](4\text{-pyridinealdoxime})_8\cdot 3H_2O$ (**1**), $(Fe_{0.92}Co_{0.08})_2[Nb(CN)_8](4\text{-pyridinealdoxime})_8\cdot 3H_2O$ (**2**), $(Fe_{0.71}Co_{0.29})_2[Nb(CN)_8](4\text{-pyridinealdoxime})_8\cdot 3H_2O$ (**3**), $(Fe_{0.50}Co_{0.50})_2[Nb(CN)_8](4\text{-pyridinealdoxime})_8$ $\cdot 3H_2O$ (**4**), $(Fe_{0.21}Co_{0.79})_2[Nb(CN)_8](4\text{-pyridinealdoxime})_8\cdot 3H_2O$ (**5**), and $Co_2[Nb(CN)_8](4\text{-pyridinealdoxime})_8$ $\cdot 3H_2O$ (**6**). Calcd. for **1**: Fe, 7.74; Nb, 6.43; C, 46.59; H, 3.77; N, 23.28. Found for **1**: Fe, 7.98; Nb, 6.69; C, 46.71; H, 3.72; N, 23.39. Calcd. for **2**: Fe, 7.11; Co, 0.65; Nb, 6.43; C, 46.57; H, 3.77; N, 23.28. Found for **2**: Fe, 7.07; Co, 0.67; Nb, 6.40; C, 46.57; H, 3.66; N, 23.33. Calcd. for **3**: Fe, 5.49; Co, 2.36; Nb, 6.43; C, 46.53; H, 3.77; N, 23.25. Found for **3**: Fe, 5.68; Co, 2.45; Nb, 6.44; C, 46.44; H, 3.69; N, 23.39. Calcd. for **4**: Fe, 3.86; Co, 4.07; Nb, 6.42; C, 46.49; H, 3.76; N, 23.23. Found for **4**: Fe, 3.81; Co, 4.13; Nb, 6.32; C, 46.40; H, 3.69; N, 23.39. Calcd. for **5**: Fe, 1.62; Co, 6.42; Nb, 6.41; C, 46.43; H, 3.76; N, 23.20. Found for **5**: Fe, 1.56; Co, 6.47; Nb, 6.39; C, 46.71; H, 3.71; N, 23.12. Calcd. for **6**: Co, 8.13; Nb, 6.41; C, 46.39; H, 3.75; N, 23.18. Found for **6**: Co, 8.26; Nb, 6.54; C, 46.37; H, 3.71; N, 23.18.

3.2. Measurements

Elemental analyses for C, H, and N were carried out by standard microanalytical methods while those for Fe, Co, and Nb were analyzed by inductive plasma mass spectroscopy. FT-IR spectra were recorded on a FT-IR4100 spectrometer (JASCO, Tokyo, Japan). X-ray powder diffraction was measured on a Ultima-IV powder diffractometer (Rigaku, Tokyo, Japan). Rietveld analyses were performed using PDXL program (Rigaku, Tokyo, Japan). Magnetic susceptibility and magnetization measurements were carried out using a MPMS superconducting quantum interference device (SQUID) magnetometer (Quantum Design, San Diego, CA, USA).

4. Conclusions

In this work, we synthesized and characterized ternary metal cyanido-bridged metal assemblies of $(Fe_xCo_{1-x})_2[Nb(CN)_8](4\text{-pyridinealdoxime})_8 \cdot zH_2O$. The magnetic measurements reveal that all of the Fe-containing systems present a spin-crossover phenomenon. In particular, $(Fe_{0.21}Co_{0.79})_2[Nb(CN)_8](4\text{-pyridinealdoxime})_8 \cdot zH_2O$ exhibits a coexistence of a spin-crossover phenomenon and a magnetic phase transition with T_c of 12 K and a large H_c of 3100 Oe. Additional investigations on the photomagnetic effect are in progress.

Supplementary Materials: The following are available online at www.mdpi.com/2304-6740/5/4/63/s1. Cif and cif-checked files of **1–6**. Figure S1: SEM images and particle size distributions of **1–6**; Figure S2: Co fraction (x) dependence of spin-crossover transition temperature of **1–5**.

Acknowledgments: The present research was supported in part by a JSPS Grant-in-Aid for specially promoted Research (Grant Number 15H05697). We also recognize the Cryogenic Research Center, The University of Tokyo, and Nanotechnology Platform, which are supported by MEXT.

Author Contributions: Shin-ichi Ohkoshi conceived and designed the project. Kenta Imoto and Shinjiro Takano performed the experiments. Kenta Imoto and Shinjiro Takano analyzed the data. Shin-ichi Ohkoshi and Kenta Imoto wrote the paper.

Conflicts of Interest: The authors declare no conflict of interest.

References

1. Gütlich, P.; Goodwin, H.A. *Spin Crossover in Transition Metal Compounds*; Springer: New York, NY, USA, 2004.
2. Halcrow, M.A. *Spin Crossover Materials: Properties and Applications*; John Wiley & Sons: Chichester, UK, 2013.
3. Gütlich, P.; Hauser, A.; Spiering, H. Thermal and Optical Switching of Iron (II) Complexes. *Angew. Chem. Int. Ed. Engl.* **1994**, *33*, 2024–2054. [CrossRef]
4. Decurtins, S.; Gütlich, P.; Köhler, C.P.; Spiering, H.; Hauser, A. Light-induced excited spin state trapping in a transition-metal complex: The hexa-1-propyltetrazole–iron (II) tetrafluoroborate spin-crossover system. *Chem. Phys. Lett.* **1984**, *105*, 1–4. [CrossRef]
5. Hauser, A. Intersystem Crossing in Fe(II) Coordination Compounds. *Coord. Chem. Rev.* **1991**, *111*, 275–290. [CrossRef]
6. Real, J.A.; Andres, E.; Munoz, M.C.; Julve, M.; Granier, T.; Bousseksou, A.; Varret, F. Spin Crossover in a Catenan Supramolecular System. *Science* **1995**, *268*, 265–267. [CrossRef] [PubMed]
7. Halder, G.J.; Kepert, C.J.; Moubaraki, B.; Murray, K.S.; Cashion, J.D. Guest-dependent spin crossover in a nanoporous molecular framework material. *Science* **2002**, *298*, 1762–1765. [CrossRef] [PubMed]
8. Renz, F.; Oshio, H.; Ksenofontov, V.; Waldeck, M.; Spiering, H.; Gütlich, P. Strong field iron (II) complex converted by light into a long-lived high-spin state. *Angew. Chem. Int. Ed.* **2000**, *39*, 3699–3700. [CrossRef]
9. Gaspar, A.B.; Seredyuk, M.; Gütlich, P. Spin crossover in metallomesogens. *Coord. Chem. Rev.* **2009**, *253*, 2399–2413. [CrossRef]
10. Ould Moussa, N.; Molnar, G.; Bonhommeau, S.; Zwick, A.; Mouri, S.; Tanaka, K.; Real, J.A.; Bousseksou, A. Selective Photoswitching of the Binuclear Spin Crossover Compound {[Fe(bt)(NCS)$_2$]$_2$(bpm)} into Two Distinct Macroscopic Phases. *Phys. Rev. Lett.* **2005**, *94*, 107205. [CrossRef] [PubMed]
11. Papanikolaou, D.; Margadonna, S.; Kosaka, W.; Ohkoshi, S.; Brunelli, M.; Prassides, K. X-ray Illumination Induced Fe(II) Spin Crossover in the Prussian Blue Analogue Cesium Iron Hexacyanochromate. *J. Am. Chem. Soc.* **2006**, *128*, 8358–8363. [CrossRef] [PubMed]
12. Bertoni, R.; Cammarata, M.; Lorenc, M.; Matar, S.; Létard, J.-F.; Lemke, H.-T.; Collet, E. Ultrafast Light-Induced Spin-State Trapping Photophysics Investigated in Fe(phen)$_2$(NCS)$_2$ Spin-Crossover Crystal. *Acc. Chem. Res.* **2015**, *48*, 774–781. [CrossRef] [PubMed]
13. Trzop, E.; Zhang, D.; Pineiro-Lopez, L.; Valvarde-Munoz, F.J.; Munoz, M.C.; Palatinus, L.; Guerin, L.; Cailleau, H.; Real, J.A.; Collet, E. First Step towards a Devil's Staircase in Spin-Crossover Materials. *Angew. Chem. Int. Ed.* **2016**, *55*, 8675–8679. [CrossRef] [PubMed]

14. Kahn, O.; Martinez, C.J. Spin-transition polymers: From molecular materials toward memory devices. *Science* **1998**, *279*, 44–48. [CrossRef]
15. Molnar, G.; Salmon, L.; Nicolazzi, W.; Terki, F.; Bousseksou, A. Emerging properties and applications of spin crossover nanomaterials. *J. Mater. Chem. C* **2014**, *2*, 1360–1366. [CrossRef]
16. Gütlich, P.; Link, R.; Steinhäuser, H. Mössbauer-Effect Study of the Thermally Induced Spin Transition in Tris(2-picolylamine)iron(II) Chloride. Dilution Effect in Mixed Crystals of $[Fe_xZn_{1-x}(2\text{-pic})_3]Cl_2 \cdot C_2H_5OH$ (x = 0.15, 0.0029, 0.0009). *Inorg. Chem.* **1978**, *9*, 2509–2514. [CrossRef]
17. Ganguli, P.; Gütlich, P.; Müller, E.W. Effect of metal dilution on the spin-crossover behavior in $[Fe_xM_{1-x}(phen)_2(NCS)_2]$ (M = Mn, Co, Ni, Zn). *Inorg. Chem.* **1982**, *21*, 3429–3433. [CrossRef]
18. Haddad, M.S.; Federer, W.D.; Lynch, M.W.; Hendrickson, D.N. An explanation of unusual properties of spin-crossover ferric complexes. *J. Am. Chem. Soc.* **1980**, *102*, 1468–1470. [CrossRef]
19. Martin, J.-P.; Zarembowitch, J.; Bousseksou, A.; Dworkin, A.; Haasnoot, J.G.; Varret, F. Solid State Effects on Spin Transitions: Magnetic, Calorimetric, and Mössbauer-Effect Properties of $[Fe_xCo_{1-x}(4,4'\text{-bis-}1,2,4\text{-triazole})_2(NCS)_2] \cdot H_2O$ Mixed-Crystal Compounds. *Inorg. Chem.* **1994**, *33*, 6325–6333. [CrossRef]
20. Tayagaki, T.; Galet, A.; Molnar, G.; Munoz, M.C.; Zwick, A.; Tanaka, K.; Real, J.A.; Bousseksou, A. Metal Dilution Effects on the Spin-Crossover Properties of the Three-Dimensional Coordination Polymer Fe(pyrazine)[Pt(CN)$_4$]. *J. Phys. Chem. B* **2005**, *109*, 14859–14867. [CrossRef] [PubMed]
21. Krivokapic, I.; Chakraborty, P.; Enachescu, C.; Bronisz, R.; Hauser, A. Low-Spin→High-Spin Relaxation Dynamics in the Highly Diluted Spin-Crossover System $[Fe_xZn_{1-x}(bbtr)_3](ClO_4)_2$. *Inorg. Chem.* **2011**, *50*, 1856–1861. [CrossRef] [PubMed]
22. Chakraborty, P.; Enachescu, C.; Walder, C.; Bronisz, R.; Hauser, A. Thermal and Light-Induced Spin Switching Dynamics in the 2D Coordination Network of $\{[Zn_{1-x}Fe_x(bbtr)_3](ClO_4)_2\}_\infty$: The Role of Cooperative Effects. *Inorg. Chem.* **2012**, *51*, 9714–9722. [CrossRef] [PubMed]
23. Zheng, S.; Siegler, M.-A.; Costa, J.-S.; Fu, W.-T.; Bonnet, S. Effect of Metal Dilution on the Thermal Spin Transition of $[Fe_xZn_{1-x}(bapbpy)(NCS)_2]$. *Eur. J. Inorg. Chem.* **2013**, *2013*, 1033–1042. [CrossRef]
24. Paradis, N.; Chastanet, G.; Palamarciuc, T.; Rosa, P.; Varret, F.; Boukheddaden, K.; Létard, J.-F. Detailed Investigation of the Interplay Between the Thermal Decay of the Low Temperature Metastable HS State and the Thermal Hysteresis of Spin-Crossover Solids. *J. Phys. Chem. C* **2015**, *119*, 20039–20050. [CrossRef]
25. Baldé, C.; Desplanches, C.; Létard, J.-F.; Chastanet, G. Effects of metal dilution on the spin-crossover behavior and light induced bistability of iron(II) in $[Fe_xNi_{1-x}(bpp)_2](NCSe)_2$. *Polyhedron* **2017**, *123*, 138–144. [CrossRef]
26. Kahn, O.; Gatteschi, D.; Miller, J.S.; Palacio, F. *NATO ARW Molecular Magnetic Materials*; Kluwer Academic Publishers: London, UK, 1991.
27. Dunbar, K.R.; Heintz, R.A. Chemistry of Transition Metal Cyanide Compounds: Modern Perspectives. *Prog. Inorg. Chem.* **1997**, *45*, 283–391. [CrossRef]
28. Miller, J.S.; Drillon, M. *Magnetism–Molecules to Materials*; Wilely-VCH: Weinheim, Germany, 2005.
29. Train, C.; Gruselle, M.; Verdaguer, M. The fruitful introduction of chirality and control of absolute configurations in molecular magnets. *Chem. Soc. Rev.* **2011**, *40*, 3297–3312. [CrossRef] [PubMed]
30. Ohkoshi, S.; Tokoro, H. Photomagnetism in Cyano-Bridged Bimetal Assemblies. *Acc. Chem. Res.* **2012**, *45*, 1749–1758. [CrossRef] [PubMed]
31. Ferlay, S.; Mallah, T.; Ouahès, R.; Veillet, P.; Verdaguer, M. A room-temperature organometallic magnet based on Prussian blue. *Nature* **1995**, *378*, 701–703. [CrossRef]
32. Hatlevik, Ø.; Buschmann, W.E.; Zhang, J.; Manson, J.L.; Miller, J.S. Enhancement of the Magnetic Ordering Temperature and Air Stability of a Mixed Valent Vanadium Hexacyanochromate(III) Magnet to 99 °C (372 K). *Adv. Mater.* **1999**, *11*, 914–918. [CrossRef]
33. Holmes, S.M.; Girolami, G.S. Sol−Gel Synthesis of $KV^{II}[Cr^{III}(CN)_6] \cdot 2H_2O$: A Crystalline Molecule-Based Magnet with a Magnetic Ordering Temperature above 100 °C. *J. Am. Chem. Soc.* **1999**, *121*, 5593–5594. [CrossRef]
34. Ohkoshi, S.; Mizuno, M.; Hung, G.J.; Hashimoto, K. Magnetooptical Effects of Room Temperature Molecular-Based Magnetic Films Composed of Vanadium Hexacyanochromates. *J. Phys. Chem. B* **2000**, *104*, 9365–9367. [CrossRef]

35. Verdaguer, M.; Bleuzen, A.; Marvaud, V.; Vaissermann, J.; Seuleiman, M.; Desplanches, C.; Scuiller, A.; Train, C.; Garde, R.; Gelly, G.; et al. Molecules to build solids: High T_C molecule-based magnets by design and recent revival of cyano complexes chemistry. *Coord. Chem. Rev.* **1999**, *190*, 1023–1047. [CrossRef]

36. Ohkoshi, S.; Einaga, Y.; Fujishima, A.; Hashimoto, K. Magnetic properties and optical control of electrochemically prepared iron–chromium polycyanides. *J. Electroanal. Chem.* **1999**, *473*, 245–249. [CrossRef]

37. Herrera, J.M.; Marvaud, V.; Verdaguer, M.; Marrot, J.; Kalisz, M.; Mathonière, C. Reversible Photoinduced Magnetic Properties in the Heptanuclear Complex $[Mo^{IV}(CN)_2(CN–CuL)_6]^{8+}$: A Photomagnetic High-Spin Molecule. *Angew. Chem. Int. Ed.* **2004**, *43*, 5468–5471. [CrossRef] [PubMed]

38. Ohkoshi, S.; Tokoro, H.; Hozumi, T.; Zhang, Y.; Hashimoto, K.; Mathonière, C.; Bord, I.; Rombaut, G.; Verelst, M.; Cartier dit Moulin, C.; et al. Photo-induced magnetization in copper octacyanomolybdate. *J. Am. Chem. Soc.* **2006**, *128*, 270–277. [CrossRef] [PubMed]

39. Mahfoud, T.; Molnar, G.; Bonhommeau, S.; Cobo, S.; Salmon, L.; Demont, P.; Tokoro, H.; Ohkoshi, S.; Boukheddaden, K.; Bousseksou, A. Electric-Field-Induced Charge-Transfer Phase Transition: A Promising Approach toward Electrically Switchable Devices. *J. Am. Chem. Soc.* **2009**, *131*, 15049–15054. [CrossRef] [PubMed]

40. Bleuzen, A.; Marvaud, V.; Mathonière, C.; Sieklucka, B.; Verdaguer, M. Photomagnetism in Clusters and Extended Molecule-Based Magnets. *Inorg. Chem.* **2009**, *48*, 3453–3466. [CrossRef] [PubMed]

41. Pajerowski, D.M.; Andrus, M.J.; Gardner, J.E.; Knowles, E.S.; Meisel, M.W.; Talham, D.R. Persistent Photoinduced Magnetism in Heterostructures of Prussian Blue Analogues. *J. Am. Chem. Soc.* **2010**, *132*, 4058–4059. [CrossRef] [PubMed]

42. Ozaki, N.; Tokoro, H.; Hamada, Y.; Namai, A.; Matsuda, T.; Kaneko, S.; Ohkoshi, S. Photoinduced magnetization with a high Curie temperature and a large coercive field in a Co–W bimetallic assembly. *Adv. Funct. Mater.* **2012**, *20*, 2089–2093. [CrossRef]

43. Ohkoshi, S.; Arimoto, Y.; Hozumi, T.; Seino, H.; Mizobe, Y.; Hashimoto, K. Two-dimensional metamagnet composed of cyano-bridged $Cu^{II}–W^V$ bimetallic assembly. *Chem. Commun.* **2003**, *22*, 2772–2773. [CrossRef]

44. Kato, K.; Moritomo, Y.; Takata, M.; Sakata, M.; Umekawa, M.; Hamada, N.; Ohkoshi, S.; Tokoro, H.; Hashimoto, K. Direct Observation of Charge Transfer in Double-Perovskite-Like RbMn[Fe(CN)$_6$]. *Phys. Rev. Lett.* **2003**, *91*, 255502. [CrossRef] [PubMed]

45. Tokoro, H.; Matsuda, T.; Nuida, T.; Moritomo, Y.; Ohoyama, K.; Dangui, E.D.L.; Boukheddaden, K.; Ohkoshi, S. Visible-light-induced reversible photomagnetism in rubidium manganese hexacyanoferrate. *Chem. Mater.* **2008**, *20*, 423–428. [CrossRef]

46. Vertelman, E.J.M.; Lummen, T.T.A.; Meetsma, A.; Bouwkamp, M.W.; Molnar, G.; Loosdrecht, P.H.M.; Koningsbruggen, P.J. Light- and Temperature-Induced Electron Transfer in Single Crystals of RbMn[Fe(CN)$_6$]·H$_2$O. *Chem. Mater.* **2008**, *20*, 1236–1238. [CrossRef]

47. Tokoro, H.; Nakagawa, K.; Imoto, K.; Hakoe, F.; Ohkoshi, S. Zero thermal expansion fluid and oriented film based on a bistable metal-cyanide polymer. *Chem. Mater.* **2012**, *24*, 1324–1330. [CrossRef]

48. Kosaka, W.; Nomura, K.; Hashimoto, K.; Ohkoshi, S. Observation of an Fe(II) Spin-Crossover in a Cesium Iron Hexacyanochromate. *J. Am. Chem. Soc.* **2005**, *127*, 8590–8591. [CrossRef] [PubMed]

49. Arai, M.; Kosaka, W.; Matsuda, T.; Ohkoshi, S. Observation of an Iron(II) Spin Crossover in an Iron Octacyanoniobate-Based Magnet. *Angew. Chem. Int. Ed.* **2008**, *47*, 6885–6887. [CrossRef] [PubMed]

50. Ohkoshi, S.; Imoto, K.; Tsunobuchi, Y.; Takano, S.; Tokoro, H. Light-induced spin-crossover magnet. *Nat. Chem.* **2011**, *3*, 564–569. [CrossRef] [PubMed]

51. Ohkoshi, S.; Takano, S.; Imoto, K.; Yoshikiyo, M.; Namai, A.; Tokoro, H. 90-degree optical switching of output second-harmonic light in chiral photomagnet. *Nat. Photonics* **2014**, *8*, 65–71. [CrossRef]

52. Imoto, K.; Takemura, M.; Nakabayashi, K.; Miyamoto, Y.; Orisaku, K.K.; Ohkoshi, S. Syntheses, crystal structures, and magnetic properties of Mn–Nb and Co–Nb cyanido-bridged bimetallic assemblies. *Inorg. Chim. Acta* **2015**, *425*, 92–99. [CrossRef]

53. Ohkoshi, S.; Iyoda, T.; Fujishima, A.; Hashimoto, K. Magnetic properties of mixed ferro-ferrimagnets composed of Prussian blue analogs. *Phys. Rev. B* **1997**, *56*, 11642–11652. [CrossRef]

54. Ohkoshi, S.; Abe, Y.; Fujishima, A.; Hashimoto, K. Design and Preparation of a Novel Magnet Exhibiting Two Compensation Temperatures Based on Molecular Field Theory. *Phys. Rev. Lett.* **1999**, *82*, 1285–1288. [CrossRef]

55. Shannon, R.D. Revised Effective Ionic Radii and Systematic Studies of Interatomic Distances in Halides and Chalcogenides. *Acta Cryst.* **1976**, *A32*, 751–767. [CrossRef]

56. Herrera, J.M.; Franz, P.; Podgajny, R.; Pilkington, M.; Biner, M.; Decurtins, S.; Stoeckli-Evans, H.; Neels, A.; Garde, R.; Dromzée, Y.; et al. Three-dimensional bimetallic octacyanidometalates $[M^{IV}\{(\mu\text{-CN})_4Mn^{II}(H_2O)_2\}_2 \cdot 4H_2O]_n$ (M = Nb, Mo, W): Synthesis, single-crystal X-ray diffraction and magnetism. *Comptes Rendus Chim.* **2008**, *47*, 6885–6887. [CrossRef]

MDPI

St. Alban-Anlage 66

4052 Basel, Switzerland

Tel. +41 61 683 77 34

Fax +41 61 302 89 18

http://www.mdpi.com

Inorganics Editorial Office

E-mail: inorganics @mdpi.com

http://www.mdpi.com/journal/inorganics

www.ingramcontent.com/pod-product-compliance
Lightning Source LLC
Chambersburg PA
CBHW051849210326
41597CB00033B/5831